Readings from
American Scientist

The Solar S...
Its Strange Objects

Edited by

Brian J. Skinner

Yale University

Member, Board of Editors
American Scientist

WILLIAM KAUFMANN, INC.   LOS ALTOS, CALIFORNIA

Front cover: Color picture of Mars reconstructed from three frames—
each through a different filter—shuttered by the Viking 1 Orbiter on
June 18, 1976. Courtesy of Jet Propulsion Laboratories, Pasadena, CA.

Library of Congress Cataloging in Publication Data
Main entry under title:

The Solar system and its strange objects.

  (Earth and its inhabitants)
  Bibliography: p.
  Includes index.
  1. Solar system—Addresses, essays, lectures.
2. Cosmology—Addresses, essays, lectures.
I. Skinner, Brian J., 1928–   II. American
scientist.   III. Series.
QB501.S63   523.2   81-21156
ISBN 0-913232-84-X   AACR2

ISBN 0-913232-84-X

Contents

Introducing

Earth and Its Inhabitants

A new series of books containing readings originally published in *American Scientist*.

The 20th century has been a period of extraordinary activity for all of the sciences. During the first third of the century the greatest advances tended to be in physics; the second third was a period during which biology, and particularly molecular biology, seized the limelight; the closing third of the century is increasingly focused on the earth sciences. A sense of challenge and a growing excitement is everywhere evident in the earth sciences—especially in the papers published in *American Scientist*. With dramatic discoveries in space and the chance to compare Earth to other rocky planets, with the excitement of plate tectonics, of drifting continents and new discoveries about the evolution of environments, with a growing human population and ever increasing pressures on resources and living space, the problems facing earth sciences are growing in complexity rather than declining. We can be sure the current surge of exciting discoveries and challenges will continue to swell.

Written as a means of communicating with colleagues and students in the scientific community at large, papers in *American Scientist* are authoritative statements by specialists. Because they are meant to be read beyond the bounds of the author's special discipline, the papers do not assume a detailed knowledge on the part of the reader, are relatively free from jargon, and are generously illustrated. The papers can be read and enjoyed by any educated person. For these reasons the editors of *American Scientist* have selected a number of especially interesting papers published in recent years and grouped them into this series of topical books for use by non-specialists and beginning students.

Each book contains ten or more articles bearing on a general theme and, though each book stands alone, it is related to and can be read in conjunction with others in the series. Traditionally the physical world has been considered to be different and separate from the biological. Physical geology, climatology, and mineral resources seemed remote from anthropology and paleontology. But a growing world population is producing anthropogenic effects that are starting to rival nature's effects in their

magnitude, and as we study these phenomena it becomes increasingly apparent that the environment we now influence has shaped us to be what we are. There is no clear boundary between the physical and the biological realms where the Earth is concerned and so the volumes in this series range from geology and geophysics to paleontology and environmental studies; taken together, they offer an authoritative modern introduction to the earth sciences.

Volumes in this Series

The Solar System and Its Strange Objects Papers discussing the origin of chemical elements, the development of planets, comets, and other objects in space; the Earth viewed from space and the place of man in the Universe.

Earth's History, Structure and Materials Readings about Earth's evolution and the way geological time is measured; also papers on plate tectonics, drifting continents and special features such as chains of volcanoes.

Climates Past and Present The record of climatic variations as read from the geological record and the factors that control the climate today, including human influence.

Earth's Energy and Mineral Resources The varieties, magnitudes, distributions, origins and exploitation of mineral and energy resources.

Paleontology and Paleoenvironments Vertebrate and invertebrate paleontology, including papers on evolutionary changes as deduced from paleontological evidence.

Evolution of Man and His Communities Hominid paleontology, paleoanthropology, and archaeology of resources in the Old World and New.

Use and Misuse of Earth's Surface Readings about the way we change and influence the environment in which we live.

Introduction: The Solar System and Its Strange Objects

The immensity of the universe and the millions of stars visible from Earth will always raise questions and fire imaginations. But no question is more intriguing than the question of life existing elsewhere in the universe—and not only life, but intelligent, human-like life. The answer to this problem does not lie entirely within the realm of science fiction. The possible chemistries by which life forms might develop are limited by the inherent properties of atoms. Biological chemistry must be carbon-based and probably cannot differ too greatly from the chemistry found on Earth. Starting with that premise and using Earth as a model to estimate times for development and evolution, it is possible to discover many constraints that limit the number of places in the universe where life may exist. The first three articles in this collection—by Trimble, Huang, and Pollard—discuss various aspects of this absorbing problem.

Trimble reviews briefly and lucidly the history of the entire universe, of which the appearance and evolution of life is only one part; her paper is, therefore, an excellent place to start. Huang and Pollard restrict their discussions more closely to the actual characteristics of planets that could have life forms. Pollard also presents the interesting argument that continental drift, by keeping animal populations apart for long periods, has provided separate evolutionary experiments here on Earth. Since humans and other hominids did not originate in America or Australia, even though mammals were present, Pollard suggests that our species is probably a random, one-time end-product of an evolutionary line in a single, specific environment.

While the question of forms of life elsewhere in the universe is a matter of speculation, theories concerning the structure and evolution of galaxies, of stars they contain, of the formation of heavy chemical elements within stars, and of the origin of the planets that circle stars are becoming less and less speculative. Because Earth is a small member of a large family of planets, and because it is now believed that planets circling stars are a usual part of stellar evolution, it seems appropriate to include four papers that deal with these larger phenomena of stars and galaxies.

Greenstein's paper, first published in 1961, is still one of the most lucid discussions of the origin of heavy chemical elements as a result of the stellar evolutionary cycle. Without nuclear processes that combine hydrogen and helium into heavier elements, there could be no heavy, rocky planets like Earth. But the heavy elements born in the core of an aging star must somehow be recycled to become part of a new solar system. Wheeler's paper, which discusses supernovae, addresses the question of how the recycling occurs. It is humbling to realize that every heavy atom in Earth was formed in a now deceased and completely recycled star. In studying galaxies, astronomers observe the endless processes of star formation, evolution, and decay. A balanced view of the solar system can best be gained through a larger awareness of the aggregate of all the other suns in the millions of galaxies in the universe. The papers by Larson and Geller have therefore been included to present our current understanding of larger-scale, galactic-sized structure.

Many years ago, the planets were thought to be simply bright spots of light in the sky and seemed to be as distant as stars. We now know better. As a result of observations through high-resolution telescopes and of space exploration, we now have photographs of some of the planets. To those of us who can only learn about distant places through photographs, the surface of the moon is probably more familiar than central Antarctica or the Gobi Desert. Even the geology of the moon and Mars is better known than the geology of some places on Earth. Therefore, seven papers discussing planets form the largest section in the book. The flood of data from wide-ranging spacecraft is so great that, unfortunately, no paper can stay up-to-date for long. Nevertheless, each of the seven is a recent and succinct discussion of some of the most important discoveries in planetology.

The synthesis of data bearing on the geologic evolution of the terrestrial planets by Head, Wood, and Mutch provides a fine overview of planetary features. As for Earth's nearest neighbor, Moon, continuing studies of lunar rocks brought back by the astronauts add daily to our understanding of lunar chemistry. Yet our understanding of the origin and development of the larger features on the

1

moon has changed little since the astronaut landings, and discussions of the discoveries made during the Apollo missions are as valid today as they were when they were written. The paper by Muehlberger and Wolfe analyzes the Apollo 17 mission, the last of the lunar landing missions, and in so doing synthesizes and summarizes results from the previous lunar landings. We now know that even though the moon is physically close to Earth, it is very different; it has been tectonically quiet for over 3 billion years and is devoid of familiar patterns of erosion and weathering because it has no atmosphere.

But the moon is small compared to Earth. Much better comparisons can be drawn from Venus and Mars, and unmanned space probes have now landed on both. Venus is often called a sister planet to Earth because the two bodies are about the same size. But they hardly seem like sisters when their atmospheres are compared. Venus is enveloped by dense clouds of carbon dioxide, which act as a thermal blanket and produce temperatures as high as 600°K on its surface. The clouds also prevent us from making direct visual observations of the surface from orbiting vehicles. We have, therefore, had to develop much of our understanding of the surface properties of Venus by inference and calculation, and the paper by Lewis demonstrates how this can be done. By contrast, the atmosphere of Mars is thin and tenuous, and the pitted surface and polar caps of the planet are readily visible; thus, we have a host of spectacular and sometimes puzzling images of its surface. In his overview Carr points out that Mars, like Earth, is both a volcanically and tectonically active planet, and furthermore, that the Martian surface has been affected by the erosive action of wind, water, and ice. There are many differences between Mars and Earth, though; Earth's lithosphere is moved and recycled by plate tectonics, whereas the rigid Martian lithosphere is fixed, which means the products of erosion accumulate and remain essentially unaltered on the surface.

The giant planets, Saturn and Jupiter, are very different from their small terrestrial relatives. They present massive gassy exteriors to passing space probes, and although they presumably have rocky cores buried deep inside, the cores are completely inaccessible. Therefore, the articles on Jupiter by Smoluchowski and on the rings of Saturn by Pollack deal with features that are completely different from those of the rocky, terrestrial planets. Nevertheless, Jupiter and Saturn account for so much of the mass of the planetary family and play such a key role in planetary dynamics and in our understanding of the origin and development of all the planets, that no book of readings would be balanced without discussions of these giants.

The sun and all of the planets were formed from a primitive solar nebula, but each planet has been changed and modified so much by internal thermal processes during the 4,600 million years since its birth that many questions about the birth process remain obscure. Some objects in the solar system have changed less, however. Meteorites, which are discussed by Mason, were formed as a result of the same condensation events as the planets, and they have either not changed since birth or have changed in such different ways from planets that they provide valuable information not available from the planets themselves. Studies of meteorites are an essential element in any coherent program of studies of the solar system. Comets remain more enigmatic than meteorites, probably because we have never "captured" one for study, but they, too, may be essentially unchanged fragments that date from the birth of the solar system. With a proposal for a space probe to visit and sample a comet before the end of the century now under consideration, Feldman's discussion of the composition of comets can be read both as a summary and as a prediction that may soon be tested.

Tektites, unlike comets and meteorites, are far from primitive. As King's article indicates, they are small glassy objects found in several places on Earth. At each location the tektites have the same radiometric ages, but the ages differ greatly from one location to another. We must conclude that each group of tektites owes its origin to a special formative event, which raises the questions of what the events were and where they occurred. After considerable debate, it is now generally, but not universally, believed that tektites are terrestrial objects, small splashes of molten rock thrown out of large craters at times of meteorite impacts.

To those who live here, Earth is a special and very different planet. But imagine for a moment that you are a visitor from another part of space and are observing Earth for the first time as your spaceship passes through the solar system. What would you see and what conclusions might you draw? The surface would certainly differ from the surfaces of the other terrestrial planets because of the atmosphere, hydrosphere, and biosphere. If you lingered long enough, seasonal changes would be evident, and curious anthropogenic features (such as cities) would mark Earth as the home of organized, large, land-based animals (people). Networks of land-based transportation systems connecting the cities would appear, and the regular geometric patterns evident on much of the land surface would indicate food production through intensive agriculture. It would soon occur to an intelligent observer that Earth must support an unusually large population of big animals for such a small planet. Large populations of big animals, wherever they occur, use a lot of food; huge cities and complex transportation systems use vast quantities of energy and mineral resources. The intelligent observer, unaware that Earth's huge population and bulging cities are recent phenomena, would no doubt conclude that Earth must have a stable population and enjoy an efficient, waste-free system for using and controlling its necessarily limited resources.

We certainly have not yet reached the sensible and desirable situation deduced by the space traveler; but the day is fast approaching when we will have to do so if human civilization is to continue. One of the most important new developments in this respect is our ability to see

Earth as a space traveler can—though remote-sensing vehicles that orbit Earth and gather data that relate to weather, crops, and other processes and features crucial to our daily activities. Data are radioed back to Earth to be processed quickly and efficiently. Large-scale features not previously visible can be seen and used in mineral exploration, while subtle, short-term changes, such as might occur in crops and forests, can help farmers and foresters reduce crop failures. Remote sensing holds tremendous promise for solutions to problems of global management. Accordingly, the three final papers in the volume address questions concerning the use of Earth-orbiting vehicles for remote sensing; they bring the volume to a close on a high note, for satellite observations of Earth itself may, in the long run, be the most important result of all space exploration.

Suggestions for Further Reading

Chapman, C.R., *The Inner Planets* (New York: Charles Scribners and Sons, 1977).

Dole, S.H., *Habitable Planets for Man,* 2nd ed. (New York: American Elsevier, 1970).

Jastrow, R., and M.H. Thompson, *Astronomy: Fundamentals and Frontiers,* 3rd ed. (New York: John Wiley and Sons, 1977).

King, Elbert A., *Space Geology. An Introduction* (New York: John Wiley and Sons, 1976).

Kopal, Z., *The Moon in the Post-Apollo Era* (Dordrecht, Holland: D. Reidel, 1974).

Masursky, H., G.W. Colton and Farouk El-Bas, editors, *Apollo over the Moon. A View from Orbit,* NASA SP-362 (Washington, D.C.: Scientific and Technical Information Office, National Aeronautics and Space Administration, 1978).

Mutch, Thomas A., R.E. Arvidson, J.W. Head, III, K.L. Jones and R.S. Saunders, *The Geology of Mars* (Princeton, New Jersey: Princeton University Press, 1976).

Short, N.M., P.D. Lowman, Jr., S. C. Freden and W. A. Finch, Jr., *Mission to Earth: Landsat Views the World,* NASA SP-360 (Washington, D.C.: Scientific and Technical Information Office, National Aeronautics and Space Administration, 1976).

Taylor, S.R., *Lunar Science: A Post-Apollo View* (New York: Pergamon, 1975).

Wasson, J.T., *Meteorites: Classification and Properties* (New York: Springer, 1974).

Wood, J.A., *The Solar System* (Englewood Cliffs, New Jersey: Prentice-Hall, 1979).

Authoritative and up-to-date reviews, summaries, and analyses of many of the topics discussed in this volume can be found in volumes published by Annual Reviews, Inc., Palo Alto, California 94306. Articles of special interest will be found in the annual volumes of *Annual Review of Earth and Planetary Sciences,* commencing with Volume 1, 1973, and *Annual Review of Astronomy and Astrophysics,* commencing with Volume 1, 1962.

PART 1 _Life in the Universe_

Virginia Trimble

Cosmology: Man's Place in the Universe

In which we review the history of the Universe and explore the relationships between its properties and the presence of life

What I want to try to do in the next few pages is to review the history of the Universe from the earliest times for which we have any evidence down to the present day, with special emphasis on how conditions favorable for life seem to have arisen, and then to explore the extent to which this history is dependent upon the Universe having roughly the properties it does, and finally to inquire into the implications of varying those properties.

A Cook's tour of the universe and its early history

Let's start by taking a look (Table 1) at the scales of the things we will be discussing. Notice that the human scales in each case are close to the geometric means of the astronomical and atomic scales. Thus, we should not be surprised to find that our presence here is dependent both on

Virginia Trimble is grateful to the Aspen Center for Physics, where much of this was written, for hospitality, and to the Alfred P. Sloan Foundation for a Research Fellowship (1972–74). The author's thoughts on the Universe and its contents have inevitably been influenced by more people than can conveniently be mentioned, but those from whom she first heard some of the ideas discussed here, and whom she would therefore like to thank especially, include Dave Arnett, William A. Fowler, Jim Gunn, Philip Morrison, Paula Moddel, Bohdan Paczyński, Martin Rees, Bill Saslaw, Starling Trimble, and (last only in this deliberately alphabetical list) Joe Weber. Address: Department of Physics, University of California, Irvine CA 92717 (January to June); Astronomy Program, University of Maryland, College Park MD 20742 (July to December).

the large-scale phenomena of astronomy and on the details of atomic physics.

The largest phenomenon of all is, of course, the Universe itself. It is important to be sure we agree about what we mean by "the Universe" and the various other terms we will be using. The earth and eight other planets, about 34 moons, and a variety of smaller objects are in gravitationally bound orbits around a star called the sun. We refer to this grouping as the Solar System. It has a total mass of about 2×10^{33} grams (virtually all in the sun, though most of the angular momentum is in the planets), a diameter of about 2×10^{15} cm, and an age of about 5×10^9 yr. The sun is a perfectly typical star, having a mass of 2×10^{33} g (the solar mass, abbreviated $M_\odot$, is often used as a unit for other stars), an electromagnetic radiation energy output of 4×10^{33} ergs/sec (one solar luminosity, $L_\odot$), a spectrum approximately that of a 5700 K black body, a radius of 7×10^{10} cm (1 $R_\odot$), and a composition by weight (at least in its outer, visible layers) about 73% hydrogen, 25% helium, and 2% everything else (about half of it carbon and oxygen).

The sun, in turn, is one of about 2×10^{11} stars that are gravitationally bound in a rotating, roughly spherical system (although the most conspicuous members are concentrated in a plane considerably flatter than the proverbial pancake) called the Milky Way Galaxy (or just the Galaxy). It has a mass of at least 3×10^{44} g (but see Ostriker et al. 1974 for evidence that it may be ten times more massive than this) and a diameter of about 10^{23} cm. It is at least 10^{10} yr old.

The Milky Way, in turn, is bound in a small cluster of about 30 galaxies (all but one much less massive than ours) called the Local Group. It is not certain whether higher-order structures are gravitationally bound, but there does seem to be some clustering of the clusters (Hauser and Peebles 1973). The clusters range from small ones like the Local Group up to much richer ones containing thousands of galaxies and having masses of 10^{15} $M_\odot$. Completely isolated galaxies are probably rare (Tifft and Gregory 1976). The properties of the medium between the galaxies (except within the rich clusters, where a hot intracluster gas is often a strong source of X rays; Kellogg et al. 1973) are very poorly known. The average density could be anywhere from 0 to 10^{-5} particles cm^{-3}, the intergalactic medium comprising anywhere from 0 to 90% of the total average density over large regions of space. If the density is high, the matter must also be rather hot ($\sim 10^6$ K) or exceedingly clumpy to prevent detection. A preponderance of the evidence (as summarized, e.g., by Gott et al. 1974) now seems to favor an intergalactic density at the low end of the possible range.

The clusters of galaxies (or perhaps the superclusters) appear to be distributed at random through space, with separations such that they contribute an average density of at most 10^{-31} g cm^{-3} (Ostriker et al. 1974). There is no detectable falloff of the density of clusters of galaxies out to the largest distances at which they can be seen with present telescopes. This is about 10^{28} cm or 3000 Mpc (Megaparsecs; one parsec = distance at which an object has a *parallax* of one *second* of arc = 3×10^{18} cm), corresponding to a light travel time of several billion years. We probably observe quasi-stellar objects (quasars) at much larger distances, but their properties are so poorly understood that they add very little to our knowledge of the large-scale structure of the Universe.

The volume surveyed is sometimes called the observable universe, and it is the region for which we have direct observational evidence. Spectra of the vast majority of galaxies within this region and outside the Local Group show red shifts which are proportional to their distances from us. These are normally interpreted as Doppler shifts, implying that all the objects within the observable universe are receding from one another at speeds proportional to their separations. The proportionality constant is generally called the Hubble constant. Its value (at the present time in the history of the Universe and of astronomical research) is about 57 km/sec/Mpc (Sandage and Tammann 1975). This proportionality (Hubble's Law) is our chief evidence that the Universe is expanding.

Within the framework of some reasonable theory of gravity, like General Relativity, we can extrapolate beyond the observable region and try to learn something about the entire four-dimensional space-time volume that can, in principle, be connected to us by light signals. The word *Universe* properly refers to this entire volume and, in this sense, is not much more than 50 years old, dating back to the realization that certain bright, fuzzy patches in the sky are, in fact, galaxies like our own (Curtis 1919; Hubble 1925).

Efforts to model the Universe go back just about as far and always involve a variety of simplifying assumptions. The simplest possible set of such assumptions has proved remarkably successful. We assume (1) that General Relativity is the right theory of gravity (probably without the arbitrary additional repulsive kind of gravity, the cosmological constant, introduced by Einstein to permit a static universe), (2) that the expansion implied by Hubble's Law is isotropic and would be seen to be isotropic by any observer moving with the galaxies, (3) that the Universe is homogeneous on a sufficiently large scale, and (4) that pressure is presently unimportant and that the

mass-energy of the Universe is now mostly in the form of matter rather than radiation or other zero-rest-mass particles. Under these assumptions, the Einstein field equations yield a two-parameter family of models, called the Friedmann models, and the problem of deciding what the Universe is like reduces to finding values for the two parameters. These turn out to be the present value of the Hubble constant, H_0, which we know quite well, and the present value of the local average density of mass-energy (in all forms), ρ_0, which may be uncertain by a factor of 100. Given values of these, we can answer a variety of interesting questions, like: How old is the Universe? Is it finite or infinite in volume? Will the present expansion continue forever or will gravity cause the galaxies to slow down and eventually fall back together? Roughly, a low-density ($\rho_0 < 10^{-30}$ g cm^{-3}) universe has infinite volume, is 16–20 billion years old, and will expand forever; while a high-density universe ($\rho_0 > 10^{-30}$ g cm^{-3}) has finite volume, is less than 16 billion years old, and will eventually (in a hundred billion years or so) turn around and recontract. With many ifs, buts, maybes, and other caveats, evidence now available seems to indicate that our universe is a low-density one (Gott et al. 1974). Notice that if the universe is to be a high-density one, then $\gtrsim 90\%$ of the mass-

energy is neither visible nor in galaxies.

Under the same assumptions, there are some questions that we cannot answer or even ask in a meaningful fashion. One of these is Where is the center of the Universe? The assumed homogeneity and isotropy of the expansion imply that all the matter was arbitrarily close together a finite time ago in the past, so that the center of the expansion exists only in four dimensions and is something like the instant of creation; while the geometry of space-time within the framework of General Relativity is such that space is either infinite (and so can have no center) or uniformly curved, so that all points are equivalent (rather like the curved surface of the earth, only in three dimensions). It will become clear shortly that "what came before the present universe?" is another of these unanswerable questions.

With this background, we can now say that the earliest event for which we have any evidence is a time about 15–20 billion years ago when the Universe was much hotter and denser than it is at present. The evidence for the time scale comes from (1) running the Hubble expansion backwards in time and asking how long ago would all of the galaxies have been arbitrarily close together ($H_0 = 50$ km/

Table 1. Scales of phenomena being considered

Atomic scale	Human scale	Astronomical scale
	T I M E	
Nuclear decays 10^{-14} *seconds*	Attention span of physics undergraduates *1 minute = 60 seconds*	Age of the Universe 6×10^{17} *seconds*
	M A S S	
Hydrogen atom 2×10^{-24} *grams*	Typical Sigma Xi member *140 lb = 6.5×10^4 grams*	Solar mass 2×10^{33} *grams*
	L E N G T H	
Diameter of atomic nucleus 10^{-13} *centimeters*	Height of dean at prestigious university *18 feet = 546 cm*	Distance from sun to next star *1 parsec = 3×10^{18} cm*
R A T E O F E N E R G Y O U T P U T		
Atomic decay 10^{-3} *erg/second*	Output of large electricity generating plant *200 megawatts = 2×10^{15} ergs/second*	Luminosity of the sun 4×10^{33} *ergs/second*

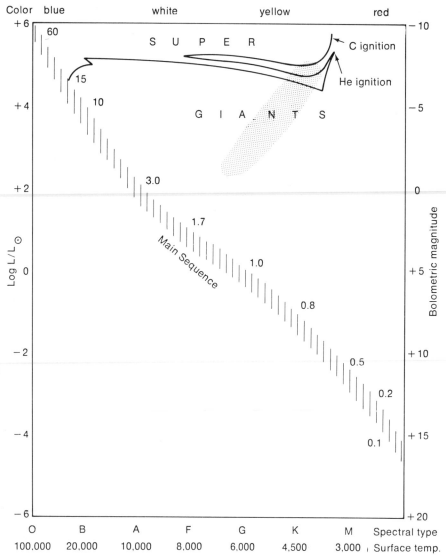

Figure 1. Hertzsprung-Russell (HR; color-magnitude) diagram for a representative group of stars young enough that the entire Main Sequence is still populated. Notice the curious scales always used by astronomers. The vertical scale is the logarithm of total luminosity in solar units, or magnitudes which are logarithms (base $100^{1/5}$) of the reciprocal of the luminosity in fairly arbitrary units. The horizontal scale is surface temperature (though neither exactly linear or logarithmic) or color or spectral class (an ancient and honorable way of dividing up the stars which somewhat predates the realization that temperature is the most important determinant of spectral line intensities) and runs backwards. Masses in solar masses are given at representative points along the main sequence. The evolution of a typical massive star through the supergiant region is shown, along with the points at which helium and carbon burning start. Most of the time is spent close to the main sequence and in the red supergiant region. The stars in the stippled region of the diagram are generally or always variable in luminosity with regular periods of 3–30 days. They are called Cepheid variables, after the prototype, Delta Cephei, and are important distance indicators for nearby galaxies, because their periods are correlated with their total luminosities.

picture of the Universe backwards about 20 billion years, we see it at a temperature $T \gtrsim 10^{10}$ K and a density $\rho \gtrsim 1$ g cm^{-3}. Under these circumstances, many kinds of matter and radiation come into equilibrium, and the relative numbers of various kinds of particles (protons, neutrons, electrons, positrons, neutrinos, photons, and perhaps others) depend only on T. As the Universe expands and cools, unstable particles decay or annihilate, and others undergo nuclear reactions, resulting in about 25% He^4 and traces of H^2 (deuterium), He^3, and Li^7, as well as about 75% ordinary hydrogen (Wagoner et al. 1967) in the standard cosmological models.

Unfortunately, this hot, dense phase (sometimes called the Big Bang) also wiped out any evidence of what (if anything) went before. Hence the question What happened before the Big Bang? belongs to the realm of pure speculation (philosophy?) rather than that of physics. It is rather like putting a car into a steel blast furnace and asking the trickle of molten metal that comes out whether it was a Pinto or a VW before. You just can't tell, because the evidence has been destroyed.

Galaxies and stars

Coming out of the hot, dense early universe we therefore see some radiation (which continues to cool, down to 3 K by the present time) and some matter, in the form of hydrogen and helium. Luckily this is not all that happened, because the chemistry of H and He is not very interesting! The matter at this stage was not perfectly smooth but was concentrated in lumps. This is also fortunate for us, because, as we have already seen, the average density of matter in the Universe is exceedingly low. Thus, in the absence of local concentrations of matter, the average hydrogen atom would not have encountered another hydrogen atom for the last 10 billion years or so, and would be very lonely. The cause of the lumps is not well understood, though they are not unexpected, since, when the Universe was very young, there had not yet been time for interactions and smoothing to have occurred across large distances. But they must have been there, because we see galaxies and clusters now. There has been

sec/Mpc = 1.67×10^{-17} sec^{-1}, or $1/H_0$ = 2×10^{10} yr), (2) the ages of the oldest stars in our Galaxy, probably 12–18 billion years, and (3) the ages of the radioactive elements in the Solar System, which tell us that the earth and meteorites solidified about 4.65 billion years ago, and that synthesis of these elements had been going on for 7 to 13 billion years before that (Gott et al. 1974). The evidence for the high temperature and density comes (1) again from running the Hubble expansion backwards, conserving mass-energy and the numbers of various kinds of particles, including photons, and (2) from the existence of two relics of the hot, dense phase. These relics are an isotropic background of microwaves having a blackbody spectrum corresponding to a present temperature of 2.7 K (Peebles 1971) and the seemingly universal presence of helium, with an abundance of 20–30% by weight (Trimble 1975). Thus, if we run a

some success in calculating how the lumps must have grown and developed into proto-galaxies (Jones 1976).

The evolution of a galaxy is largely a matter of the exchange of material between stars and an interstellar gaseous medium and the nuclear processes that occur in stars. Many different types of galaxies are observed (of which the most clearly defined are Ellipticals, with their brightest stars distributed throughout a spheroidal volume, and Spirals, with their brightest stars concentrated in spiral-shaped arms in a plane; see any elementary astronomy textbook for pretty pictures). They come with different total masses, luminosities, kinds of stars, colors, amounts of gas, shapes, abundances of heavy elements, and spatial distributions of these and other properties. Serious efforts to understand the evolution of galaxies go back only about 20 years, and the field is changing so rapidly now that it is hard to say more than that it looks as if we may be able to account for the observed ranges of properties and their correlations in terms of a rather small number of initial conditions in the lump, e.g. total mass, size, angular momentum, and degree of turbulence (Trimble 1975; Audouze and Tinsley 1976).

A protogalaxy becomes a galaxy when some appreciable fraction of its gas has been converted into stars. The most distant galaxies, seen as they were 3–5 billion years ago, do not look very different from the nearby ones, but the quasars may represent some early stage of the galaxy formation process, which appears to have been largely completed within a few billion years after the initial hot, dense phase of the Universe. We do not, in other words, see any obviously young galaxies nearby.

The process of star formation from interstellar gas, on the other hand, has continued to the present time in our own and most other galaxies (though at widely varying rates). We can almost see it happen before our eyes, at least in the sense that we see some bright, naked-eye stars (for instance the brightest ones in Orion) that did not yet exist as separate bodies at the time of our remote Zinjanthropan ancestors. We do not have an adequate theory of star formation

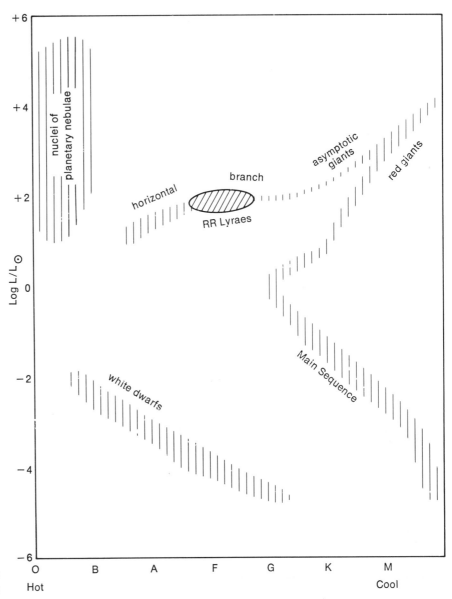

Figure 2. Hertzsprung-Russell diagram for a cluster of stars old enough (~10^{10} yr) that stars like the sun (~1 $M_\odot$) are leaving the Main Sequence. The evolutionary track of a single star would go from the main sequence up into the red giant region along the heavily populated track, down to the horizontal branch (when He is ignited), back up to the red giant region along the asymptotic branch, then horizontally (and very rapidly) across from right to left as a planetary nebula is shed, and finally down into the white dwarf region. The stars in the region labled RR Lyraes are variables (also named for their prototype RR Lyrae) with periods less than about a day.

(lots of people would say we have none at all), but we can learn quite a lot about it by looking.

In our region of the Galaxy (often called the solar neighborhood), young stars are (almost?) invariably found in groups and in the presence of denser-than-average clouds of gas and dust, in the spiral arms of the Galaxy. The very youngest stars are typically still hidden behind the remnants of the clouds from which they formed and are seen only as infrared sources. We are led, then, to a picture in which a typical dense (10^{3-5} cm^{-3}), massive

(~10^5 $M_\odot$) interstellar cloud (these are observed as sources of radio line emission) is shocked by a collision with another cloud, with the expanding shell from one of the supernovae (which we will meet later) or with the density wave which is believed to be responsible for the characteristic spiral shape of our own and many other galaxies. The shock starts the cloud collapsing under its own weight. As it contracts, excess angular momentum forces it to fragment into star-sized pieces. This means masses from about 0.05 to 100 $M_\odot$, the lower limit being set by the requirement

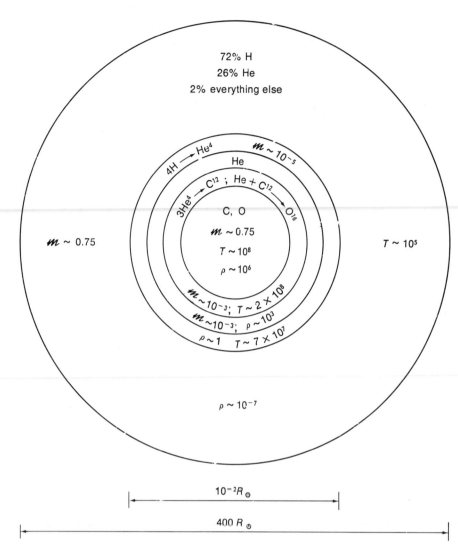

Figure 3. Interior structure of a 1.5 $M_\odot$ (solar type) star shortly before it sheds its planetary nebula and becomes a white dwarf. Masses of the various zones are given in solar masses, temperatures in Kelvins, and densities of the zones in g cm^{-3}. Only the primary constituents and nuclear reactions are indicated. Notice the great disparity in scale between the inner and outer regions.

The differences are (1) the stars can start with pure H and convert two protons to neutrons (via the weak interaction) to make He, while we must start with substances (deuterium, tritium, lithium) in which this has already been done; (2) the star does it at much higher temperature and density than is contemplated in the lab; and (3) the star will take 10^{6-12} years to get all the energy out, while we hope for rather faster operation.

The energy liberated by hydrogen-burning keeps the pressure inside the star constant and stops its contraction. The star will remain in hydrostatic equilibrium until its hydrogen fuel is exhausted. The hydrogen-burning phase of stellar life is called the Main Sequence stage, from the star's position on a diagram of brightness vs. surface temperature (Figs. 1 and 2). Such plots are called Hertzsprung-Russell (HR) diagrams and are of considerable assistance in understanding stellar evolution. Notice the one-to-one relationship between stellar mass and position on the Main Sequence. This, in turn, implies a relation between mass and lifespan, since the fuel supply depends on mass and the rate of fuel consumption depends on luminosity (which scales as about M^3 on the Main Sequence, as was first understood by Eddington 1926). A star's structure remains stable until hydrogen has been exhausted in about the inner 10% of its mass. This implies a main sequence lifetime of nearly 10^{10} yr for a star like the sun (70% hydrogen in the inner 2×10^{32} grams, yielding 7×10^{18} ergs/gram, and being used to supply 4×10^{33} ergs/sec). But a 20–100 $M_\odot$ star at the upper end of the main sequence can last only 10^{6-7} yr, while stars of less than about 0.75 $M_\odot$ have not had time to leave the main sequence in the age of the Universe. Almost 90% of the stars we see are on the Main Sequence, accounting for the name and implying that it is the longest-lived phase.

Most astronomers and physicists feel that we have a good quantitative understanding of the main sequence phase (despite the continuing non-appearance of the expected solar neutrinos; Bahcall and Davis 1976), but the ratio of speculation to "hard" theoretical (and observational) facts will gradually increase as we move away from the main sequence. The

that the center of the piece eventually become hot enough for nuclear reactions to occur and the upper limit by the tendency for radiation pressure to blow material off a star that is too massive and bright. We cannot predict how many stars of each mass will form from a particular cloud (this is one of the senses in which we have no theory of star formation), but observation shows that, over the history of our own and most other galaxies, more than half the mass has gone into stars less massive than the sun. Thus small fragments are favored over large ones, and the smallest stars are the most common.

The fragments are called proto-stars and continue to contract under their own self-gravity (on a time scale that depends on their mass, amounting to

about 0.1% of the time they will spend undergoing nuclear reactions) until their centers reach a temperature of about 10^7 K. Because the very first generation of stars must have been nearly pure hydrogen and helium, a variety of processes that now occur during the contraction phase (including, probably, those that lead to the formation of planets) will not have happened. At a central temperature near 10^7 K, hydrogen begins to fuse to form helium (either directly, or using carbon, nitrogen, and oxygen as catalysts, when they are available and at slightly higher temperature) with the liberation of about 7×10^{18} ergs/gram.

Stars have clearly solved the problem of controlled nuclear fusion, which we are now attacking in the laboratory.

evolution of normal stars as a function of mass has been studied by numerous groups. Three recent series of papers by W. D. Arnett, I. Iben, and B. Paczyński (referenced in Trimble 1975) cover all the phases we will be discussing here.

Exhaustion of hydrogen in the stellar core introduces a discontinuity in composition and mean molecular weight. As a result, the equilibrium structure becomes a very extended one (Hoyle and Lyttleton 1942, 1949). The outer layers of the star rapidly expand and cool, while the core once again contracts under its own weight, liberating gravitational potential energy to keep the star shining. The resulting star is called a red giant (or red supergiant, in the case of the most massive stars which become even bigger and brighter) for obvious reasons.

Our sun will become a red giant in about another five billion years. When it does, it may become so large that its outer layers engulf the earth. In any case, its greatly increased luminosity is expected to raise the temperature of the earth to the point where the oceans boil away.

The increasing temperature of the contracting stellar core soon raises the temperature of the surrounding hydrogen enough that it begins to fuse to helium, again liberating nuclear energy. About as much hydrogen is burned in the red giant phase as was burned on the main sequence, but the star is about 10 times as bright, so the phase lasts only about 10% as long. Red giants (Betelguese and Antares are examples) are thus rarer in the sky than main sequence stars.

For all but very tiny stars ($\lesssim$.3 $M_\odot$, which have not yet had time to leave the main sequence anyway), the core eventually becomes hot enough for nuclear reactions involving helium to occur. The onset of helium burning occurs explosively in stars like the sun (because the cores are so dense the matter is partly degenerate, thus pressure does not immediately increase when the reactions drive the temperature up) causing a readjustment of the stellar structure. The star lands in the horizontal branch region of the HR diagram (Figs. 1 and 2), while more massive stars merely remain in the red giant region during helium burning. Both phenomena are

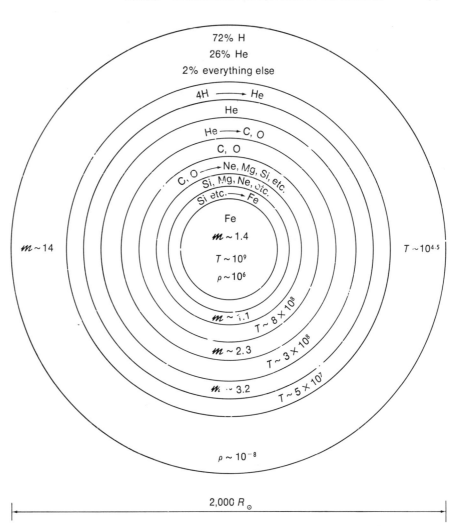

Figure 4. Interior structure of a massive (22 $M_\odot$) star shortly before it becomes a supernova. Only the primary constituents are shown, and those rather schematically. Masses of the various zones are in solar masses, temperatures in Kelvins, and densities in g cm^{-3}. Notice again that more than half the mass and virtually all of the volume is still occupied by the hydrogen-rich envelope, whose composition has been relatively unaffected by nuclear processing in the star.

seen—the lower-mass helium burners as a collection of stars strung out horizontally across the HR diagrams of old star clusters, in which stars ~1 $M_\odot$ are leaving the main sequence, and the high-mass helium burners as a clump on the red giant branch of the HR diagrams of young clusters, in which ~2–10 $M_\odot$ stars are now leaving the main sequence. The helium-burning phase is still shorter than the red giant phase, because the star remains quite bright and helium-burning produces fewer ergs per gram than hydrogen burning.

The products of helium burning are carbon and oxygen, in roughly equal amounts. This is clearly of some importance, since we and other terrestrial living creatures are in large measure made of them. The approximately equal production of

carbon and oxygen is the result of a delicate balance between the electromagnetic repulsion of the interacting helium atoms and the detailed nuclear structure of the products (Fowler et al. 1975) and, in the absence of some deeper understanding of the forces involved, must be regarded as an extremely fortunate coincidence from our point of view.

The exhaustion of helium at the center of the star, like that of hydrogen earlier, results in the core contracting to liberate gravitational potential energy and the outer layers again expanding. Helium soon begins to burn in a thin shell around the inert carbon-oxygen core (Figs. 3 and 4). During this double-shell-burning phase, low-mass stars again ascend to the red giant region of the HR diagram (where, in old clusters, we see them as a scattering of stars to the blue side of the normal red giant branch, the so-called asymptotic giant branch), more massive ones remaining as red supergiants. During this phase, convection can bring the products of nuclear reactions in the stars themselves up to their surfaces. Many asymptotic giant stars show excess carbon (and s-process products) in their atmospheres. In addition, in second and later generation stars, which had some heavy elements to begin with, several reactions involving He, C, N, and Ne liberate some neutrons. These are captured, primarily by iron and heavier nuclei, building up many (but not all) of the elements and isotopes between iron and lead (atomic numbers 27–82); by the so-called s process (the slow addition of neutrons, interspersed with beta decays).

From here on, the evolution of a star depends importantly on its mass and takes one course for stars $\lesssim 6\,\mathrm{M_\odot}$ and another for stars $\gtrsim 6\,\mathrm{M_\odot}$ with a possible narrow intermediate range in which carbon burning starts so explosively (again because the core is degenerate) that the star is completely disrupted. The mass at which this transition occurs has been discussed often and from many viewpoints (e.g. Woolf 1974).

All stars shed some material throughout their lives. We see a continuous solar wind of particles leaving our sun, and the rate of mass loss is observed to be much larger in some bright, red stars. Thus a star may reach the end of helium shell burning with only $\frac{1}{2}$–$\frac{3}{4}$ of its original mass (or even less, if it is in a binary system). But it is the original mass that counts, the center of the star continuing to evolve almost unaware of the surface losses. The lower mass stars, like our sun, never get hot enough for any further nuclear reactions to occur after helium burning. Rather, various instabilities in the shells gradually lift off the outer layers of the star, leaving behind the very hot, very dense core. The lost layers are heated and ionized by light from the remaining core and become responsible for one of the more frequently photographed phenomena of astronomy, the planetary nebulae (again see any elementary text for pictures). The name reflects their appearance in a small telescope, not any imagined connection with planets.

The remaining core has a density of about $10^6\,\mathrm{g\,cm^{-3}}$ and a radius of only about 10^4 km (the size of the earth). It is initially very hot even at the surface (10^{5-6} K) since it has just finished nuclear reactions, sometimes resulting in a detectable source of soft X rays (Hearn et al. 1976), but cools off in a matter of millions of years to some tens of thousands of degrees. The nebulae dissipate within 10^4 years (judging from their measured expansion rates), leaving behind faint (because they are tiny) stars, moving in the HR diagram from the region marked "nuclei of planetary nebulae" down toward the white dwarf region. The matter in these stellar cores is degenerate, thus they can neither radiate away the kinetic energy of the electrons nor contract any further. Degenerate matter is subject to the uncertainty principle; thus to compress it you must localize the particles further, raising the uncertainties in their momenta and, therefore, their kinetic energies. You must add energy to degenerate matter in order to compress it. Only the thermal energy of the nuclei is available and is gradually radiated away over billions of years, leaving the star as a gradually cooler and fainter white (and eventually yellow, red, and black) dwarf. This will be the final state of our sun. As seen from earth (which will, of course, be thoroughly frozen), it will not look much brighter than the full moon does now. The answer to Robert Frost's question therefore appears to be fire first and ice later.

Notice that the white dwarfs retain most or all of the carbon and oxygen produced in low-mass stars. Thus, we are going to have to follow the evolution of more massive stars to see the origins of the heavy elements found in the sun, the earth, and ourselves. In stars above about $6\,\mathrm{M_\odot}$, the carbon-oxygen core gets hot enough for further nuclear reactions to occur. The burning of carbon and oxygen produces relatively abundant elements like neon, magnesium, silicon, and sulfur. As one source of fuel is exhausted, further contraction of the core causes further heating until another fuel can be tapped. Finally, the burning of elements near silicon produces a stellar core which is mostly iron and nickel, the star as a whole rather resembling an onion (see Fig. 4). Energy is liberated in each of the nuclear reactions, though not as much as in hydrogen or helium

burning, but most of it is carried away by neutrinos (rather than photons), which do not contribute to keeping the star shining. These phases are, therefore, very short lived ($<10^4$ yr), and the HR diagram is no longer a particularly useful tool for following the evolution.

Once the star develops an iron core of about $1\,\mathrm{M_\odot}$, it is in serious trouble. The nucleus of iron is the most tightly bound of all atomic nuclei; hence no reaction involving it can liberate any energy. The core of the star just continues to contract and get hotter and denser until several processes begin to occur (more or less simultaneously) which drain energy away from the core, causing it to collapse suddenly and catastrophically. The two main processes are (1) the breaking up of Fe back into He nuclei by high energy gamma rays and (2) the reaction of protons with electrons to form neutrons and neutrinos. Both of these soak up energy, and the latter especially also reduces the effective sizes of the particles in the stellar core. The core therefore collapses until a new source of pressure (the neutrons becoming degenerate) stops it at a radius of about 10 km. The product is called a neutron star, and the process liberates some 10^{53} ergs of gravitational potential energy (comparable with the total radiation output of the star over its entire previous life).

This liberation of energy and its distribution through the star result in a variety of violent phenomena, including (probably), (1) a burst of neutrinos, (2) a burst of gravitons, (3) the rapid addition (r process) of neutrons to iron from the core to build heavy elements up to at least plutonium and maybe further, (4) some very high-temperature nuclear reactions in the intermediate layers which yield a variety of rare, intermediate-weight isotopes, (5) a rapid brightening of the outer layers of the star until it may outshine its entire galaxy for some weeks (we call this a supernova when we see it, which happens at a rate of once every few hundred years in the solar neighborhood, or about every 30 years in a large galaxy), (6) the acceleration of some material from the surface of the star to relativistic speeds, after which we call the particles cosmic rays, (7) the expulsion of one-to-many solar masses of material from the star at speeds of 10^{3-4} km/sec (we see such expanding clouds around old supernovae in the solar neighborhood), and (8) the storing of energy in rapid rotation of the neutron star, which then gradually feeds it out again over the next 10^{6-7} yr in forms which cause us to see it as a pulsar. The association of neutron star formation with supernovae was first suggested by Baade and Zwicky (1934) and confirmed by the detection of a pulsar in the Crab Nebula (Staelin and Reifenstein 1968), which is the gaseous remnant around the site where the Chinese reported seeing a supernova explosion in 1054 C.E. From our point of view, the most interesting thing a supernova does is to distribute the heavy elements made by nuclear reactions in the massive star back into the interstellar medium.

I have said nothing so far about the evolution of pairs of stars—binary systems—which constitute at least half the stars in the solar neighborhood. The presence of a close companion influences a star's life profoundly (see, e.g., Paczyński 1971 for a review of some of the effects) particularly in the later stages. It may even lead to an occasional massive star collapsing past the neutron star stage down inside its own Schwarzschild horizon to form a black hole. (Possibilities for black hole formation and their properties have been reviewed many times, e.g. Thorne 1974.) No one has yet followed a massive star in a close binary system all the way from the main sequence to supernova explosion, and the effects on nucleosynthesis are unknown but thought to be small. The core of the star should continue on its merry way even after the outer layers are stripped off and given to the companion star.

In the case of normal supernovae, we observe the shedding of the outer layers, and careful spectroscopic studies have found extra amounts of various heavy elements in them (Kirshner and Oke 1975; Peimbert and van den Bergh 1971; Peimbert 1971), representing the products of nuclear reactions in the stars. These then enrich the general interstellar medium, so that second and subsequent generations of stars that form will contain a component of elements besides hydrogen and helium (about 2% in our sun and 3–4% in stars now forming), in proportions which are reasonably well understood theoret-

ically (Trimble 1975). The presence of heavy elements in turn implies the possibility of planets like the earth.

Planets and life

If we are interested in life in the universe, one of the first things we will ask a star is whether or not it has planets, preferably with solid surfaces and preferably located at such a distance from the parent star that water is a liquid at least some times and in some places. Unfortunately, direct observation cannot answer this question for any star but the Sun. In the past, theories of the formation of the solar system often involved events, like the close passage of two stars, which would be so rare that we could easily be the only planetary system in the Galaxy. Recent theoretical studies (see, e.g., *Scientific American*, Sept. 1975) have, however, come around to a view in which leaving behind some material in planets is a natural part of condensing a star out of the interstellar medium. On this basis, planets should be quite common. There is some indirect supporting evidence (van de Kamp 1975). Careful study of the motions of several nearby stars across the plane of the sky has revealed evidence of a roughly sinusoidal wiggle, typically with a period of years and an amplitude of a small fraction of a second of arc, such as would be produced by the star and a planet with a mass 1–10 times that of Jupiter orbiting around their mutual center of mass. And if the presence of Jupiters can be taken to imply the presence of Earths, then many or most stars (at least those of roughly solar type) may well have planets with solid surfaces.

The nature of the atmospheres of these planets will also be important for the possibility of life. The earth's orginal atmosphere was not its present oxidizing one, but rather a reducing one (made up of things like CO, CO_2, NH_3, CH_4, H_2, and H_2O at various stages) according to several lines of evidence. These include the kinds of gases found on other planets and released by volcanoes, the reduced nature of the minerals deposited until about a billion years ago, and the necessary conditions for the origin of life itself.

The constituents of this primitive atmosphere are interesting from several points of view. First, they are

made of the commonest chemically reacting elements (H, O, C, and N in order of abundance). Second, they contain the elements that occur most abundantly in terrestrial living creatures. Third, they are more or less the things you need in order to do the kind of experiment suggested by Urey and carried out by Miller (1953; and many others since) in which substances like methane, ammonia, formaldehyde, and carbon dioxide are dissolved in water and energy added (in the form of ultraviolet radiation, electric currents, mechanical shocks, etc.). Under these circumstances, many chemical reactions occur, whose products (if the right inorganic reactants are made available) include a wide variety of simple organic molecules—amino acids, sugars, bases, phosphates—and some of their polymers (Stephen-Sherwood and Ord 1973). And fourth, these and a wide variety of other simple organic and inorganic molecules are now known to be widely distributed through the interstellar medium (Herbst and Klemperer 1976). Molecules detected include interstellar CO, CN, NH_3, H_2S, H_2O, formaldehyde and hydrocyanic acid, formic acid, methyl and ethyl alcohol, and HC_5N, which has the same molecular weight as glycine, the simplest amino acid, and about 30 others. These molecules and the presence of amino acids (in a racemic mixture of left- and right-handed forms, which leads us to believe they are nonbiological) in several meteorites (Jungclaus et al. 1976, and references therein) seem to indicate that both the raw materials and prebiological organic molecules will be widely distributed in suitable environments, like the surfaces of planets.

No laboratory experiment has yet taken the complete step from the raw materials to a self-reproducing molecule. Perhaps this is just as well; the first biochemist who makes a self-reproducing molecule may find out that it eats biochemists. Ways to approach self-reproduction have been discussed by Calvin (1975) in these pages. In any case we (or at least the infidels among us) do know that the step has been taken at least once, for here we are; and by the time you read this, laboratory experiments or the Viking Mars lander may have shown that it can be taken more than once. In the meantime, we can only say that there does not seem to be any reason to suppose that the earth is special or that chemical life is not likely to have appeared many times over the history of the galaxy. The evolution of terrestrial life and its effects on the atmosphere and other components of the environment have been reviewed in this journal by Cloud (1974). Suffice it to say that, once you have self-reproduction, then the entire mechanism of Darwinian selection and mutations takes over, and the evolution from a primeval slime mold to a local politician seems practically inevitable. We therefore expect that life will eventually become intelligent life in many cases.

But this, in turn, raises another question; if there is this sort of inevitability in the progress from the hot, dense early universe to the formation of galaxies, stars, planets, and life, then there should be lots of other civilizations floating around space. Where are they? Of course, there are people who think they know the answer to this question—large silver ships drop out of the sky, land in their gardens, and take them for rides. These people generally have other problems as well. Others have tackled the question in a slightly different way. They start with the total number of stars in the galaxy, eliminate those that are too short-lived, too cool, or too close to another star to provide comfortable environments, guesstimate factors for the probabilities of planets, the origin of life, and the development of civilization, by which they mean not the quartets of Mozart or the dialogues of Plato, but rather radio astronomy—that is, the ability to communicate across interstellar distances. Many such efforts (Shklovskii and Sagan 1966; Ponnamperuma and Cameron 1974; Sagan 1973; and references therein) end up with an estimate that a civilization may have appeared every $10^{2 \, {}^{+2}_{-1}}$ years over the history of the Galaxy. A carbon-based biochemistry is assumed, but there is no presumption (and little likelihood!) that the products would look like us. Other kinds of biochemistry may possibly increase the numbers of civilizations above these estimates.

The question of whether we have millions of neighbors, thousands, or none then reduces to the question of how long a civilization lasts. Possible answers range from 10^2 years (the time over which we seem capable of advancing(?) from the first radio broadcasts to self-destruction) to 10^7 years (the time scale for biological evolution from one species to another among the higher mammals) on up to 10^{10} years (the main sequence lifetime of the host star). These answers correspond to our having no company at all, other civilizations within a thousand parsecs or so, or near neighbors within 100 or even 10 parsecs, corresponding to round-trip light travel times comparable with human lifetimes.

Some effort has been made to test this last, most optimistic, hypothesis by pointing large radio telescopes in the directions of nearby, solar-type stars and listening for a while (see Sagan and Drake 1975 for some of the details). No positive results have been reported. This is not particularly discouraging, because the sensitivity was such that nothing could have been seen unless a comparably powerful telescope, operating in a radar, broadcasting, mode, was pointed directly at us. The range of such experiments will be considerably extended when the Very Large Array, presently under construction in New Mexico, comes into operation in the 1980s. We could do still better with present technology, to the point of being able to detect even a relatively weak radio planet like the earth out to a distance of several hundred parsecs (though the cost would be comparable with the cost of the Apollo program or a very small war). This is, therefore, perhaps a good time to start thinking about the consequences for society (including zealous companionship in scientific research!) of the possible discovery that ours is merely one of many civilizations, and (given the time scales involved, inevitably) rather a backward one at that. The sociological implications may be worth worrying about.

Universal constants and their importance

Whether we are alone or one of thousands or millions of civilizations in the Galaxy, we seem to be a fairly natural product of the total history of the Universe. It is, therefore, of some interest to ask whether any universe would have done, or is ours special? This is quite different from asking whether the earth and sun are special, because there are lots of stars and one can do some kind of statistics. The Universe, on the other hand, is by definition unique in our experience. One can, however, characterize the Universe in terms of a fairly small number of properties, which lead to the history we have discussed, and ask what would have happened if one or more of these properties had been different from what we see it to be. We will require eight parameters—four small-scale ones and four large-scale ones (Table 2). The four small-scale properties are the four forces of nature, the four ways in which one bit of matter or radiation can interact with another. Understanding the internal structure of the so-called elementary particles (protons, neutrons, mesons, etc.) may require a fifth force, or even an entirely different way of looking at the problem (Heisenberg 1976), but we cannot say anything very concrete about its size or possible variations at the moment. These four forces are the gravitational, electromagnetic, nuclear (or strong), and weak interactions, in order of decreasing familiarity. Table 2 lists them in order from strongest to weakest.

Most of the dire consequences of changing the forces mentioned in Table 2 require changes of an order of magnitude or more and thus do not rule out a factor of two decrease in G over the past 10 billion years (Van Flandern 1976), though this can probably be ruled out in other ways (Roxburgh 1976). But changing the electromagnetic force (the charge on the electron) by even a factor of three would mean that water could not be a liquid at any temperature, while very small changes in the ratio of the electromagnetic to nuclear force would cause helium burning to produce either all carbon or all oxygen rather than a mix of the two.

Similarly, looking at the Universe on the largest scale, we find four important parameters. Two of these are the numbers that told us which of the Friedmann models we live in. They are H_0 and ρ_0 and measure the age and average density of the Universe. A third is the entropy of the Universe, which can also be expressed as the ratio of the number density of photons to the number density of heavy particles (protons and neutrons). This defines the present temperature (which we measure with some precision) and, therefore, the temperature and density at which helium was formed in the Big Bang. Finally, we have the homogeneity and isotropy, which we originally put in as one of our cosmological assumptions, but which are also observed to be approximately true and seem to be necessary for galaxy formation (Collins and Hawking 1973). The changes in these properties required to produce the dire consequences are often several orders of magnitude, but the constraints are still nontrivial, given the very wide range of numbers involved. Efforts to avoid one problem by changing several of the constants at once generally produce some other problem. Thus we apparently live in a rather delicately balanced universe, from the point of view of hospitality to chemical life.

Implications

It seems, in other words, that the Universe must be more or less the way it is just because we are here. This anthropic principle ("cogito ergo mundus talis est") has been further explored by Carter (1974). It is of some interest to try to understand what it might mean that our Universe should fall within a rather narrow range, favorable to life, for each of several parameters. It may just mean that God (or the Initial Conditions, depending on your point of view) has been very careful. This is a possible answer—it may even be the right one—but it is not a scientific answer,

at least in the narrow sense that it does not lead one to make any further observations, do any further experiments, or carry out any further calculations.

There are, of course, other possibilities. It could be, for instance, that ours is merely one of many universes, and that it is only those very few with particularly favorable properties that ever develop living creatures who ask these strange questions. There are two senses in which this could be so. If our Universe is, in fact (contrary to the preponderance of the evidence, but by no means impossibly) of the sort that will eventually turn around and recontract, one might imagine a (finite or infinite) series of cycles of expansion and contraction, each constituting a separate universe and each having its own values of the fundamental constants. This violates our initial assumption that general relativity is the right theory of gravity, because, within its framework, when you once achieve singular conditions (like infinite density) out of nonsingular ones (i.e. after the first recontraction), you can never get out again (Hawking and Ellis 1973). Perhaps this should not worry us too much, though, because general relativity is a classical (non-quantum) theory, and at sufficiently high densities, quantum mechanical effects will become important (Zeldovich 1971) and must profoundly change

the nature of gravitation (Misner et al. 1973).

The alternative, if our Universe is an infinite, ever-expanding one, is that a number of four-dimensional space-times (universes) might be imbedded in five (or higher) dimensional space, existing simultaneously, from the point of view of a five (or higher) dimensional observer. There is no easy way to picture five (or higher) dimensional space with a three (or lower) dimensional mind, and it is probably hopeless to try.

A third possibility is that, at some time in the distant future, when we have understood "all" the physics of the universe, it will be obvious that the various quantities must have the relationships they do. There is some hint that this might be so in the "large numbers." These are several simple combinations of the constants in dimensionless form that, depending on just how you write them, are all of order unity, 10^{40} or 10^{80}. For instance, the radius of the observable universe ($c/H_0 \sim 10^{28}$ cm) divided by the classical electron radius ($e^2/m_e c^2 \sim 10^{-12}$ cm) is about 10^{40}. And the ratio of the electromagnetic to gravitational forces between an electron and a proton ($e^2/Gm_e m_p$) is 2×10^{39}, and the number of particles in the observable universe, N_0, is about 10^{80}, i.e. $(10^{40})^2$. Alternatively, these can be written as $8\pi G\rho_0/3H_0^2 \sim 1$ (this is

Table 2. Universal forces and properties and consequences of varying their values

Force	Phenomena controlled	Consequences of lowering	Consequences of raising
Nuclear or strong	Structure and reactions of atomic nuclei	Early universe converts all matter to heavy elements; no source of energy for stars.	No nuclear reactions at all or none past helium; no heavy elements made, so no chemistry.
Electromagnetic	Structure and interactions of atoms and molecules	Electrons not bound to atoms; no chemical reactions possible.	Electrons inside nuclei; no chemical reactions possible.
Weak	$\nu_e + n \leftrightarrow p + e^-$ and other beta decays	No hydrogen burning possible; no source of heat or heavy elements.	Early universe converts all matter to helium. No energy sources for stars.
Gravitational	Structure and dynamics of planets, stars, and galaxies	Stars don't get hot enough for nuclear reactions to occur.	Nuclear reactions so rapid that star lifetimes very short.

Property of Universe	Value	Consequences of lowering	Consequences of raising
Rate of expansion	$H_0 \sim 55$ km/sec/Mpc	Matter all comes out of early universe in dense configurations.	Galaxies can't form; matter ends up spread out uniformly.
Average density now	$\rho_0 = 10^{-32} - 10^{-30}$ g cm^{-3}	Galaxies can't form; atoms very lonely.	Early universe turns all matter into heavy elements; no energy sources.
Entropy or temperature	$\eta = \eta_\gamma/\eta_{baryon} \sim 10^{-9}$ or $T = 2.7$ K	Early universe turns all matter into heavy elements; no stellar energy source.	Galaxies can't form due to radiation pressure.
Isotropy and homogeneity	$\Delta T/T \sim 10^{-3}$	Galaxies can't form in anisotropic expansion.	

equivalent to saying that the universe is not far from the boundary between ever-expanding and reconstracting models) and $(c/H_0)/(e^2/m_e c^2) \sim N_0^{1/2} \sim 10^{40}$. The largest number of constants appears in $j(j + 1)\hbar c/e^2 = ln(j(j + 1)\hbar c/Gm^2)$ (J. W. Follin, Jr., pers. comm. 1976), where, in all of these expressions, H_0 is the present value of Hubble's constant, m_e, e, and j are the rest mass, charge, and spin of the electron ($j = \frac{1}{2}$), G is the constant of gravity, c is the speed of light, ρ_0 is the present average density of the universe, $\hbar = h/2\pi$ and h is Planck's constant, and $m = m_e m_\mu/(m_e + m_\mu)$ and m_μ is the rest mass of the muon.

There are other numbers (involving stars and such) that come out $\sim N_0^{1/4}$ and $N_0^{3/4}$. A variety of interpretations of these large numbers have appeared from the time of Dirac (1938) and Eddington (1946) to the present (Rees 1972; Wheeler 1974). Some of the numbers may only express conditions for the formation of stable stars with reasonable lifetimes. Others, especially $G\rho_0/H_2^0 \sim 1$ (which implies that G must change if it is to be true for all time) and its variants, have been made the cornerstone of whole new, nonrelativistic cosmologies (typically not in very good accord with observations). Perhaps when we have learned enough physics, we will understand that these relationships must hold and why.

Finally, it may be that the complexity and nonpredictability which we call intelligence (in ourselves, and orneriness in our adolescent children) is an inevitable result of just having enough particles interacting. By way of analogy (R. P. Feynman, pers. comm. 1975), consider a water molecule, whose structure, energy levels, and so forth can be calculated with some precision by the methods of quantum mechanics. But nothing in that calculation would ever lead us to predict waterfalls. The waterfall is a result of very many particles interacting in ways we cannot, in practice, predict or calculate. Similarly, perhaps whenever there are enough particles interacting, no matter what laws or forces govern their behavior, a sort of complexity results which we would acknowledge as a fellow intelligence. One might, therefore, imagine a universe in which the early, hot state had turned everything into neutron star material, and the ex-

ceedingly compact creatures living there would contemplate a universe like ours and claim (in very low voices, presumably) that it couldn't possibly have any intelligent life in it because the density was too low.

When this paper is delivered as a talk to Sigma Xi chapters, I generally stop at this point and ask if there are any questions, comments, rude remarks, or other feedback. I have tried to work some of the commonest of these into the text, but the reader who would like to know more about "the Universe" will find a variety of other interesting questions—and a few answers—in the references cited.

References

Audouze, J., and B. M. Tinsley. 1976. *Ann. Rev. Astron. Astrophys.* 14: 43–80.

Baade, W., and F. Zwicky. 1934. *Phys. Rev.* 45: 138 and *Proc. Nat. Acad. Sci.* 20: 254.

Bahcall, J., and R. D. Davis. 1976. *Science* 191: 264.

Calvin, M. 1975. *Am. Sci.* 63: 169.

Carter, B. 1974. In *Confrontation of Cosmological Theory with Observational Data*, IAU Symposium No. 63, ed. M. S. Longair, p. 291. Dordrecht: D. Reidel.

Cloud, P. 1974. *Am. Sci.* 62: 54.

Collins, C. B., and S. W. Hawking. 1973. *Astrophys. J.* 180: 317.

Curtis, H. D. 1919. *J. Wash. Acad. Sci.* 9: 212.

Dirac, P. A. M. 1938. *Proc. Roy. Soc.* A 165: 199.

Eddington, A. S. 1926. *The Internal Constitution of the Stars.* Cambridge University Press.

———.1946. *Fundamental Theory.* Cambridge University Press.

Fowler, W. A., G. Caughlan, and B. A. Zimmerman. 1975. *Ann. Rev. Astron. Astrophys.* 13: 69.

Gott, J. R., J. E. Gunn, D. Schramm, and B. M. Tinsley. 1974. *Astrophys. J.* 194: 543.

Hauser, M. G., and P. J. E. Peebles. 1973. *Astrophys. J.* 185: 757.

Hawking, S. W., and G. F. R. Ellis. 1973. *The Large Scale Structure of Space-Time.* Cambridge University Press.

Hearn, D. R., J. A. Richardson, H. V. D. Bradt, G. W. Clark, W. H. G. Lewin, W. F. Meyer, J. E. McClintock, F. A. Primini, and S. A. Rappaport. 1976. *Astrophys. J.* 203: L21.

Heisenberg, W. 1976. *Phys. Today,* March, p. 32.

Herbst, E., and W. Klemperer. 1976. *Phys. Today,* June, p. 32.

Hoyle, F., and R. A. Lyttleton. 1942. *Mon. Not. Roy. Astron. Soc.* 102: 218.

———.1949. *Mon. Not. Roy. Astron. Soc.* 109: 614.

Hubble, E. 1925. *Publ. Amer. Astron. Soc.* 5: 261.

Jones, B. 1976. *Rev. Mod. Phys.* 48: 107.

Jungclaus, G., J. R. Cronin, C. B. Moore, and G. V. Yuen. 1976. *Nature* 261: 126.

van de Kamp, P. 1975. *Ann. Rev. Astron. Astrophys.* 13: 295.

Kellogg, E., S. Murray, R. Giaconni, H. Tannenbaum, and H. Gursky. 1973. *Astrophys. J.* 185: L13.

Kirshner, R. P., and J. B. Oke. 1975. *Astrophys. J.* 200: 574.

Miller, S. L. 1953. *Science* 117: 528.

Misner, C. W., K. S. Thorne, and J. A. Wheeler. 1973. *Gravitation.* Ch. 44. W. H. Freeman.

Ostriker, J. P., P. J. E. Peebles, and A. Yahil. 1974. *Astrophys. J.* 193: 4.

Paczyński, B. 1971. *Ann. Rev. Astron. Astrophys.* 9: 183.

Peebles, P. J. E. 1971. *Physical Cosmology.* Princeton University Observatory.

Peimbert, M. 1971. *Astrophys. J.* 170: 261.

Peimbert, M., and S. van den Bergh. 1971. *Astrophys. J.* 167: 223.

Ponnamperuma, C., and A. G. W. Cameron. 1974. *Interstellar Communication.* Houghton Mifflin.

Rees, M. J. 1972. *Comm. Astrophys. Sp. Phys.* 4: 179.

Roxburgh, I. W. 1976. *Nature* 261: 301.

Sagan, C. 1973. *Communication with Extraterrestrial Intelligence.*

Sagan, C., and F. D. Drake. 1975. *Sci. Amer.* 232 (5): 80.

Sandage, A. R., and G. Tammann. 1975. *Astrophys. J.* 202: 583.

Shklovskii, I. S., and C. Sagan. 1966. *Intelligent Life in the Universe.* San Francisco: Holden Day.

Staelin, D. H., and E. C. Reifenstein. 1968. *Science* 162: 1481.

Stephen-Sherwood, E., and J. Oró. 1973. *Space Life Sci.* 4: 5.

Tifft, W. G., and S. A. Gregory. 1976. *Astrophys. J.* 205: 696.

Thorne, K. S. 1974. *Sci. Am.* 231 (6): 32.

Trimble, V. 1975. *Rev. Mod. Phys.* 47: 877.

Van Flandern, T. G. 1976. *Sci. Am.* 234 (2): 44.

Wagoner, R., W. A. Fowler, and F. Hoyle. 1967. *Astrophys. J.* 148: 3.

Wheeler, J. A. 1974. *Am. Sci.* 62: 683.

Woolf, N. J. 1974. In *Late Stages of Stellar Evolution,* IAU Symposium No. 66, p. 143. Dordrecht: D. Reidel.

Zeldovich, Ya. B. 1971. *Comm. Astrophys. Sp. Phys.* 3: 179.

Occurrence of Life in the Universe

Su-Shu Huang

The necessary requirements for the occurrence of life, especially in an advanced form, are here discussed in the light of our present knowledge of the stars and their evolution.

Man has overcome the confines of the earth's gravitational field. In a few years he will learn with certainty about the existence or non-existence of life on the other planets in the solar system. However, because of the vast distances between the stars, he is still a long way from being able to detect empirically the occurrence of life outside the solar system. This does not prevent him from attacking the problem in a "theoretical" way with our present knowledge of astrophysics, biology, and chemistry as a guide.

In order to study the problem of life in the astronomical universe, we have to consider two independent questions: (1) the possibility of formation of planets around stars, and (2) the possibility of occurrence of life on such planets. The first question is connected with the mode of star formation and will not be discussed here. Suffice it to say that, according to current understanding, planet formation around a star is by no means rare [1]. The second problem is of "astro-biological" interest and is the main concern of this note.

Granted that all kinds of stars have an equal chance of possessing planets, we ask: Is there any way of knowing which kinds of stars favor the existence of life on their planets? This question can be reasonably answered, we find, with our present knowledge.

It is well-known that most atoms do not possess the property of coagulating into large molecules which are the building blocks of living matter. It is principally carbon atoms that have this property. So, if life exists on other worlds, it must depend upon this remarkable property of carbon atoms. In other words, the chemistry of organic

A physicist trained in China, Su-Shu Huang changed his interest to astrophysics at the University of California on his arrival in this country as a Guggenheim Fellow in 1947. The author here examines the general conditions governing the occurrence of life in the universe without going into the problem of star formation. Life in the universe, the author finds, is a direct consequence of the mode of star formation. Address: Berkeley Astronomical Department, University of California, Berkeley, CA 94720.

substances should be universal, although actual morphological forms of life may be different in different places. Therefore, we can only expect life to exist on planets where the temperature lies within certain limits. It is from this fact that we draw the following conclusions.

Time-scales of biological and stellar evolution

Let us first define t_b as the time of biological evolution which brings the chaotic inanimate atoms into an orderly form of intelligent life. In the case of biological evolution on the earth, t_b is of the order of a few billion years.

Next, we consider the evolution of stars which are but a large mass of gaseous particles held together by their mutual gravitational attraction. The luminosity of a star (i.e., its rate of energy output in the form of radiation) is maintained either by thermonuclear reactions of its constituent particles or by gravitational contraction. In the very beginning, when the star is not hot enough to start thermonuclear reactions, its luminosity is derived from gravitational contraction. This is the first phase of stellar evolution. When the temperature of the star rises as a result of gravitational contraction, thermonuclear reactions, converting hydrogen into helium, take place in the core of the star. From then on the energy output of the star is exactly balanced by the energy liberated in these reactions. Therefore, the star maintains for a long time its constant energy output and constant radius until the hydrogen in the core is exhausted. We call this a main sequence star. The luminosity and effective temperature of the star on the main sequence depend upon its mass. A star remains in the main-sequence stage for a much longer time than in the early contracting stage. After the exhaustion of hydrogen in the core, the star moves away from the main sequence and again evolves rapidly with changing luminosity, forming the third stage of evolution. Now, in order to have life on one of its planets, the central star must be on the main sequence; otherwise, the rapidly evolving stars of changing luminosity will destroy life. Thus, if we define the time of stellar evolution, t_s, as the time that a star remains on the main sequence, the first condition for expecting life in an advanced form on any planet of a star is

$$t_s \geq t_b$$

Otherwise, life will be destroyed (by evolution of the star) before it reaches a climax. Now, t_s can be computed for main-sequence stars of different masses. It varies from 10^7 years for early O-type stars to more than 10^{11} years for M-type stars [1].

We now ask what is the value of t_b. The earth is about 4.5×10^9 years old, while paleontological evidence shows the existence of life on the earth over an interval of 1×10^9 years. We may tentatively take $t_b = 3 \times 10^9$ years. If so, the time-scales, t_s, of all the early-type stars (O, B, A) are less than t_b. Hence, we would not expect life of an advanced order to develop on planets associated with these kinds of stars.

One may question the wisdom of using 3×10^9 years for t_b in general, because this value is only based on a single case—that of the earth. However, we can argue that the time-scale, t_b, cannot be greatly shortened because, according to current ideas in biology, the natural selection and evolution of organisms is a result of mutations which are of a random nature and are therefore slow. Here one may ask: because of intense ultraviolet radiation, and possibly also of high-energy corpuscles from early-type stars, should one expect a higher mutation rate for the living organisms on their planets than on the earth and, thereby, an accelerated evolution of organisms? In order to answer this question we must point out that most mutations are harmful. It follows that, in order to work for natural selection, mutations must be rare. If mutations were too frequent, there would be a considerable chance of different mutations occurring in one and the same individual. The injurious ones would dominate the advantageous ones because the former occur more frequently. Thus, the species would perish instead of being improved by an over-activity of the mutation process [2]. Consequently, the value of t_b cannot be greatly shortened. In this way we exclude O, B, A, and maybe early F stars as the energy sources which could make life in its advanced forms flourish on their planets. For O and B stars, t_s is of the order of 10^7–10^8 years; we therefore wonder whether life even in its most primitive forms is possible on their planets.

The habitable zone of a star

Let the luminosity of a star be L. This energy supplies necessary heat to the living organism. The heat received by the living beings on a planet must be neither too large nor too small. Otherwise they will perish. In consequence, we can define a habitable zone around each star such that the rate of energy received per unit area facing the star, i.e., $L/(4\pi R^2)$ where R is the distance from the star, lies between an upper limit

$$\epsilon_1 = \frac{L}{4\pi R_1{}^2}$$

and a lower limit

$$\epsilon_2 = \frac{L}{4\pi R_2{}^2}$$

where ϵ_1 and ϵ_2 are constants independent of the nature of the star itself, while R_1 and R_2 are defined by these two equations. A planet is habitable if its distance R from the star satisfies the condition that

$$R_1 < R < R_2.$$

If all planets of a star are formed nearly in one fundamental plane (like those in the solar system) the habitable zone is a concentric circular ring having an area, A, equal to

$$A = \pi(R_2{}^2 - R_1{}^2) = \frac{L}{4}\left(\frac{1}{\epsilon_2} - \frac{1}{\epsilon_1}\right)$$

which is proportional to the luminosity, L, of the star. If the planets could be formed throughout the entire space surrounding the star, the habitable zone would consist of a concentric spherical shell having a volume, V, equal to

$$V = \frac{4}{3}\pi(R_2{}^3 - R_1{}^3) = \frac{L^{3/2}}{6\pi^{1/2}}\left(\frac{1}{\epsilon_2{}^{3/2}} - \frac{1}{\epsilon_1{}^{3/2}}\right)$$

which is proportional to $L^{3/2}$. In any case, the habitable zone increases with the luminosity of the star. The more luminous the star, the greater is the probability that a planet falls into its habitable zone and can thereby support life. Therefore, stars on the lower part of the main sequence (like late-K and M type dwarfs) have a small chance of possessing planets within their habitable zones. For example, a K-5 main sequence star is only $1/10$ as luminous as the sun. Thus, its habitable zone is only $1/10$ as large as the sun's if all planets of the star are confined to a plane. The habitable zone is only $1/10^{3/2}$ as large if its planets are not confined to a plane. Since the luminosity decreases rapidly when we move from K-5 down the main sequence, the habitable zones of late-K and M stars become very small indeed.

The chance of finding life on planets associated with intrinsically faint stars will be even smaller if we consider the fact that planetary orbits are usually not circular but are elliptical. For, if only a small part of its eccentric orbit falls outside the habitable zone, the planet will become uninhabitable.

Of course, we cannot rule out completely the possibility of finding life on planets revolving around faint main-sequence stars. But, the probability of finding life there is extremely small. Since there are a great number of faint dwarf stars in our galaxy and perhaps also in other galaxies, chances are that a few may possess planets with living beings. If the latter do exist, they have a long time for their biological evolution because the faint dwarfs remain for an extremely long time on the main sequence. Consequently, living organisms there can evolve to a very high form.

From the arguments given here and in the preceding section we conclude that, if we want to look for life of a high order outside the solar system, the likely places to look are in the main-sequence F (preferably late-F), G, and K (preferably early-K) stars. If they should possess planets, there is a good chance that life flourishes on some of them. It is interesting to note that our sun, which is a main-sequence G2 star, does support life abundantly on at least one of its planets, fully in agreement with the present conclusion.

Dynamical and other considerations

Since we know that about one half of the stars in the solar neighborhood are binary and multiple systems, it would be of interest to examine whether such combinations of stars favor the occurrence of life on their planets. The answer is clear-cut. They have damaging effects on the

Table I. Near-By Stars on Whose Planets (if any) Living Beings Have a Better-Than-Average Chance to Develop.

Star	Distance (in light years)	Spectral Type	Luminosity, L
Sun		G2	1.00
ε Eridani	10.8	K2	0.34
τ Ceti	11.8	G4	0.38

development of life for two reasons, one dynamical, the other physical.

Dynamically, a planet around one component of a binary or multiple system is always perturbed by its companions. Its chance of remaining within the habitable zone of its central star on a time scale greater than t_b, and therefore of supporting life on it, is greatly reduced. For binary systems of large separations (visual binaries), quasi-stable orbits in the immediate neighborhood of each component are possible. This can be seen, for example, from the fact that the orbit of the moon around the earth is reasonably stable in spite of the perturbation by the sun. However, the actual time-scale during which a planet remains within the habitable zone of its central star is a problem involving three bodies and is difficult to estimate. For close binaries (eclipsing and spectroscopic), the habitable zone, which must surround both components, most likely does not contain dynamically stable orbits unless the mass of one component is very much greater than that of the other or their separation is very small.

Physically if either one of the components in a close binary system is of early type (O, B, A), no life is expected for the same reason as given in the earlier discussion, no matter what kind of star the other component is. For binaries of large separation, this restriction does not apply and we can consider both components separately. However, even in binaries of large separation, the presence of a white dwarf as a component may make the system an unlikely abode for living beings. If, as astronomers nowadays suspect, a star has to undergo some catastrophic change in luminosity (like in a nova explosion, etc.) before it finally degenerates into a white dwarf, such changes in luminosity, possibly with accompanying high-energy corpuscles, would destroy all life that is nearby. Then another time-interval of t_b has to elapse before life may appear again on the planets of the system where the catastrophic change took place.

Finally, one may ask whether the orbit of a star around the galactic center has any bearing on the life-supporting property of its planets. The sun moves with most of its neighbors approximately in a circular orbit around the galactic center, so that there is very little change in its immediate surroundings in each revolution (of about 2×10^8 years). If a star (like a high-velocity star) moves in an eccentric orbit in the galaxy, has the change in its environment any harmful effect on the maintenance of life? To attempt to answer this question would be speculative and is therefore outside the scope of the present note.

Examinination of nearby stars

The conclusions already reached are valid wherever the chemical principle, or more fundamentally the quantum mechanics, applies. That the latter applies everywhere in the universe can be seen from the spectral lines the distant galaxies emit, for the spectral lines are predicted by quantum mechanics. In fact, both the spectral lines emitted by atoms and the formation of molecules are quantum mechanical phenomena of the same valency electrons. Thus, the previous discussions can give us an upper limit to the percentage of stars with which living beings may be associated. Because most stars in space are M-dwarfs, many are close binaries, and a large number are not main-sequence stars, we set the upper limit at 5 per cent for stars near the solar neighborhood. The exact percentage of stars that acutally support life depends upon the mode of star and planet formation and cannot be estimated objectively.

Before concluding this note let us examine the life-supporting ability of the nearby stars. Within 5 parsecs (about 16.7 light years) of the solar system, there are 42 stars (including the sun) [3], of which most are faint M-dwarfs and late-K stars. Many are binary and multiple systems, some with white dwarf companions. If we further exclude from consideration the high velocity stars which are only temporary visitors in the solar neighborhood, there are left only three stars (listed in Table I) on whose planets life may be found. Among these three, the sun has the highest luminosity and therefore has the best chance for supporting life on its planets. If both ε Eridani and τ Ceti are truly single stars, there is also a good chance of finding their planets (if any) abundant with life. On several previous occasions Struve has mentioned that τ Ceti is one of the stars in the immediate neighborhood of the solar system that may possess planets inhabited by living beings because of its close resemblance to the sun.

Finally, we inquire into the problem of life on planets (if any) of our nearest neighbor, α Centauri. This is a triple system, the two more massive components (G4, K1) of which revolve around each other in an orbit with a semi-major axis of only about 20 astronomical units. The eccentricity of this binary orbit is quite large. Thus, it is difficult to imagine that its planets can remain in the habitable zone for a long time. Consequently we do not expect any well-evolved life on any planet of our nearest neighbor in space.

Acknowledgment: I thank Professor O. Struve for valuable discussions on this interesting subject of life in the astronomical universe.

References

1. O. Struve, Stellar Evolution, Princeton University Press, 1950.

2. E. Schrödinger, What Is Life?, Macmillan Co., 1941.

3. P. van de Kamp, *Sky and Telescope, 14,* 498, 1955.

William G. Pollard

The Prevalence of Earthlike Planets

Although a rough estimate can be made of how many Earthlike planets might exist in the universe, the possibility that life has actually evolved on them cannot be calculated even in principle

One of the most significant results of the space program of the United States has been the new perspective on Earth provided for all mankind by the magnificent color pictures of the planet brought back from the Apollo missions. From space, where Earth can be seen as one among many astronomical objects, nothing else can compare with its sparkling beauty. By contrast, what the space program has shown us of Mars, Venus, and the moon serves to heighten our appreciation for our abode in the universe on this fragile planet.

It is almost certain that no other planet in our solar system now supports the phenomenon of life. The question still remains, however, as a persistent field of speculation, as to how many other stars in the galaxy or the universe as a whole may support a planet like Earth. In science fiction and fantasies like *Star Wars,* such stars are numerous. Even in the scientific community, a blue-ribbon panel established by the National Aeronautics and Space Administration has designed a major international program to search for extraterrestrial intelligence (SETI) (Morrison et al. 1977), and a branch of

Dr. Pollard, a theoretical physicist, was Executive Director of Oak Ridge Associated Universities (formerly Oak Ridge Institute of Nuclear Studies) from its incorporation in 1946 until August 1974. Since his retirement in 1976 he has served as a consultant to the ORAU Institute for Energy Analysis on questions relating to solar and nuclear energy, biomass fuels, and industrial conservation. Much of the information for the present study comes from the recent (1978) collection Protostars and Planets, *edited by T. Gehrels. Address: 191 Outer Drive, Oak Ridge, TN 37830.*

science (so far without content) called exobiology has been designated and widely accepted. In view of such great interest it seems desirable to seek an estimate of the fraction of stars that might have associated with them an Earthlike planet and the probability that such a planet would be capable of supporting a prolonged evolutionary history of living organisms. That is the purpose of the study described in this article.

The most primitive bodies in the solar system are the carbonaceous and ordinary chondritic (stony) meteorites. Considerations based on the abundance of argon-36 and argon-40 in the atmosphere of Venus and on the ratio of potassium to uranium in Earth and the moon suggest that both Earth and Venus have a composition similar to these meteorites (P. R. Bell, pers. comm.). The surface material on the moon and Mars is quite different, but their bulk composition could also be chondritic. In seeking the probability of Earthlike planets associated with other stars, we shall look for those that might be expected to have a composition similar to the carbonaceous chondrites that are found in an inner region like that extending out to the asteroid belt in the solar system. This assumption serves two essential purposes for this study.

First, a planet with a composition similar to that of the carbonaceous chondrites would have the possibility of having large land masses similar to Earth's continents. Only when such land masses have been invaded by macroscopic multicellular life forms will we consider a planet to be Earthlike. A planet possessing only microscopic unicellular organisms in its hydrosphere, although intensely interesting biologically, is not con-

sidered in this study to be sufficiently Earthlike to qualify.

The second important consequence of this composition is the presence in carbonaceous chondrites of a variety of reduced carbon and nitrogen compounds such as those found in interstellar gas and dust clouds. A planet formed from a coalescence of such bodies would have stored within it the materials essential for the subsequent production of molecular species required for the evolution of life.

Even more essential than Earthlike land masses is the presence of sizable bodies of liquid water throughout the evolutionary history of the planet. A full evolutionary development of complex organelles and organisms is not conceivable apart from an ample continuous marine environment. Moreover, cosmic abundances would make CO_2 a major constituent of the secondary atmosphere of any planet of chondritic composition. Liquid water is essential for the conversion of atmospheric CO_2 into solid carbonates. The contrast between CO_2 in the atmospheres of Mars and Venus and in that of Earth gives emphasis to this point.

In order for life to develop on another planet, there must be some form of molecular-information coding and expression. A coding scheme simple enough to be capable of extensive elaboration and duplication would probably involve complementary pairing similar to that of base pairs in nucleic acids on Earth. Among possible molecular species that might be available on another planet, there does not seem to be any simple substitute for purines and pyrimidines. A selection other than adenine-thymine or -uracil and guanine-cytosine is conceivable, but there is no evident alternative to this class of bases. The operation of such a coding scheme requires long chains to which the bases are easily attached, and again for this function there seems to be no simple or readily available alternative to sugar-phosphate chains of unlimited length.

In addition, to express and utilize the information contained in the code, amino acids are probably essential because of their unique ability to form peptide linkages of indefinite extent. On another planet, however, a different selection of an "alphabet" of some 15 to 25 amino acids in place of the 20 on Earth might well evolve, with corresponding novelties in their protein analogs. These considerations suggest that if life is to evolve on another planet it must at a minimum begin with an adequate supply in its hydrosphere of phosphoric acid together with purines, pyrimidines, amino acids, and simple sugars like ribose.

In the last stages of the formation of a carbonaceous chrondritic planet, crustal melting from cratering by large meteorites combined with internal heating and volcanic activity releases various C, N, O, and H molecular species, in addition to H_2O, into a growing planetary atmosphere and hydrosphere. It is known experimentally that ultraviolet radiation from the central star will photochemically convert such species into sugars, purines, pyrimidines, and amino acids. This process could be supplemented by electrical discharges in the atmosphere and radioactivity in the hydrosphere. Ultraviolet radiation of wavelength $\lambda < 2200$ Å will break hydrogen bonds in CH_4, NH_3, HCN, and H_2O and is at the threshold for the O–CO bond. It follows that a planet of the assumed composition and possessing a liquid hydrosphere could evolve life provided its central star radiated with sufficient intensity in the ultraviolet below 2200 Å.

A final consideration concerns the length of time required for evolution. Table 1 gives the key events in the biological history of Earth and the approximate time—measured in units of 10^9 years, or *aeons*—at which they occurred. The time scale for a similar sequence of key events on another planet might be either compressed or lengthened.

Table 1. Key events in the evolution of life on Earth

Time (aeons)	Key events
0	Gravitational collapse to proto-sun
0.2	Formation of Earth and other planets
0.7	Formation of hydrosphere with dissolved nucleic acids and amino acids
1.5	Prokaryote cells formed
2.0	Mg + porphyrin produced chlorophyl and chloroplasts
	Anaerobic photosynthesis
2.3	Fe + porphyrin produced cytochrome c and mitochondria
	Aerobic photosynthesis
	Precipitation of soluble feric Fe from oceans
2.5	Beginning of O_2 in atmosphere
3.3	Eukaryote cells formed
4.0	O_2-rich atmosphere
	O_3 ultraviolet shield layer
4.1	Metazoa formed in oceans
4.3	Invasion of land masses by macroscopic flora, then fauna
4.7	Present

SOURCES: Schopf 1978 and Hart 1978

pressed or lengthened. A compression by 35%, from 4.7 to 3 aeons, would seem an extreme case biologically, considering the number of steps that must be achieved in sequence. Therefore, for a rough estimate of the prevalence of Earthlike planets on which macroscopic multicellular flora and fauna may have covered land masses fairly extensively we will seek stars formed between 3 and 4 aeons ago or longer.

Characteristics of the central star

The number of stars in our galaxy is immense—of the order of 1 or 2 times 10^{11}. Including all galaxies in the universe raises the total number above 10^{20}. These stars, however, have a great variety of physical states, and the majority would be entirely unsuited to support an Earthlike planet able to maintain water in the liquid state for 3 or 4 aeons.

The characteristics of the nearest 100 stars within 6.5 parsecs (about 21 light-years) of the sun have been tabulated by Allen (1973). Their masses, when known, range between $0.035M_\odot$ and $2.31M_\odot$ ($M_\odot$ is the mass

of the sun). Their colors, as measured by the difference between their blue and visual magnitudes (B–V), range from 0 to 2.12, corresponding to a range of surface temperatures from 10,800K to 2,600K. There are 7 white dwarfs and 4 subdwarfs; most of the remaining 89 are on the main sequence. Of the 100, 51 are members of visual binaries or triplets, and 72 are members of stellar systems when spectroscopic binaries and companions not visibly observed are included. Only single stars on the main sequence are suitable for an Earthlike planet, since there are no stable periodic orbits in a binary system, and even the quasi stable orbits involve large variations in the distance from one or the other of the two stars during each period, resulting in extreme thermal cycling (Goudas 1963).

We denote by f_{MS} the fraction of any large total group of N stars that are on the main sequence and by f_S the fraction that are single or not a member of a system of stars. For the 100 nearest stars, $f_{MS} = 0.89$ and $f_S = 0.28$. An analysis by Abt (1978) of the frequency of binaries along the main sequence for solar-type F3–G2 primaries gives 54% with companions of mass $0.07–1.2M_\odot$ and 13% with nonvisible "black dwarf" companions of mass $0.01–0.07M_\odot$. If the remainder are all counted as single stars, then $f_S = 0.33$. For the purpose of this article we shall use $f_{MS} = 0.9$ and $f_S = 0.3$.

Some characteristics of a single star that have relevance for its ability to support an Earthlike planet in orbit around it are its lifetime in residence on the main sequence, its luminosity and spectral characteristics, and its mass and chemical composition.

The average luminosity L of a star during its residence on the main sequence varies as the nth power of its mass M, with n about 5 near the mass of the sun and decreasing to about 3 for stars of more than 9 solar masses and for those of less than half the solar mass. Table 2 gives values of $L/L_\odot = (M/M_\odot)^n$, with $n = 5$, except at the two extremes, where a value of 4.5 has been used for the largest mass and of 4.0 for the smallest.

The lifetime of a star on the main sequence, t_{MS}, can be taken very roughly as proportional to its initial store of hydrogen as measured by its mass divided by the rate of hydrogen

Table 2. Conditions necessary for central star to be able to support an Earthlike planet

$M/M_\odot$	$L/L_\odot$	MS lifetime (aeons)	Surface temp (K)	λ_{max} Å	$f(\lambda < 2{,}200)/$ $f_\odot (\lambda < 2{,}200)$
2.0	23.0	0.9	9,000	3,200	20.0
1.5	7.6	2.0	7,400	3,900	6.9
1.2	2.5	4.8	6,300	4,600	2.3
1.1	1.6	6.8	6,000	4,800	1.5
1.0	1.0	10.0	5,700	5,100	1.0
0.9	0.6	15.0	5,300	5,500	0.5
0.8	0.3	>15.0	5,000	5,800	0.3
0.5	0.06	>15.0	4,000	7,200	0.02

burning as measured by its luminosity. Taking the main-sequence lifetime of the sun as 10 aeons, this gives $t_{MS} \sim 10(M/M_\odot)/(L/L_\odot)$. Lifetimes so calculated are given in Table 2. During its residence on the main sequence, a star increases its luminosity by 50% or more, depending on the stellar model employed, as a result of the replacement of hydrogen by helium. This increase for the sun since its entry onto the main sequence is generally taken to be about 25%. To avoid an excessive planetary temperature increase, the main-sequence lifetime of the central star should probably be at least double the assumed age of an Earthlike planet. For an assumed age of 3 aeons, this would require $t_{MS} > 6$ aeons or a maximum mass of 1.15 $M_\odot$. For the more probable age of 4 aeons, $t_{MS} > 8$, and the maximum mass is 1.06 $M_\odot$. Thus stars with $M > 1.1 M_\odot$ are probably excluded for an Earthlike planet that has had a sufficiently long evolutionary history.

The fourth and fifth columns of Table 2 give the approximate surface temperatures and wavelength $\lambda_{max} = 2.9 \times 10^8/T$ Å at maximum radiative intensity. Because of the probable role of ultraviolet with wavelength $\lambda < 2200$ Å in the production of organic compounds required for the evolution of some form of life on an associated planet, it is of interest to compare the fraction of the total luminosity of a star with lower wavelength to that for the sun. The ratio of these fractions is given in the last column of Table 2. Both this fraction and the lowered photosynthetic efficiency implied by the value of λ_{max} suggest that stars of mass <0.8 $M_\odot$ might not be suitable for an Earthlike planet. Later when the tidal couple between a planet and its central star are considered, it will

be seen that stars of mass <0.9 $M_\odot$ are probably also excluded. Thus we take the mass range 0.9 $M_\odot < M < 1.1$ $M_\odot$ as the probable limit on a central star supporting an Earthlike planet.

A further restriction on suitable stars is based on the population class to which they belong. Stars in globular clusters and in the central nucleus of the galaxy belong mostly to population II, which is characterized by a low abundance of elements heavier than hydrogen and helium. The more suitable population I stars are found in the galactic disc and were formed late in the history of the galaxy from matter cycled through the hot interiors of massive earlier stars.

The composition of the sun is representative of the composition of the original solar nebula out of which the planets were formed and of the composition of population I stars generally. The components of the solar nebular disc can be divided into three

Table 3. Composition of solar nebular disc by classes

Class	Components	Relative mass	
I	H_2	259	} ~360
	He	104	
II	CH_4	1.53	} ~7
	NH_3	0.52	
	H_2O	3.17	
	Ne	1.45	
III	Mg	0.21	} 1
	Si	0.23	
	S	0.13	
	Fe	0.38	
	Na, Al, Ca	0.05	

broad classes, with the relative mass of the third class, out of which the chondrites and terrestrial planets were formed, taken as unity (Table 3). For the formation of an Earthlike planet, a portion of the nebular disc must be swept clear of the much more massive class I and II components so that the coalescence into planets can be accomplished almost entirely from class III matter. The requirements for such an outcome depend on the dynamics of the formation of the star in the process of gravitational collapse and will be discussed below. First, however, a rough estimate will be made of the fraction of single population I stars with masses between 0.9 and 1.1 $M_\odot$.

A very crude estimate of the value of this fraction (f_{IM}) can be obtained by assuming the age of the galactic disc to be 12 aeons, the rate of star formation in the disc to have been constant throughout its history, and counting only stars formed less than 6 aeons ago as having sufficient class III matter to make the formation of an Earthlike planet of sufficient mass possible. With these admittedly rather dubious assumptions, the fraction of stars whose lifetime on the main sequence is greater than 12 aeons and which were formed at least A but not more than 6 aeons ago is $(6 - A)/12$. For stars with $6 < t_{MS} < 12$, this fraction is $(6 - A)/t_{MS}$ or $((6 - A)/10)(M/M_\odot)^4$.

The observed stellar mass spectrum has been fitted by Scalo (1978) to the empirical relation $dN \propto M^{-\gamma}dM$, with $\gamma = 1.94 + 0.94 \log (M/M_\odot)$. From this relation and the two preceding fractions, the fraction f_{IM} of stars in the mass range of $0.9 M_\odot < M < 1.1 M_\odot$ formed between 6 and A aeons ago of all stars with masses between $0.1 M_\odot$ and $50 M_\odot$ is easily obtained. The result is $f_{IM} \approx 0.010$ for $A = 4$ aeons and $f_{IM} \approx 0.015$ for $A = 3$ aeons. This means that the fraction of stars in this mass range belonging to population I that are single and on the main sequence—i.e. $f_{MS}f_{S}f_{IM}$—and that could, if they had developed a solarlike planetary system, have an Earthlike planet 4 aeons old is about 3 in 1,000. The fraction for such a planet 3 aeons old is about 4 in 1,000.

In the gravitational collapse from which the solar system was formed, the major portion of the angular momentum remained in the solar nebula, which became deployed in a large equatorial disc (Fig. 1). Initially the composition of the disc was the same as that of the central body. But during the T Tauri phase of the collapse and subsequent evolution of the sun onto the main sequence, an inner zone was swept clear of the volatile class I and II components, leaving behind only nonvolatile refractory dust of the class III component. This zone, labeled A in Figure 1, has a radius of about 4 astronomical units (a.u.) in

This zone consisted mainly of class II and III components, out of which Uranus and Neptune were formed.

A candidate star for an Earthlike planet would have had to have undergone a similar axially condensed mode of gravitational collapse leading to a similar disc with differentiated zones. If such a star belonged to population II, it would have very little class II and practically no class III matter in its nebula. Its disc, if formed, would have zones A and C

f_{DA} will turn out to have a value between 0.2 and 0.8, though a much smaller value remains a possibility.

The planet and its orbit

The formation of planets comparable to the minor planets of the solar system (those in region A of Fig. 1) is a highly stochastic process, the details of which are not well understood. An estimate of the distribution in number, mass, and radial distance of similar planets formed in the nebular

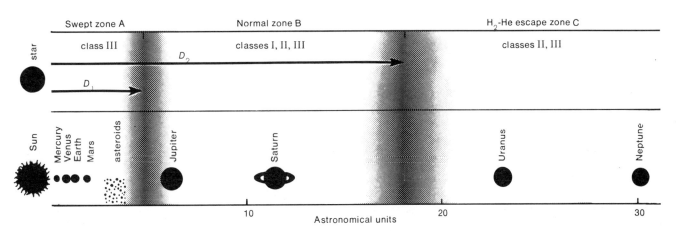

Figure 1. In the gravitational collapse from which stellar systems that might support Earthlike planets are formed, the major portion of the original mass falls into the central body (the sun, in the case of the solar system), and

the major portion of angular momentum stays in the stellar nebula, deployed in a large equatorial disc, represented here, in which the remaining matter is distributed into three zones. These zones are shown for the sun and

for another population I star formed in a similar way. (Scale of planets and distances are approximate; Pluto would be roughly 15 cm off the right-hand edge of the diagram.)

the solar system and D_1 in a comparable stellar planetary system. This sweeping of class I and II elements could have been the result of intense flares of solar wind, possibly combined with magnetic coupling of ions with the sun's rotating magnetic field. Eventually the terrestrial planets and asteroids coalesced out of this dust (see Fig. 1).

Beyond the swept zone, the composition of the disc was normal and included all the class I, II, and III matter approximately in the proportions shown in Table 3, except for some escape of H_2 and He from the gravitational field of the disc. Within this "Normal" zone (B in Fig. 1), the planets Jupiter and Saturn were formed. The weak gravitational field in the outer reaches of the disc allowed a more rapid escape of class I components, which formed zone C beyond a radius of 14–16 a.u. from the sun, or D_2 from another similar star.

empty and it could only end up with Jupiterlike planets in zone B. In order for a population I star to produce an Earthlike planet, it would in the process of disc formation have had to have produced a swept zone A of sufficient radius D_1 to contain enough matter for at least one planet of sufficient mass.

Although we would like to estimate the fraction f_{DA} of single stars forming a solar-type disc, D, around a central star with such a swept zone A, this is not possible at the present level of the theory of gravitational collapse, as Bodenheimer and Black (1978) have shown. Initial conditions of angular momentum and magnetic field in the cloud prior to collapse are important determinants of the outcome. Some details of the sun and solar disc may prove to be anomalous and related to the initiating supernova explosion, which was certainly involved (Cameron 1978). It seems likely that

discs of other stars is beyond the reach of our present understanding of the processes and mechanisms involved in the conversion of the gas and dust in the original nebula into gross planetary objects. The nature of this problem and the approaches to its resolution are reviewed by Hartmann (1978) and Wetherill (1978). The best that can be done is to discuss some of the parameters and their ranges for a planet once it has been formed and to define the conditions it must meet in order to be Earthlike. The probabilities for actualizing these characteristics must remain indeterminate for the present.

In order to have achieved an extensive land-based biosphere, a planet would have to be sufficiently massive to have retained a moderate atmosphere, including an appreciable O_2 fraction, for several aeons. This rules out small bodies such as the moon and Mercury and requires a mass somewhat greater than that of Mars. A lower limit of one-fourth the mass of the Earth seems reasonable. The amount of material in region A of a star even with large D_1 probably precludes the accretion of as much as two earth masses ($m_\oplus$) on any one planet. Thus the mass range for an Earthlike planet can be taken to be $0.25\,m_\oplus < m < 2m_\oplus$.

For other characteristics of the planet and its orbit, we take

R = mean distance of planet from star ($R_\oplus = 1$ a.u.)
e = eccentricity of orbit ($e_\oplus = 0.017$)
ω = rotational or spin angular velocity ($\omega_\oplus = 7.3 \times 10^{-5}$ rad/sec)
θ = spin-orbit angle ($\theta_\oplus = 23.5°$)

The rate of heat input to the planet from the central star is $\pi r^2(1 - \overline{A})$ $(L/4\pi R^2)$, where r is the radius of the planet, $\overline{A}$ its spectrally averaged albedo, and $L/4\pi R^2$ the radiation intensity at the top of the atmosphere (the solar constant in the case of Earth). $\overline{A}$ will depend on the area of snow, ice, and cloud cover, the relative area of land and sea, vegetation, etc. Excluding the "greenhouse" effect of the planetary atmosphere, the rate of heat loss from the planet is $4\pi r^2 \epsilon \sigma T^4$, where ϵ is the emissivity at blackbody temperature T and σ is the Stephan-Boltzmann constant. For a planet in a circular orbit with large ω and $\theta = 0$, daily temperature variations are small, and the balance between heat input and output gives an average temperature $\overline{T}$ of

$$\overline{T}^4 = (1 - \overline{A})L/16\pi\epsilon\sigma R^2 \quad (1)$$

For small values of ω, daily variations in temperature are large, and seasonal variations become significant in elliptical orbits of large eccentricity or large spin-orbit angles θ. For such cases the mean temperature given by Eq. 1 is not a very informative measure of the actual situation.

Under the assumptions adopted here, an Earthlike planet in another stellar system would have a hydrosphere and an equilibrium partial pressure of CO_2 approximately that in Earth's atmosphere. For Earth, the CO_2 greenhouse effect raises the average temperature obtained from Eq. 1 by 30–35K, and we may assume that it would do about the same on a candidate planet. Assuming also a similar albedo and emissivity, Eq. 1 can be used to relate the mean orbit radius of

another planet to the mass of its central star for the same mean temperature as that of the Earth

$$R = R/R_\oplus =$$
$$(L/L_\odot)^{1/2} = (M/M_\odot)^{5/2} \quad (2)$$

since $R_\oplus = 1$ a.u., and we have used $L/L_\odot = (M/M_\odot)^5$ in the restricted mass range now under consideration. For a star of mass $1.2\,M_\odot$, this gives $R = 1.57$ a.u., which is just slightly beyond the orbit of Mars, and for a star of mass $0.8\,M_\odot$, $R = 0.57$ a.u., which is half way between the orbits of Mercury and Venus. For values of R somewhat less than those given by Eq. 2, water would be mostly in the vapor phase, and for values somewhat greater, mostly in the solid. However, the complexities of atmospheric and hydrospheric convection, CO_2 equilibrium solubility, and changes in albedo and emissivity make an estimate of the range of R difficult. A recent computer simulation carried out by Hart (1978) for Earth showed a runaway greenhouse at an early stage in Earth history for $R_\oplus = 0.95$ a.u. and a runaway glaciation about 2 aeons ago for $R_\oplus = 1.01$ a.u. These results suggest that an Earthlike planet must have an orbit radius within 1–4% of that given by Eq. 2.

Characteristics of the orbit other than its mean radius will govern the magnitude of daily and seasonal variations in surface temperature. For rotational or spin angular velocities $\omega < 2 \times 10^{-5}$ rad/sec (>87-hour day), the daily temperature variation would be extreme and probably unacceptable. For a value of ω equal to the orbital angular velocity, the same face of the planet would be illuminated at all times, and any hydrosphere the planet may have possessed would have been completely converted to ice and snow on its dark side.

The spin orbit angle θ results in a hemispheric variation between summer and winter insolations (instellarations in another system), I_s and I_w, of $I_s/I_w = (1 + \sin\theta)/(1 - \sin\theta)$. For Earth and Mars this ratio is 2.3, and for $\theta = 30°$ it is 3. For a planet like Uranus with $\theta \sim 90°$, the winter insolation is zero and the summer insolation twice that at equinox. As θ increases, the width of the moderate temperature band centered on the equator decreases.

The eccentricity of the orbit also produces seasonal variations in the

insolation. The ratio of the insolation at perihelion to that at aphelion is $I_p/I_a = (1 + e)^2/(1 - e)^2$. This ratio is 1.03 for Venus, 1.07 for Earth, 1.45 for Mars, and 2.3 for Mercury. For an eccentricity $e = 0.27$, it is 3, and the time spent by the planet between aphelion and a position at which a line to the star is normal to the major axis is twice that spent between this point and perihelion. Larger eccentricities than this would result in increasingly severe environmental constraints on evolutionary potentials.

Some combinations of eccentricity and spin axis inclination could, however, produce suitable environments on a portion of the planet. In an orbit with $e = 0.27$ and $\theta = 30°$ with the spin axis parallel to a plane through the major axis normal to the plane of the orbit, the insolation on one hemisphere is a maximum at equinox and decreases only 14% in going from equinox to either perihelion or aphelion. That on the other hemisphere would increase by a factor of 2.6 in going from the spring equinox to perihelion and decrease by a factor of 0.3 in going from the fall equinox to aphelion, with this value remaining below 50% of that at equinox for half a year and then rising and falling by a factor of 6 during the next half year. On such a planet, life could evolve under very favorable environmental conditions on one hemisphere, while being excluded from the other.

A final aspect of planetary orbits is the deceleration of the spin angular velocity due to tidal action. The theory of such action has been reviewed in detail by Jeffries (1970). For Earth the rate of decrease of rotational angular velocity due to tides raised in the oceans by the sun and moon together is given by

$$\dot{\omega} = -\frac{N_m + N_s}{C} =$$
$$-7.9 \times 10^{-6} \text{ rad/sec/aeon}$$

(p. 314), where N_m is the tidal couple with the moon, N_s that with the sun, and C is the moment of inertia of Earth. The couple with the moon is stronger than that with the sun and $N_m = 5.1\,N_s$ (p. 299, Eq. 14). For comparison with another stellar system, we seek only that part $\dot{\omega}_s$ of $\dot{\omega}$ due to the couple with the sun, or

$$\dot{\omega}_s = -N_s/C =$$
$$-1.3 \times 10^{-6} \text{ rad/sec/aeon} \quad (3)$$

The tidal couple with Earth's oceans due to the sun is proportional to the height H of the tide, and H is taken to be proportional to $M_\odot/R^3$. Using this proportionality in Jeffries's Eq. 13 (p. 299) gives

$$N_s \propto r_\oplus^4 M_\odot^2/R_\oplus^6$$

where $r_\oplus$ is the radius of Earth. Taking $C = \beta m r^2$ gives

$$\dot\omega_s \propto -\frac{r_\oplus^2 M_\odot^2}{\beta m_\oplus R^6}$$

For a planet in another stellar system, the ratio of this deceleration to that for Earth is thus

$$\frac{\dot\omega}{\dot\omega_s} = \frac{\beta_\oplus}{\beta}\left(\frac{r}{r_\oplus}\right)^2\left(\frac{m_\oplus}{m}\right)\left(\frac{M}{M_\odot}\right)^2\left(\frac{1}{R}\right)^6 \quad (4)$$

since $R_\oplus$ is 1 a.u.

Since under the assumption used here, an Earthlike planet would have a hydrosphere and land areas, its tides would also be ocean tides, and this ratio should be directly applicable to it. For its moment of inertia we may assume that $\beta \simeq \beta_\oplus$. Using $r^3 \propto m/\rho$, with ρ the density of the planet, and Eqs. 2 and 3, we obtain from Eq. 4

$$\dot\omega = -1.3 \times 10^{-6}\left(\frac{m_\oplus\rho_\oplus^2}{m\rho^2}\right)^{1/3}\left(\frac{M}{M_\odot}\right)^{-13} \quad (5)$$

For a planet of the same mass and density as Earth, the term $(m_\oplus\rho_\oplus^2/m\rho^2)^{1/3}$ is unity, while for one of mass one-fourth that of Earth and the density of Mars, its value is 2. For this range of values of this factor, the tidal decelerations of spin angular momentum are given in Table 4 for stars of several masses. Since it is desired to maintain $\omega > 2 \times 10^{-5}$ rad/sec throughout a planetary history of at least 4 aeons, it is evident from this table that stars less massive than $0.85 M_\odot$ are excluded. A lower limit of $0.9 M_\odot$ is a safe choice and is, of course, the one already adopted. Evidently this tidal constraint is stronger than that applied earlier based on the wavelength λ_m at peak intensity and the fraction of the spectrum with $\lambda < 2200$ Å. Even a planet of low mass and density with sufficient initial angular velocity to avoid locking-in at some small multiple of the orbital angular velocity would experience increasing temperature swings between day and night after 4 or 5 aeons of evolutionary development.

Considering the variety of possibilities for the formation of planets in the A region of population I stars, a reasonable judgment might be that the chance of forming a planet of adequate mass in an orbit whose mean radius is within close limits of that given by Eq. 2 and with an acceptable combination of ω, e, and θ is probably not as great as one in 200 or less than one in 5,000. We designate the fraction of such stars possessing a planet of sufficient mass in an orbit, O, satisfying all these conditions by f_{mO} and guess that its value is no greater than 0.005 and possibly no less than 0.0002.

Since this is the last of the fractions for which any physically measurable estimates are possible, it will be useful to collect and define all those that have been introduced and estimated to this point.

Estimated fraction of all stars on the main sequence: $f_{MS} = 0.9$

Estimated fraction of main-sequence stars that are not members of a stellar system, i.e. single stars: $f_S = 0.3$

Estimated fraction of single, main-sequence stars belonging to population I, whose mass lies between 0.9 and 1.1 $M_\odot$, of age greater than 4 aeons: $f_{IM} = 0.01$. (If an age of 3 aeons is considered adequate for evolutionary development, the estimate is 0.015.)

Estimated fraction of single, main-sequence, population I stars of suitable mass and age that in the process of their formation produced an extended equatorial disc with a swept zone A containing only class III matter of sufficient extent for an Earthlike planet as measured by the radius D_1: $f_{DA} = 0.2$–0.8

Estimated fraction of the stars defined above having a planet of mass between 0.25 and 2 $m_\oplus$ in an orbit of radius and eccentricity R and e, and having spin angular velocity and orientation ω and θ, that could have been Earthlike for 4 aeons: $f_{mO} = 0.005$–0.0002

Given these fractions, the number of stars with an Earthlike planet, N_p, in a total population of N stars is

$$N_p = N f_{MS} f_S f_{IM} f_{DA} f_{mO} \quad (6)$$

If the estimates made here are anywhere near correct, N_p is between 10^{-5} and 10^{-7} of all stars. At the local density of stars in our galaxy, this would mean that there may be between 10 and 1,000 Earthlike planets per cubic kiloparsec, or between one and 100 within a distance of 1,000 light-years.

The course of evolution

Throughout this study we have attempted only to bring an Earthlike planet to a stage at which a wide invasion of land areas by macroscopic, multicellular flora and fauna would be a possibility. Whether life would develop at all on such a planet has not been evaluated, and indeed it is entirely possible that there are planets 5 aeons old very much like Earth which have been devoid of life throughout their history. The attempt has been only to define an environment and an adequate time span under which life could have evolved as it did on Earth. But so far no consideration has been given to the question of what evolutionary course might have been followed after this stage had been reached on another planet. The center of interest in such a study as this lies, however, in the probability that from such a stage evolution will ultimately lead to the production of intelligent, self-aware, hominidlike species capable of cultural and technological development. We turn in this concluding section to a consideration of this question.

The basic problem in any such attempt is the historic character of evolution which, like all history, involves a complex interplay of chance and accident in a continuously developing fabric of innumerable and complicated interconnections. And like all history, the history of life is inherently unpredictable. It is often said that the selective advantage of intelligence and language will ensure its achievement in evolution. But natural selection can operate only on

Table 4. Deceleration of spin angular velocity by tidal friction

$M/M_\odot$	$-\dot\omega$ (rad/sec/aeon)
0.90	0.5–1.0×10^{-5}
0.85	1.1–2.2×10^{-5}
0.80	2.4–4.7×10^{-5}
0.75	5.5–11×10^{-5}
0.70	13–27×10^{-5}

the material at hand, not on unrealized potentialities. This essentially indeterminate and nonrepeatable aspect of evolution has been cogently argued by George Gaylord Simpson (1964).

The achievement of the initial stage for our Earthlike planet is no guarantee that it will eventually lead to any particular species or type. The fraction of such planets with hominids is not calculable in principle because most of the decisive events in the history of life or human history have an a priori probability of zero (e.g. the a priori probability for the emergence of Alexander the Great or Napoleon).

The best we can do is to rely on an experiment carried out on Earth by continental drift. During the first invasion of land areas by life in the Devonian, 0.4 aeons ago, the land was consolidated into a single continental mass called Pangaea. Land life evolved on this common continuous land mass for the next 0.2 aeons. At this stage a major portion of the mass, Gondwanaland, broke off to the south, and in the mid-Triassic, 0.18 aeons ago, when reptiles were rapidly evolving into the first dinosaurs, Australia and Antarctica broke from Gondwana and thereafter pursued their own course of evolution independent of the rest. We can think of them as a sample of another planet, which we shall call Planet A, in another stellar system, as "Earthlike" as any such planet could be.

The course of mammalian evolution on Australia followed the marsupial branch rather than the placental as it did elsewhere. Until man migrated to Australia from Southeast Asia within the last 50,000 years, the course of evolution on Planet A showed no signs of leading to any kind of hominid. Given another 0.1 to 0.2 aeons of evolutionary development, such a species could conceivably have emerged if no men had migrated there from the outside. On the other hand, Parker (1977) has argued rather persuasively that such an outcome from an initial stock of marsupial-type mammals is unlikely.

Later, at the end of the Jurassic period 0.13 aeons ago, at the peak of the giant dinosaurs, South America began to separate from Africa. We can think of this area—Planet S—as belonging to another stellar system and constituting the land mass of a fully Earthlike planet in that system. At this time placental mammals were beginning an increasingly rapid development. On Planet S they continued their development, which included a primate branch. The New World monkeys, however, are very different from the primates of Africa, which we now designate Planet E, for Earth. There are a number of species of New World primates, but they are all small animals that have never left the trees. Ignoring the recent arrival of *Homo sapiens* in South America by migration from Mongolia, it is possible to imagine a future evolutionary path from this native primate stock that could have led in another 0.1 to 0.2 aeons to hominids. All we can say with assurance, however, is that it was not headed that way when *Homo sapiens* arrived on the scene and drastically altered its future course.

Thus continental drift has produced for us three Earthlike "planets." Only on one of them, Planet E, did evolution lead to man. Planet A and Planet S experimented with their own distinctive and fascinating evolutionary histories, but neither of these histories shows much promise of taking a direction that might in time have produced hominids.

In this study we have attempted to identify as precisely as possible the constraints involved in the production elsewhere in the universe of a planet having a constitution, environment, and age adequate for a full-scale evolution of living systems to the stage of widespread occupation of land areas by macroscopic multicellular flora and fauna. Given these conditions, we have no way of knowing whether life would develop at all on such a planet and certainly no way of knowing whether it would ever produce self-conscious beings like ourselves, capable of language, rational thought, manipulative skills, and high technology. Life on such a planet could use a modified molecular coding scheme and a different amino acid alphabet. Its evolutionary history could be strikingly different from that on Earth, and its hominids, if they developed at all, very different creatures from ourselves.

There is a deeply ingrained conviction in the great majority of mankind, to which the appeal of science fiction and fantasy bears witness, that the universe is so constituted that if an opportunity exists for life to originate, it will be actualized, and if an opportunity exists for hominids to evolve, that too will be actualized. Whatever may be the basis for such convictions, it clearly must be sought outside the domain of science. The most this study has been able to establish is that even the opportunity for such achievements occurs quite rarely among the vast profusion of forms in which matter is consolidated in the universe.

References

Abt, H. A. 1978. The binary frequency along the main sequence. In *Protostars and Planets*, ed. T. Gehrels, pp. 323–38. Univ. of Arizona Press.

Allen, C. W. 1973. *Astrophysical Quantities*, 3rd ed. pp. 234–37. Univ. of London, The Athlone Press.

Bodenheimer, P., and D. C. Black. 1978. Numerical calculation of protostellar hydrodynamic collapse. In *Protostars and Planets*, ed. T. Gehrels, pp. 288–322. Univ. of Arizona Press.

Cameron, A. G. W. 1978. Physics of the primitive solar nebula and of giant gaseous protoplanets. In *Protostars and Planets*, ed. T. Gehrels, pp. 453–87. Univ. of Arizona Press.

Goudas, C. L. 1963. Three dimensional periodic orbits and their stability. *Icarus* 1:1–18.

Hart, M. H. 1978. The evolution of the atmosphere of the earth. *Icarus* 33:23–39.

Hartmann, W. K. 1978. The planet forming state: Toward a modern theory. In *Protostars and Planets*, ed. T. Gehrels, pp. 58–73. Univ. of Arizona Press.

Jeffries, H. 1970. *The Earth: Its Origin, History and Physical Constitution*, 5th ed. Cambridge Univ. Press.

Morrison, P., J. Billingham, and J. Wolfe, eds. 1977. The search for extraterrestrial intelligence: SETI. NASA SP-419. Washington: Superintendent of Documents, Government Printing Office.

Parker, P. 1977. An ecological comparison of marsupial and placental patterns of reproduction. In *The Biology of Marsupials*, ed. B. Stonehouse and D. Gilmore. London: Macmillan.

Scalo, J. M. 1978. The stellar mass spectrum. In *Protostars and Planets*, ed. T. Gehrels, pp. 265–87. Univ. of Arizona Press.

Schopf, J. W. 1978. The evolution of the earliest cells. *Sci. Am.* 239:111–38.

Simpson, G. G. 1964. *This View of Life: The World of an Evolutionist*. Harcourt Brace and World. The non-prevalence of humanoids. 1964. *Science* 143:769–75.

Wetherill, G. W. 1978. Accumulation of terrestrial planets. In *Protostars and Planets*, ed. T. Gehrels, pp. 565–98. Univ. of Arizona Press.

PART 2 *Suns, Galaxies, and the Origin of Elements*

Stellar Evolution and the Origin of the Chemical Elements

Jesse L. Greenstein

The beauty of our universe and its immensity in space and time give astronomy an extraordinary hold on the popular imagination. This fascination is not less for the professional working in an observatory, that silver bubble on a mountain top, earthbound, but with its vision directed outwards.

In most of the physical sciences, the questions of meaning can be laid aside, fortunately for daily progress. In astronomy, however, the telescope looks out through such enormous reaches of space and time, and through such an unexpected universe, that one group of the deeper questions, that of origins, becomes the daily subject to which the methods of this physical science are applied. Cosmological speculation appears early in all theologies; most bibles start with "In the beginning. . . ," a story of creation. Unfortunately, our daily proximity to deep problems does not inevitably lead to deep answers. Astronomers, cosmologists, and physicists are bound by the current state of human knowledge and by the limits of the knowable. Nevertheless, they are impelled by strong internal forces, complex psychological derivatives of the childish need to know "where, when, how, and why." The ideal textbook picture of science as a logical, hypothetico-deductive system leaves out these human motives, leaves out the importance of creative intuition, and most importantly, leaves out the beauty of the world we observe.

How significant these inner compulsions have been in the development of scientific cosmology and cosmogony is a fascinating topic in the history of science. My subject is one particular, limited, world view, clothed with the usual detail of physical science, speculation, extrapolation from a few facts, and using mathematics, physics, and astronomy. But fundamentally, it is a detective story. The question is not "who did it?" but "what was done and when?" The composition of living matter turns out to be a better sample of the universe than is the dead earth. Is there a reason for this composition? How are atoms born, and how do they evolve?

Dr. Jesse L. Greenstein is of the Mount Wilson and Palomar Observatories in Pasadena, California. He emphasizes that, in astronomy, "the telescope looks out through such enormous reaches of space and time, and through such an unexpected universe" that questions of origins become "the daily subject to which the methods of this physical science are applied." The article abundantly illustrates the processes of synthesis of the chemical elements, both light and heavy, through many varieties of stars.

There have been many recent attempts to give all-embracing answers. Jeans and Eddington popularized the first applications of relativity theory, of atomic and nuclear physics, to the structure of the stars and to questions of the time scale of the universe. Lemaître and Gamow then explored the connection between the expansion of the universe and the formation of the atoms in a possible primeval explosion. Later, Hoyle and others introduced the concept of the steady-state universe, continuous creation, and stressed the formation of the elements in the stars. It is unlikely that we have exhausted the possible cosmogonies, or that we have even realized the complexity of the problem. Nevertheless, limiting the possible range of fantasy, one major advance has occurred in the last ten years; the growth of our knowledge of the nature of the stars, of stellar evolution, of nuclear energy sources and nuclear processes, has begun to provide us with real clues. There is tangible evidence concerning events that occurred billions of years ago, with examples of similar events now occurring. We still make wild extrapolations, but now within a range limited by some facts. How close are we to the truth? The history of science gives us little comfort; guesses as to the future are seldom validated, seldom even remembered.

The astronomical background

Let me start with a brief outline of some major relevant facts describing the universe as we know it. Atoms occur in some ninety stable elements and several hundred isotopes. To a first approximation, the relative abundances of these atoms, with understandable differences, are the same everywhere in the earth, in the average meteorite, in the sun and the stars of our own and other galaxies [1,2,3] (Fig. 1). Hydrogen, H^1, is overwhelmingly the most abundant element, about seventy per cent by weight, helium, He^4, next, about twenty-eight per cent, and the other heavier elements increasingly rare. The complicated general abundance pattern of the heavy elements, with many subsidiary peaks, should have a rational explanation, if we postulate that no reasonable mode of creation of matter should have to provide initially so complex a family of chemical elements and isotopes. Whether man should impose his need for rationality on the universe is not obvious, but scientists exist on the hope that nature is not tricky; progress depends on the faith that natural law may be difficult to understand, but should not be stupid. Given omnipotence, I, at least, would not create a complex universe. But, fortunately, this postulate is confirmed by observed meaningful devia-

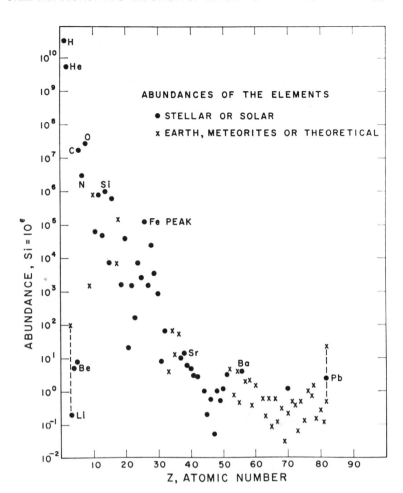

Figure 1. Composite abundance curve of the chemical elements. Circles are solar and stellar values, adapted from Goldberg, Müller and Aller [1]. Crosses are from the terrestrial, meteoritic and theoretically adjusted values given by Suess and Urey [2] and Cameron [3]. Dashed lines connect elements for which there are large differences in these determinations. Note the overwhelming abundance of H, He. Most light elements are relatively abundant, e.g. C, O, Mg, Si and there is a detached peak centered on Fe, near the region of maximum stability for heavy elements.

tions from the universal abundance curve found by study of the composition of different types of stars and also by a time-dependence of the normal composition.

Matter is largely collected in stars like the sun; in addition, interstellar matter, gas and dust, also exists in each galaxy. A typical galaxy is about 100,000 light years in diameter, and has a mass of a hundred billion suns; from one to ten per cent of the mass is interstellar gas. Galaxies apparently recede from each other with a velocity proportional to distance. The largest red-shift yet measured is nearly half the speed of light. The expansion does not occur within individual galaxies, or within clusters of galaxies, but is a property of the large scale (more than millions of light years) distribution of matter. On the basis of current estimates of the distances of the galaxies, the expansion time, measured by the reciprocal of the Hubble constant, is about 13 billion years. I omit in this treatment any discussion of the problems of the expanding universe, the time scale, or of steady-state cosmology.

Gravitational forces hold the stars together in clusters of stars (10^2 to 10^5 solar masses) or in galaxies (10^8 to 10^{12} solar masses). Rotation is almost universal, so that galaxies are flattened systems that rotate in about 10^8 years (Fig. 2). In galaxies with a relatively large amount of interstellar gas and dust, spiral arms are the loci of maximum abundance of the interstellar matter, and of newly formed stars of high luminosity. The angular momentum displayed by flattening of the galaxy, as a whole, is also present in certain types of stars. Stellar rotation varies

enormously, with equatorial speeds of 2 km/sec for the sun and up to 500 km/sec for B stars at the margin of stability. In addition, many stars are double or multiple, reflecting the tendency toward disruption of an originally single, pre-stellar condensation with excessive angular momentum and the tendency toward multiplicity in the condensation process. The tiny system of planets around our sun is another manifestation of excessive angular momentum. (Until recently, I would have called the planetary system a trivial and uninteresting fact, although I am no longer so sure!) In addition to gravitation, other important and unexpectedly large forces act on the interstellar gas. The spiral structure of the gas clouds in galaxies is probably controlled by the interstellar magnetic fields. Cosmic rays and radio-frequency noise are both intimately connected with these fields. Certain "magnetic-peculiar A stars" show very large, relatively simple, dipole fields at their surface up to 34,000 gauss. The sun, and presumably most types of stars, have localized variable fields of a few thousand gauss (the sunspots). Such a general field at the surface corresponds to an enormous field in the interior.

We will see later that stars are born by condensation from interstellar gas, and later tap nuclear energy sources which, by the conversion of mass into energy, H^1 into He^4, balance the enormous energy output. When the energy supply, dominantly hydrogen, becomes exhausted, the star "evolves" and begins to die. During the evolutionary, dying phase the star loses part of its mass to interstellar space, either violently by explosion or by slower "mass-loss," evaporation or corpuscular radia-

Figure 2. The spiral nebula, M33. The arms are rich in stars of very high luminosity, up to 10^6 $\odot$, as well as gas and dust clouds. This is a young galaxy from the point of view of star formation and evolution. It is seen nearly along the direction of its pole of rotation.

tion (the solar wind on a grander scale). Thus, waste products of nuclear reactions inside the deep interior of stars are returned to space, mixed with gases still containing the fuel, hydrogen, and recondense into a new generation of stars. Our bodies contain such waste products of long dead stars. Without one element, carbon, formed inside a star at a temperature of 150 million degrees, there would be no life in the universe.

A schematic cycle of star birth and death, together with auxiliary processes, is illustrated in Figure 3.

Except for gravitational energy released during the contraction of a star, and angular momentum available initially from galactic rotation, nuclear energy is the driving force. When it is exhausted in the stars, all activity except stellar motion ceases; for example, the magnetic fields

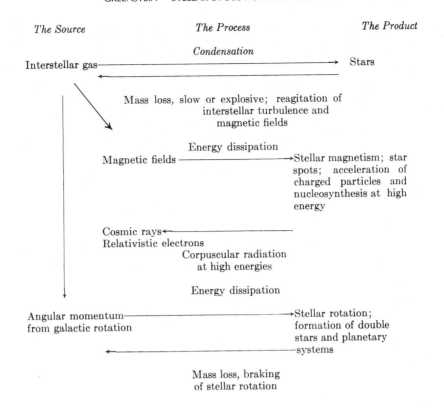

Figure 3.

and the cosmic rays are indirect results of the nuclear energy release and would eventually decay. As the one type of interaction which seems to be irreversible, and which energizes the cycles in the previous diagram, we sketch in Figure 4 the main and subsidiary lines of nuclear processes which drive all these other activities, and which synthesize the elements. Unless there are deep, unexplored cosmological effects, for example, an oscillating universe with a reversal of the direction of time or of the second law of thermodynamics, there is still only a one-way road, powered by nuclear fusion towards the heat-death of our Sun, our Galaxy and the universe we know.

The equilibrium of a star— stellar interiors

When a sufficiently large mass forms by gravitational forces, it contracts, heats up, and at central temperature of about a million degrees, taps nuclear energy sources whose energy output is a steep function of temperature. A stable configuration is reached in which the energy released is balanced by the radiation into space from the surface. The high temperature and density in the center produce a pressure gradient which balances the gravitational force. The mean temperature of a star from such stability arguments is about 5 million degrees times its mass divided by its radius, both in solar units. The star is mechanically and energetically stable as long as the energy sources last, within certain limits. Stars apparently cannot have masses much greater than about 50 times the sun because of various sources of instability, such as the pressure of radiation. Below about 0.005 solar masses, the central temperature is so low that energy release may possibly not occur; the condensation may be a solid, hydrogen-rich planet rather than a star. The leakage of radiation through a star is governed by the opacity, which depends on temperature, density, and composition.

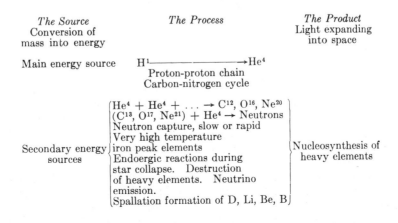

Figure 4.

Historically, two general approaches to the study of stellar interiors have been useful. In the earliest work, the main consideration was the mechanically stable gas configuration for a given mass. This depends on the composition which fixes the star's luminosity and radius. Consideration of the nuclear energy sources is not relevant, because of a fortunate accident; given the mechanical structure, the central temperature is adjustable by changing the composition within a reasonable range. The very steep temperature-dependence of the nuclear energy sources yields the correct luminosity by a proper choice of the hydrogen and helium abundances. Obviously, something is wrong with so much freedom! The more recent, detailed studies have shown that the proper method is numerical integration, step-by-step through the star, taking proper consideration of the energy sources, the pressure gradient, and the opacity and temperature gradients which regulate the energy flow and the change of composition with depth. In principle, once the mass and composition are given, the entire structure of the star is completely determined. Such a stellar model requires both a knowledge of physical theories such as nuclear fusion, the opacity of highly ionized gases and hydrodynamics, and the full resources of large computing machines.

The major results, for our purposes, of the study of stellar interiors are those connected with the properties and lifetimes of stars in:

(1) The stage of contraction from the low-density interstellar gas to the ultimately stable configuration. This is called the "contracting" of "T Tauri" stage.

(2) The stable configuration for a given mass and initial composition. This is the normal "main-sequence" stage in which a star spends most of its life.

(3) The evolving stage, that of the subgiants and red giants, which a star enters when it exhausts hydrogen in a certain fraction of its mass.

(4) The beginning of star death, when nearly all the nuclear fuel is gone and the central temperatures become extremely high.

(5) The death of stars—either through the long, slow decay of the very high-density white dwarfs, or through a fantastic explosion as a supernova.

Not all stars have gone or will go through all these stages; the above phases are characterized by very different temperatures and densities, and therefore by different nuclear processes. The consumption of the main nuclear fuel, hydrogen, at temperatures from 1 to 30 million degrees is followed near the end of the giant phase by the burning of helium at several hundred million degrees. In giant stellar explosions, supernovae, where the temperature reaches 2 to 4 billion degrees, essentially all possible reactions take place, and heavy elements like iron are formed or even reduced to protons, neutrons, alpha particles and neutrinos if the temperatures are sufficiently high.

Stellar lifetimes; star formation; evolution

If we limit ourselves to the burning of hydrogen into helium (a fairly good approximation), we compute the maximum possible lifetime of the sun on the main sequence as almost 10 billion years. The luminosity, L, is a steep power of the mass, M (both in solar units), so that the lifetime of other types of stars can be obtained from:

$$\text{Lifetime} = 10^{10} \left(\frac{M}{L}\right) \text{ years} \approx \frac{10}{L^{5/7}} \text{ billion years.}$$

Thus a highly luminous blue B star with $L = 10^5$ has a lifetime of only 3 million years. At the other extreme, a faint red M dwarf with $L = 10^{-5}$ can live over 30 trillion years. Our sun, with its average mass and moderate durability of ten billion years is now near the middle of its life. The enormous range in the nuclear lifetime of the stars and the existence now of some stars with lives of only a few million years means that not all stars were formed at the same time.

Stars are now being born from the interstellar gas; those of large mass shine brilliantly for a short while, while those of small mass condense very slowly, live longer, and may often be studied in this "contracting" phase. The interstellar gas from which the stars are born pervades the flattened, main plane of our Galaxy and is concentrated in spiral arms of gas and dust, with bright short-lived stars. The gas now represents only a few per cent of the total mass of the Galaxy. Star formation was very much more rapid in the early history of the Galaxy, billions of years ago, than it is now. A very slight counterbalance is the observed ejection of matter into interstellar space from stars at various late evolutionary stages, especially from the red giants. It is probable that the rate of star death, and therefore of ejection of mass into space, was also initially higher. There was an early period with a burst of rapid star formation in which much of the primeval gas was consumed. While some stars were still contracting, others were born, exhausted their nuclear fuel and exploded or otherwise returned part of their mass to space where it was mixed with still unused gas. This mixture condensed into further generations of stars contaminated with the waste products of nuclear processes from many stars which had traversed many varieties of evolutionary histories.

As a result, we expect that there was also an early burst of formation of heavy elements. In one cosmology, that of Hoyle, it is assumed that a galaxy begins with pure hydrogen; in some rival models of the early stage of the expansion of the universe, it is possible that isotopes of hydrogen, helium, and small amounts of a few other elements were synthesized. In another speculative cosmology, residues of matter in a different cycle of expansion and contraction might be found. Some small exchange of matter between galaxies is also now thought possible. Thus, as a reasonable starting point, independent of cosmological models, we may assume that our Galaxy began as an uncondensed, rotating spheroid of gas, largely hydrogen, with small impurities of helium, and possibly small amounts of the heavy elements. Work on galactic structure, stellar lifetimes, and evolution indicates that the rate of star formation varied roughly as the cube of the density of the residual gas. Schematically,

Figure 5. A cloud of gas and dust in our Galaxy. Note the cluster of stars recently condensed (upper right-hand corner) and the dense black clouds of dust, presumably now in gravitational contraction.

for a very rough model of this process, Figure 6 shows the behavior of the residual gas, the total number of stars formed and the rate of star formation. An arbitrary age of 20 billion years has been taken for our Galaxy, and the formation rate constants adjusted to leave 3 per cent of the original gas now, 20 billion years after T_0, the beginning of the process. Little has happened in the last 19 billion years, since 88 per cent of the stars were formed only 1 billion years after T_0, and 98 per cent by 10 billion years after T_0. (The particular model adopted for Figure 6 neglects the recycling of matter from evolving stars back to the interstellar medium.) In consequence, even though our Milky Way makes a brilliant show at night and contains 200 billion stars, it is now relatively quiescent compared to its past appearance. What is true about star formation is probably equally valid for the formation of the various chemical elements. Some 20 billion years ago, the rate of star formation was five or ten times its present value. Great numbers of massive stars of short life were born, went through the various processes of nucleosynthesis, which we will discuss later, and disappeared in the almost unimaginable fireworks of the distant past. Before the Sun was formed, almost every possible stellar temperature, density and life history had been experienced somewhere in our Galaxy. The residues are the many kinds of atoms in the universe. Some of the processes that synthesize one element will destroy others; the enormously complicated composition of our bodies, for example, requires that many different stars contributed atoms to the raw material out of which our Sun, the planets and ourselves were born.

Nuclear reactions in stars

The electrostatic repulsion of the two positively charged nuclei must be overcome for stellar thermonuclear reactions to occur. The hot gases in main-sequence stellar interiors have thermal energies, kT, of about one kilovolt. The repulsive force barrier of several million volts is measured by the product of the charges on the two nuclei, and the probability of leakage through the barrier. It depends on a factor, $Z_1Z_2/E^{1/2}$, which appears as a negative exponential. Heavy ion collisions are rare and demand very high temperatures. Actually, the high energy tail of the thermal velocity distribution permits a few collisions at higher than average energy, so that successful nuclear reactions in the stars occur mainly at from 5 to 20 kev. The collisions are random, and the yields must be obtained by integration over all velocities. The collision cross section for successful reactions is given by:

$$S_0\tau^2e^{-\tau},$$

where the parameter, τ, is:

$$\tau = 42.48 \left(\frac{Z_1^2Z_2^2A}{T}\right)^{1/3}.$$

Here, T is in 10^6 degrees, A is the reduced mass

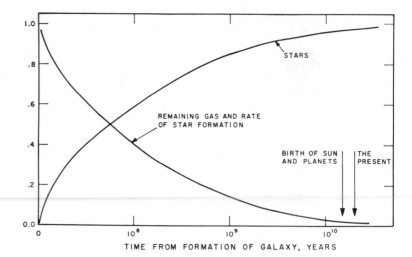

Figure 6. Schematic representation of the amount of interstellar gas left uncondensed into stars, and of the amount locked up in stars as a function of time elapsed after the formation of the Galaxy. Note the late stage at which our Sun was formed. Only a small amount of gas is now left, so that star formation is very slow.

$(= A_1 A_2 / A_1 + A_2)$, and S_0, an intrinsic nuclear reaction probability. The charge dependence is such that α-particle reactions, for example, require four times higher temperature than proton reactions (all other factors being equal); $C^{12} (\alpha, \gamma) O^{16}$ require 864 times higher T than does $D^2(p, \gamma) He^3$! Unless the parameter, τ, is not very large, the reaction rates are negligible; as a consequence, heavy-ion reactions occur only at temperatures not very far below those at which photo-disintegration effects set in.

Important factors in S_0 concern specific details of the reactions, once the two initial nuclei have been brought close together. For example, particle emission is more probable than gamma-ray emission, while beta-decay is much less so. A vast majority of collisions in the sun result only in scattering, i.e., deflection and thermal energy interchange. Among the light elements, in general, the reaction products weigh less than the two initial nuclei, so that the reaction is exoergic; reactions with nuclei heavier than iron are endoergic, except when neutrons are involved. Therefore, stellar energy generation is accompanied by nucleosynthesis only of the light elements. Heavier elements are synthesized largely by neutron capture at a stage when energy generation and nucleosynthesis are decoupled.

Starting with a contracting gas sphere heated by release of gravitational potential energy and containing only hydrogen, the first steps of nucleosynthesis and energy generation [4,5] are as follows: Two hydrogen atoms react, even at temperatures as low as a million degrees, because the electrostatic repulsion is small. But the reaction $H^1 (p, \gamma) He^2$ is impossible because He^2 is unstable, and instead $H^1 (P, \beta + \nu) D^2$ occurs, i.e., an emission of a positron and a neutrino. Such beta decay is so improbable that any given hydrogen atom in the center of the sun has a lifetime of 10^{10} years. The next step, $D^2 (p, \gamma) He^3$ is, by contrast, very rapid, taking only a few seconds, even though gamma-ray emission is slow compared to particle emission. From this argument, it seems plausible that $He^3(He^3, 2p)He^4$ should also be rapid, even though the charged particles have $Z_1 = Z_2 = 2$. Thus a star initially made of H^1 alone would build D^2, He^3, and He^4 as it generates nuclear energy, dE, by destroying mass, dM,

according to the equation

$$dE = -c^2 dM$$

where c is the velocity of light.

The intrinsic nuclear probabilities, S_0, are often resonant in character, and must be studied in the laboratory in detail to permit safe extrapolation to the low energies (≈ 10 kev) at which stellar reactions occur. Many resonant energy levels exist, and laboratory yields drop exponentially at low energies. Consequently, only the major reactions with H^1 have been explored in the laboratory below 100 kev, involving very difficult experiments with capture cross sections as low as 10^{-35} cm^2. Alpha-particle reactions have been only partially explored at the stellar energy range, and are more difficult in the laboratory. Neutron capture is usually exoergic and has been well studied at very low (laboratory-thermal) energies; laboratory cross sections are large, near 10^{-24} to 10^{-27} cm^2.

Nuclear reactions may occur elsewhere than at the center of the stars. Ionized gas in moving magnetic fields in stellar atmospheres or envelopes may suffer enormous acceleration [6] in which some particles attain energies of 100 Mev (the soft cosmic rays produced during solar flares). In addition, such high energy particles may strike heavy atoms and shatter them into nuclear fragments and neutrons. This "spallation" process has been observed when high energy cosmic rays or particles from synchrocyclotrons shatter heavy atoms in nuclear emulsions. Elements like D^2, T^3, Li^6, Li^7, Be^9, B^{10}, B^{11} as well as H^1, He^3, He^4, and neutrons are produced when an element of atomic weight 16 or more is struck by a particle of 10^8 to 10^9 ev.

Still another destructive process occurring at extreme temperatures is photo-disintegration even of stable nuclei by high-energy γ-rays such as would be present in the radiation field at temperatures above 4 billion degrees. This endoergic reversal of the fusion process destroys heavy atoms at the end of the life of certain stars. Additional endoergic processes of neutrino emission occur above 4 billion degrees and probably set the upper limit to the stellar temperatures attainable. At 500 million degrees, a typical red giant may lose 10^{24} erg/sec to neutrinos; at 2 billion, the loss is 10^{42} and at 5 billion, 10^{46} erg/sec. The latter figure is equivalent to the total solar energy production in a hundred thousand years! The neutrinos leak out of the star at the velocity of light, so that photodisintegration and neutrino losses remove energy from the hot gases very rapidly. As a result, the internal pressure is insufficient to support the weight of the overlying layers, the star collapses and heats the interior by conversion of gravitational potential energy into heat. This collapse is probably catastrophic, with a time scale of only a few seconds. Unburnt nuclear fuel is brought suddenly to high temperature, and the star explodes as a supernova. Essentially all possible nuclear reactions occur, with only the intrinsically most stable nuclei surviving. Under such conditions, elements are made in the abundance peak centered on iron.

Because of the absence of potential barriers, capture of low energy neutrons with emission of γ-rays is the simplest method of building heavy elements. Cross sections are large, show resonances, and are well correlated with the present observed abundances of the heavier elements. Each capture changes a nucleus of charge and mass Z, A into one of $Z, A+1$. If the latter is stable, the process continues until an unstable nucleus with beta-decay is reached, at which point $Z, A \rightarrow Z+1, A$. Neutron capture chains are called "slow" if the interval between successive captures of a neutron is long compared to the beta-decay time and "fast" if it is short. "Slow" chains synthesize elements on the stability line, "fast" ones synthesize neutron-rich isotopes which eventually decay (once the neutron supply is cut off) into isotopes somewhat removed from the stability line. Very "fast" neutron chains can build uranium and the trans-uranic elements. We will discuss the neutron source later.

This brief review of a few types of nuclear processes cannot fully elucidate the complex behavior of stellar matter at various stages of temperature and evolutionary age. But, obviously, there is an intimate link between energy generation and the transmutation of the elements. The goal of the theory of nucleosynthesis is to ascertain whether all the 300 stable and 50 radioactive isotopes of the 90 elements can be manufactured inside stars, or at or near their surfaces, by reasonable nuclear processes. Further details are given in articles listed in the bibliography [7,8,9,10,11]. From this current work, it appears that chains of nuclear events leading to any given isotope can be satisfactorily sketched. Ultimately, predictions of the accumulated amount of the heavier elements and comparison with observations of stars of different ages should permit study of the integrated effects of stellar energy generation and evolution in our Galaxy.

Stellar nuclear reactions and synthesis of the lighter elements

Since gravity compresses the gas in a star, the temperature must reach high values, by terrestrial standards. For a chemically homogeneous star, the correct order of magnitude of the temperature can be deduced from the requirement that the thermal energy at the mean temperature, $\bar{T}$, equals the potential energy per atom. Thus,

$$k\bar{T} = \frac{1}{5} \frac{GM m_H \mu}{R}.$$

Inserting the natural constants, writing M and R in solar units and taking the mean molecular weight, μ, near that of the sun, we find

$$\bar{T} = 4 \times 10^6 \,^\circ\text{K} \frac{M}{R}.$$

Thus, without nuclear energy release, a gas sphere like the sun must have $\bar{T} = 4 \times 10^6 \,^\circ\text{K}$ from stability considerations alone. At this $\bar{T}$, fusion reactions between D, He^3, Li, Be, and protons occur. Since the central temperature, T_c, is about three times $\bar{T}$, a gas sphere of solar mass heating up in an early contracting phase would tap nuclear energy sources at a radius several times that of the sun. The proton-proton chain would start, and even the C—N—O cycle. Both these fundamental stellar reactions involve fusion of four protons into a He^4 nucleus, with the destruction of 0.006 of the mass (25 Mev or about 4×10^{-5} erg per alpha particle formed).

Consider an ideally simple gas sphere containing only hydrogen. The steps are:

$$H^1 + H^1 \rightarrow D^2 + \beta^+ + \nu,$$
$$D^2 + H^1 \rightarrow He^3 + \gamma,$$
$$He^3 + He^3 \rightarrow He^4 + 2H^1.$$

The neutrino, ν, carries energy out and is wasted. The $\beta+$, γ-rays, and protons emitted quickly transfer their energy into heat. Once the stars have synthesized He^4, alternative chains include:

$$He^3 + He^4 \rightarrow Be^7 + \gamma,$$
$$Be^7 + \beta^- \rightarrow Li^7 + \nu + \gamma,$$
$$Li^7 + H^1 \rightarrow 2He^4.$$

The rate of energy production is set by the slow first step, $H^1 + H^1$; at the center of the sun the life of a given proton is about 10^{10} years. All other reactions are so rapid that, in a steady state, there exist only negligible concentrations of all elements except He^4. Let us continue for 10^{10} years, till the core of the star contains only He^4; since the energy source, H^1, is gone, the star would like to cool off but cannot do so. Gravitational contraction increases the temperature up to about 100 million degrees by reducing the radius of the core. Strangely enough, the star actually increases its over-all radius, becomes somewhat brighter and considerably cooler at the surface, i.e., evolves to the red-giant stage. A peculiar, slightly endoergic process starts:

$$He^4 + He^4 \rightarrow Be^8 + \gamma.$$

Since Be^8 is very unstable, the kinetic equilibrium is one in which only a tiny amount of Be^8 is ever present. On the other hand, whenever another He^4 collision occurs,

$$Be^8 + He^4 \rightarrow C^{12} + \gamma$$

and we have produced the very stable element, C^{12}, and have provided the building block necessary for the existence both of organic chemists and biologists, and for their subject of study.

Subsequent alpha-capture at high temperatures lead through the series of nuclei:

$$C^{12} + He^4 \rightarrow O^{16} + \gamma,$$

and less frequently:

$$O^{16} + He^4 \rightarrow Ne^{20} + \gamma,$$
$$Ne^{20} + He^4 \rightarrow Mg^{24} + \gamma.$$

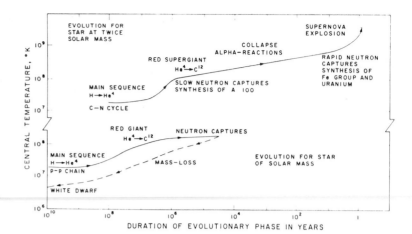

Figure 7. The central temperature attained in a star depends on its mass and on its age. These schematic curves give an indication of the time spent on possible evolutionary tracks for a star of mass 2 ⊙ (which may explode after reaching very high temperature) and also for a star of 1 ⊙ that may evolve into a white dwarf without reaching very high temperature.

We have started with H^1 only, and reached most of the abundant elements in the universe, all within one star! Once given the presence of C^{12}, the C—N—O cycle starts at temperatures of about 10 million degrees:

$$C^{12} + H^1 \rightarrow N^{13} + \gamma \rightarrow C^{13} + \beta^+ + \nu$$
$$C^{13} + H^1 \rightarrow N^{14} + \gamma$$
$$N^{14} + H^1 \rightarrow O^{15} + \gamma \rightarrow N^{15} + \beta^+ + \nu$$
$$N^{15} + H^1 \rightarrow C^{12} + He^4$$

and rarely:

$$N^{15} + H^1 \rightarrow O^{16} + \gamma$$
$$O^{16} + H^1 \rightarrow F^{17} + \gamma \rightarrow O^{17} + \beta^+ + \nu$$
$$O^{17} + H^1 \rightarrow N^{14} + He^4.$$

Thus, the important isotopes C^{13}, N^{14}, N^{15}, O^{16}, O^{17} are produced by moderate temperature fusion, once C^{12} exists. Dependent on the rate of convective or circulatory mixing, the C^{12} might be burnt with H^1 in a zone surrounding the core of the star in which it was produced. If not, the C^{12} itself might appear at the surface, and then be lost back to the interstellar gas. In either case, the group of stars formed at a later time from the gas will contain C^{12} and H^1, requisite for the C—N—O cycle, and N^{14} will be synthesized. Thus, the mean stellar composition changes secularly, with the C^{12}/H^1, He^4/H^1, N^{14}/H^1 increasing.

Logically, the production of neutrons is possible as soon as C^{12} has been synthesized, although Ne^{20} may also be important; both react with protons to produce C^{13} and Ne^{21}, and both C^{13} and Ne^{21} react with alpha particles at about 100 million degrees, i.e.:

$$C^{13} + He^4 \rightarrow O^{16} + n$$

and probably less frequently,

$$Ne^{21} + He^4 \rightarrow Mg^{24} + n.$$

While the neutrons have about 2 Mev energy, collisions reduce this very quickly to thermal energies of 10 kev; inside a star the neutrons are captured by any type of nucleus present, in proportion to its abundance and its cross section. Were hydrogen present, the major product would be D^2, which probably occurs only in stellar atmospheres; in the interior of a highly evolved star, however, the heavy elements like Fe capture most of the neutrons and synthesize the elements $A = 100$ to 200. These captures may occur on the slow time scale in red giants, over a period of 10^5–10^7 years, or on the fast scale of 10^1–10^4 seconds during a supernova explosion. While little energy is released during neutron capture, very great alterations are produced in the abundances of the heavy elements which have widely varying cross sections. Figure 7 shows schematically the run of temperature with age and therefore the time scale, together with associated nuclear reactions in stars of different mass. Not all stars (e.g., our Sun) need explode, not all follow the same evolutionary track, but averaging over a large number of different stars, we may expect all types of processes and the successful production of nuclei to be mixed back with the interstellar gas with the presently observed abundance curve. If our Galaxy started as pure hydrogen, we would expect a similar increase in He^4 and heavy elements in the interstellar gas and in young stars born from it. The degree of convective mixing between the center of the star and its atmosphere (on which spectroscopic observations can be made) depends on the opacity and the ionization of hydrogen and helium, and may vary from star to star. As a consequence of such mixing, the surface composition may in some stars reflect the interior conditions, so that spectroscopic observations can reveal nuclear reactions now active in the core. Such deviations from normal composition provide the direct evidence for comparison with the theoretical discussion.

Stellar pathology and composition anomalies

We have combined, with assumptions of varying degrees of reliability, data on the composition of the earth, meteorites, sun, normal stars, and interstellar gas to derive the schematic curve shown in Figure 1. Examination of stellar spectra shows that, after allowance has been made for differences of surface temperature, most stars have essentially the same composition [12]. In general, however, the very oldest stars confirm our most important theoretical prediction, since they have a ratio of heavy elements and metals to hydrogen much smaller than the sun. For example, the fast moving "subdwarfs," stars formed in the first billion years outside the main plane of our Galaxy, have metal abundances of 0.1 to 0.005 that of the sun (which was formed about ten billion years later). Yet, even in these old stars, the relative abundances of one metal to another are very like those in the sun. Therefore, the various fusion reactions, the neutron capture, and the high-temperature iron-peak element production had all begun early in the evolution of the Galaxy. Our ignorance of galactic and stellar

Table 1. Elements with Abundances Significantly Variable from Star to Star.

Element	Remarks
H^1	While normally the most abundant element, sometimes H is deficient by factors of 100.
D^2	Suspected present in excess during solar flares; almost certainly absent in stellar interiors.
He^3	Excessively abundant in two peculiar (magnetic?) B stars, and observed in solar cosmic ray bursts.
He^4	Normally $He^4/H^1 = 0.1$, but in many types of stars $He^4/H^1 \geq 1$. In certain white dwarfs and carbon-rich stars $He/H \approx 10^2$.
Li	Less abundant in most stars than in the earth and meteorites. In stars contracting from interstellar gas and dust clouds Li is enhanced more than 100-fold over the sun. (Surface or pre-planetary reactions.)
Be^9	Variable from star to star, enhanced in magnetic A stars (surface reaction?).
C^{12}	Present in excess by factors up to 20 in red giants, and in two white dwarfs. C/N ratio variable in some hot stars.
C^{13}	Normally $C^{13}/C^{12} = 0.01$ in the earth and stars. In some C^{12} rich stars $C^{13}/C^{12} \approx 0.1$.
N^{14}	Greatly enhanced in late stages of evolution—N/C < 1 normally, but in hot subdwarfs and a few supergiants N/C > 1.
O	Usually constant, except for weakening in peculiar magnetic A stars.
Ne	Strengthened in a few supergiants.
Si	Strengthened in some magnetic A stars.
P	Enhanced 100-fold in a peculiar B star, 3 Cen A.
S	Depleted in 3 Cen A which has excess P.
A	Enhanced in one supergiant.
Mn	Ratio to Fe abnormal in metal-poor stars. Also enhanced in magnetic stars.
Co, Ni	Possibly abnormal ratio to Fe in extremely metal-poor stars.
Ga, Kr	Greatly enhanced in 3 Cen A.
Sr	Enhanced in magnetic A stars, and in "heavy-element" stars.
Y, Zr, Nb, Mo	Enhanced in "heavy-element" stars.
Tc	All isotopes are unstable. Present in cool red giants with abnormal heavy-element content (S stars). Probably Tc^{97}, with 2 million year half-life.
Ba, La	Enhanced in "heavy-element" stars.
Ce-Dy	Same; most rare earths are also greatly enhanced in magnetic A and F stars, with Eu outstandingly strong in some magnetic A stars.

evolution conditions in the first billion years is quite complete, but the similarity of relative abundances indicates that large numbers of stars of large mass condensed, built He^4 and C^{12}, and became red giants and supernovae. The puzzlingly detailed resemblances require that the slow and fast neutron capture processes also occurred in about the same ratios as now, in vanished red giants and collapsing stars.

But let us turn to the present. Earlier I said that our problem was like a detective story—and at the end of the classic detective story there should be a reconstruction of the crime. We cannot bring the participants together in a room, but fortunately, the crime is still being committed again and again by the stars we can observe. The study of the stars of peculiar composition gives a vivid illumination of the past history of the evolution of the elements. "Peculiar" means only significantly different surface composition, since we are not yet able to study details of internal composition. In peculiar stars one or more nuclear reactions have been carried to an extreme, and the ashes of these reactions have reached the surface either by convection, during some unstable phase, or by evaporation of most of the outer envelope. High-energy reactions at the surface also yield readily observable products, although downward mixing dilutes the composition with the enormous mass below the atmosphere. An unpleasant possibility for confusion is accretion of interstellar gas, which would coat all stars with material of uniform composition! However, only in exceptionally favorable cases (low velocity passage through a dense gas cloud) is accretion appreciable.

We will consider a few critical elements in various types of stars, to see the variety of evidence on stellar nucleosynthesis. A résumé of the elements found in peculiar abundance is contained in Table 1. Not all have been analyzed quantitatively, but where factors of ten or more are involved, it is unlikely that refined stellar-atmosphere procedures will alter the general pattern. Absence of an element from Table 1 may reflect only that it is spectroscopically unobservable. If no atomic weight is given, astronomical observation cannot decide which isotope is present.

We have already discussed the nuclear processes correlated with most of the abundance differences. The H-He isotopes are directly involved in the main fusion process. He^3 is probably a spallation product as are Li and Be. The C^{12} is produced by the three-alpha process, and C^{13} by the C—N cycle. In Figure 8 we show spectroscopic evidence for the change of H/He as a result of nuclear fusion. In one pair of stellar spectra (η Leo and ν Sgr) young stars of very high luminosity are involved where the time scale for complete consumption of H is about 10^7 years. In the other pair (two white dwarfs) we see the final, integrated result of the entire history of the star. The white dwarfs are very dense (10^8 gm/cm³) so

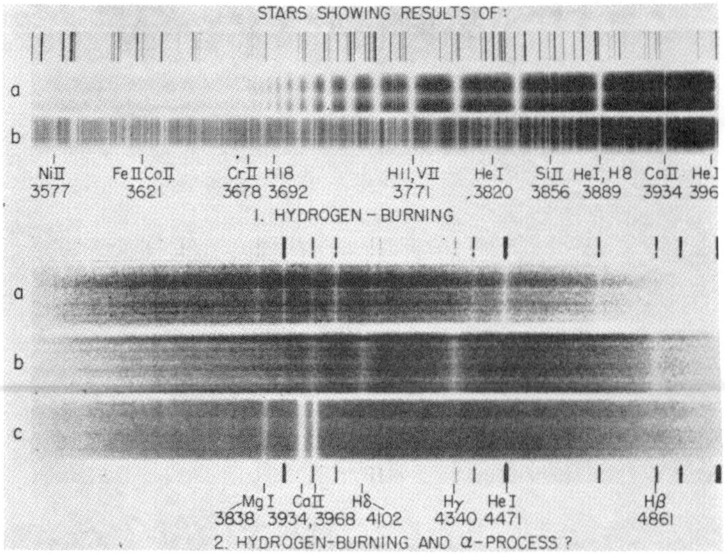

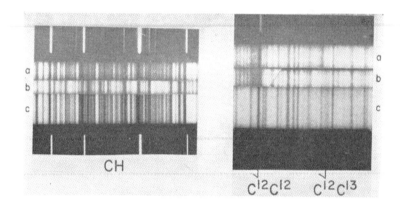

Figure 8. (Photographic negative; absorption lines appear white on a dark continuous background). Effects of conversion of H into He. The upper pair of stellar spectra shows two supergiant stars of approximately the same surface temperature; in star *a*, the hydrogen lines are strong; in star *b* H is weak, He is strong, and all metallic lines strengthened because of the low opacity of He. In the three lower spectra of white dwarfs the upper *a* shows only He, the next *b* only H; *c* has a very strange composition, showing only Ca and Mg.

Figure 9. Red giant spectra of two carbon-rich stars (*a–b*) compared with a normal red giant (*C*). (Photographic positives; absorptions dark on light background). Note that the $C^{12}C^{12}$ carbon bands are greatly enhanced in *a* and *b*, and that $C^{12}C^{13}$ is present only in *a*. The region of the CH band shows that *b* lacks CH, as well as C^{13} and is hydrogen deficient.

that nuclear reactions would proceed at an enormous, explosive rate, were there any nuclear fuel remaining. The white dwarfs must have heavy-element cores, with an atmosphere containing all of whatever lighter elements remain. Figure 8 shows one white dwarf with a hydrogen atmosphere, and another containing only helium. Figure 9 shows three red giant stars, one of normal composition (*β* Her); the other two have nearly the same surface temperatures but are rich [13] in carbon. Note that one star, HD 156074, which has strong C^{12} C^{12} molecular bands, also shows C^{12} C^{13} and the band of CH. The synthesis of C^{12} in the interior apparently was followed by the C—N cycle in an outer thermonuclear zone containing hydrogen, where C^{13} was synthesized in quantity. HD 156074 has strong hydrogen lines, in agreement with the hypothesis that the outer regions of the star are still hydrogen-rich. But in HD 182040, which has even stronger bands of C^{12} C^{12}, the C^{12} C^{13} and CH bands are missing, and there is almost no trace of hydrogen lines; this star has presumably consumed most of its hydrogen and a quantity of He^4 has been burnt into C^{12}.

The neutron capture processes have dramatic effects on the heavy-element composition. One group of stars shows high abundances of heavier elements, for example, Ba, La, and rare earths, in stars in which C is enhanced. Other cooler red giants show the lines of Tc with inten-

sities increasing when Zr/Ti is large. Such details of heavy-element abundances are now being unravelled. The changes of some heavy-element abundances from normal values are by at least a factor of 100. Other unstable elements than Tc (an element first produced in quantity by nuclear fission) may exist but will be difficult to find.

Rather than study in detail the full complexity of peculiar abundances, I would like to conclude with a particularly interesting case history, that of our knowledge [14,15,16] of the element lithium, and the related problems of deuterium and beryllium [17]. Obviously the small electrostatic repulsion factor results in a particularly large thermonuclear disintegration rate for D, Li, Be, and B (I will discuss only Li in detail). Although it appears in the path of some elementary reactions already discussed, its lifetime against proton bombardment is extremely short compared to the ages of stars. Therefore, the normal birthplace of the elements is unsatisfactory; temperatures as high as 0.5 million (for D) to 4 million degrees (for B) are disastrous. Where, then, can Li be synthesized?

The astronomical story is a typical mixture of confusion and progress. In 1951 we found that the solar abundance of Li relative to the metals was about one-hundredth of that in the earth and meteorites [14]. If the initial composition of the earth and sun were identical, mixing of solar

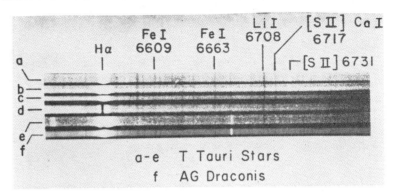

Figure 10. Stars *a* through *e* are T Tauri variable stars with strong Li lines. (Photographic positives). The bright lines indicate a disturbed high temperature envelope. Star *f*, which is not a T Tauri star, has similar bright lines, but the Li line is not present. The absorption lines in *a–e* are diffuse because of rotation.

surface and interior would reduce the Li content. Then in 1959, about 50 stars [15] cool enough to show lines of Li (which is normally highly ionized) were observed. It was found that Li varied in abundance from star to star, but the unexpected result was that they all contained even less Li than did the sun. Again convective mixing with the interior could be invoked. However, the problem was now serious; since no stars show Li excess but only deficiencies, where did the Li in the earth and meteorities originate? In fact, why is there any Li, D, Be, and B at all? The search for objects with too much Li and Be was successful, as has been indicated in Table 1. It was found that Be is also variable from star to star and is greatly enhanced in magnetic stars of type *A* (14,000–8000 °K surface temperature). Two such nearly identical stars as Sirius and Vega have Be deficiencies differing by more than 30, while several magnetic *A* stars have much more Be than does the earth.

There exists a very strange group of faint variable stars (called T Tauri stars) connected with the denser gas and dust clouds of our Galaxy. These stars are apparently in the process of contraction out of the gas, and are still in a somewhat unstable phase. They have rapid light variations, they show emission lines of hot gas clouds surrounding them, and some are connected with diffuse nebulae. Thus there still exists interaction between gas and star, sometimes of a violent kind. For example, the star may brighten by a factor of five in a few minutes, and show a blue, essentially continuous spectrum, which fades away in less than an hour—very much like a solar flare, but on an enormously greater scale. Spectroscopic observations of these faint stars are difficult, but twelve of these stars, essentially all that could be studied [16], have abundances of Li 50 to 400 times the solar value, i.e., some even higher than in the earth. This surprising agreement with terrestrial values, and the fact that T Tauri stars have been through the process of gravitational contraction, suggests that the origins of Li and of stars, and even possibly planets, have something in common. Perhaps even now, the T Tauri stars have planets forming around them as the star goes through its final birth pangs. On the basis of such arguments it might, very speculatively, become possible to predict which newly formed star will have planets 1 or 100 million years from now, the time scale for planet formation, when exploration over stellar distances becomes possible (which may represent an optimist's guess as to the technical difficulties of space travel). However, there is now experimental evidence that the formation of a star is accompanied by the synthesis of lithium and probably of the other light elements.

Such a theory has been advanced by Fowler, Hoyle, and myself [18] in a detailed discussion of both the astronomical and nuclear-physical aspects of this problem. The condensation of a star is not easy, since gravity must work against the magnetic tension of the fields as they become compressed, and against the increasing centrifugal force of stellar rotation. The interstellar medium contains some angular momentum, and on compression the conservation law results in an enormous increase of rotational kinetic energy. The star becomes unstable so that matter is expelled, carrying with it momentum and magnetic energy [19]. The total energy to be dissipated has been estimated to be as high as 10^{48} ergs, an enormous supply, part of which, perhaps 10^{45} ergs, may appear as high-energy phenomena. For example, twisted magnetic fields in a partially ionized plasma may produce high potentials and accelerate particles to relativistic energies, i.e., cosmic rays. Such "flares" at or above the surface of the newly forming star could be responsible for the strange spectra and light variations of the T Tauri stars. It is known that proton collisions above a few hundred Mev will destroy heavy nuclei by "spallation"; commonly, fragments are produced such as neutrons, deuterons, He^3 and He^4, and Li, Be, and B. Thus spallation could produce, outside the star, or in the small discoidal nebula which eventually forms the planets, the elements D, Li, Be, and B, which are otherwise so difficult to explain. On such a hypothesis, they are formed only at the surface, or in the pre-planetary nebula. The high abundance of Li in the earth or meteorites then becomes reasonable.

The argument can be provided with more detailed confirmation by considering the yields of the various light nuclei in spallation. Table 2 gives abundances in the earth and meteorites, the spallation yields and the cross sections for capture of thermal neutrons. Deuterons, and still more important, neutrons, are produced in far greater numbers than are Li^6–B^{11}. However, the final production of deuterons depends both on the $H^1(n,\gamma)D^2$ reaction, which has a cross section of 0.33 barns (10^{-24} cm²) and on the amount of hydrogen present. Note that the spallation yields vary only slightly from Li^6 to B^{11}. Why, then, is Li^6 rare compared to Li^7, and B^{10} rare compared to B^{11}? This puzzle has a solution with important consequences if we note that the cross sections for the destruction of Li^6 by $Li^6(n,\alpha)T^3$ and of B^{10} by $B^{10}(n,\alpha)Li^7$ are very large. Although the spallation yields are similar, it appears that some of the Li^6 and B^{10} formed is later destroyed by neutron capture. But if we are dealing with a gas cloud of low density, any neutron formed will either decay spontaneously or will be captured by the over-

Table 2. Production of Light Nuclei.

Isotope	Terrestrial Abundance	Spallation Yield	Cross Section Thermal Neutrons
Li^6	7.4	283	945 b*
Li^7	92.6	283	0.033 b
Be^9	20	200	0.010 b
B^{10}	4.5	689	3813 b
B^{11}	19.5	195	0.05 b

* One barn (b) = 10^{-24} cm². The abundance and yields are in units of Si = 10^6.

whelmingly abundant hydrogen if the cloud has normal stellar composition. The spallation and subsequent neutron capture must have occurred in the same place, in a relatively hydrogen-deficient solid. The most probable locale is in blocks of ice and dirt, "dirty snowballs" perhaps 10 meters in diameter, circulating at low temperature around the still half-formed sun, during the early history of the solar system. The low temperature is plausible since we believe the sun was larger, but much fainter and cooler than at present. The deficiency of hydrogen is reasonable, since in ratio to solids like Fe or Si, only a small fraction of the cosmic abundance of H could be chemically bound in non-volatile compounds like ices, frozen H_2O, CH_4 or NH_3. The "dirt" consists, presumably, of the first simple compounds of Si, Mg, Fe, and oxygen—i.e., the precursors of the rocks we know. Some of these heavier nuclei are smashed by solar cosmic rays, producing Li^6 . . . B^{11} and neutrons. The neutrons are slowed down in the solid. Some are captured by hydrogen, producing D^2, while others preferentially destroy some of the Li^6 and B^{10} already formed. Thus, detailed study of terrestrial and meteoritic abundances gives clues about events more than 5 billion years ago, when the solar system was being born. Similar events may have occurred at or near the surface of T Tauri stars, producing the observed high Li concentration. Unstable isotopes of heavier elements may also have been formed by such neutron bombardment, e.g., Pd^{107} and I^{129}, resulting in the peculiar isotopic ratios found for Ag^{107} and Xe^{129} in a few meteorites, or perhaps Al^{26} was formed, which would heat and melt the planetary material as it collected. One prediction, that may be eventually subject to experiment, is that the D^2/H^1 ratio in the major planets and in interstellar space is lower than on the earth. Still another is that the Li formed at the star's surface may have a higher Li^6/Li^7 ratio than does terrestrial Li, since the solid, hydrogen-poor stage of evolution may be necessary for the (n,α) reaction that destroys Li^6.

The case history of Li is not yet complete. Experiments in the planetary system and the sun and observations of the stars may raise new puzzles or suggest a different solution. But there is no doubt that the study of the past and of stellar evolution, using the clues provided by nuclear astrophysics, is a most fruitful and stimulating, interdisciplinary field of science. Only a few puzzles have been solved, but a reasonable road to travel has been at least surveyed.

References

1. Goldberg, L., Müller, E.A., and Aller, L.H., *Astrophys. J. Suppl. 5*, No. 45, 1960.
2. Suess, H., and Urey, H.C., *Rev. Mod. Phys., 28*, 33, 1956.
3. Cameron, A.G.W., *Astrophys. J., 129*, 676, 1959.
4. Salpeter, E.E., *Phys. Rev., 97*, 1237, 1955; *Ann. Rev. Nuclear Science, 2*, 41, 1953.
5. *Les Processus Nucléaires dans les Astres*, 1954 (Impr. Ceuterick, Louvain); also known as *Mem. Soc. Roy. Liège*, 4th series, *14*, 1953.
6. Fowler, W.A., Burbidge, G.R., and Burbidge, E.M., *Astrophys. J. Suppl., 2*, 167, 1955.
7. Greenstein, J.L., *Mem. Soc. Roy. Liège*, 4th series, *14*, 307, 1953.
8. Fowler, W.A., and Greenstein, J.L., *Proc. Nat. Acad. Sci., 42*, 173, 1956.
9. Cameron, A.G.W., *Phys. Rev., 93*, 932, 1954; *Astrophys. J., 121*, 144, 1955; Chalk River Rept. CRP-652, 1956 and CRL-41, 1957.
10. Burbidge, E.M., Burbidge, G.R., Fowler, W.A., and Hoyle, F., *Rev. Mod. Phys., 29*, 547, 1957.
11. Fowler, W.A., *Modern Physics for the Engineer*, Series II, 1961, (McGraw-Hill, New York), Chap. 9.
12. Greenstein, J.L., *Pub. Astro. Soc. Pacific, 68*, 185, 1956.
13. Searle, L., *Astrophys. J., 133*, 531, 1961.
14. Greenstein, J.L., and Richardson, R.S., *Astrophys. J., 113*, 536, 1951.
15. Bonsack, W.K., *Astrophys. J., 130*, 843, 1959.
16. Bonsack, W.K., and Greenstein, J.L., *Astrophys. J., 131*, 83, 1961; Bonsack, W.K., *ibid.*, 340, 1961.
17. Bonsack, W.K., *Astrophys, J., 133*, 551, 1961.
18. Fowler, W.A., Greenstein, J.L., and Hoyle, F., in press, *Geophys. J.*, 1961.
19. Hoyle, F., *Quart. J. Roy. Astr. Soc., 1*, 28, 1960.

After the Supernova, What?

J. Craig Wheeler

The search for black holes, neutron stars, and the site of explosive nucleosynthesis is carried on at the endpoint of stellar evolution

Man has been intrigued by exploding stars (whether or not he knew to call them that) at least as far back as the Chinese Visitor's Star of A.D. 1054 and probably into prehistory. Today we know more about this incredibly violent phenomenon, but the intrigue has only deepened. How do stars concoct to leave a pulsar remnant? What stellar explosion might make a black hole as its mark and thus lead us inexorably to consider the mysteries of those unplumbable depths? What finely tuned tinkering of stellar structure is necessary so that the explosion will produce the carbon, oxygen, nitrogen, and heavier elements we see on earth and in the sun, stars, and galaxies beyond? These questions remain unanswered, but as our knowledge improves, the plan of attack on them becomes more clear.

The conception of the nature of supernova explosions has evolved considerably since the basic theoretical scenario was outlined by Hoyle and Fowler (1960) and Fowler and Hoyle (1964) and the pioneering numerical calculations were performed by Colgate and White (1966).

J. Craig Wheeler is an assistant professor of astronomy at the Harvard College Observatory. He received a B.S. degree in physics from the Massachusetts Institute of Technology and his Ph.D. in physics from the University of Colorado in 1969. He was a postdoctoral research fellow at the Kellogg Radiation Laboratory at the California Institute of Technology. His primary research effort has been in the area of supernova dynamics and related topics. Other fields of interest to him are equations of state for dense matter and the general subject of relativistic astrophysics. The present article is based on a talk given at M.I.T. in April 1972. The author would like to acknowledge his close collaboration on many of the subjects discussed here with Drs. Z. Barkat and J.-R. Buchler. Address: Harvard College Observatory, 60 Garden Street, Cambridge, MA 02138.

This evolution of ideas is due to progress on three different fronts. Of prime importance are calculations of the late stages of stellar evolution. These provide detailed pre-supernova models so that it is no longer necessary to assume that the initial models are some "average" star. Progress has also been made in terms of the number and variety of numerical hydrodynamical calculations of the supernova explosions themselves. These calculations have given a better picture of what the mechanism of explosion can, and cannot, be.

The last and perhaps most important contribution to our understanding of supernovae comes from the increasing ability to make contact with direct observations. The prime shot in the arm here was the discovery of pulsars and their identification as rotating neutron stars. We can now search for that particular class of supernovae which satisfy the constraint that they leave a neutron star remnant. This search has promoted a considerable focusing of effort in the last two years.

Although not so dramatic, an equally important observational constraint is that of the element abundances. These observations have existed for decades in more and more refined form, but it is only now that the state of the art of nucleosynthesis calculations begins to allow full use of them. The observed abundances suggest tight constraints on the pre-supernova structure and on the nature of the explosion.

Outline of stellar evolution

When hydrogen is exhausted in the central core of a star, thermonuclear burning no longer supplies the heat to balance the loss of radiant energy at the surface. The star suffers a net loss of energy and contracts under the influence of slightly unbalanced gravitational forces. The nature of systems in static equilibrium, where gradients of thermal pressure balance gravitational forces, is such that in order to remain in equilibrium they must become hotter as they lose energy. They accomplish this feat by contracting enough to provide the energy deficit plus enough extra energy to raise the pressure to balance the stronger gravitational force in the newly contracted state. Throughout their lifetimes stars automatically manage this balance of forces and gradually heat up as they lose energy into space and become more tightly gravitationally bound. Failure to maintain this balance triggers the supernova explosion and signals the death of the star.

As stars evolve and heat up they do not contract in a uniform manner. The regions containing the heaviest elements, the ashes of the various thermonuclear burning stages, contract but the overlying layer of unburned hydrogen expands. The hydrogen layer thus becomes very loosely bound by gravity and susceptible to being driven off as the central regions become ever more dense and more tightly bound.

Once the star reaches this dense-core, rarefied-envelope stage, the envelope ceases to have much effect on the evolution of the star. The envelope, which may contain an appreciable fraction of the mass of the star, determines the appearance of the star but the dense core controls the evolution. For this reason the relevant mass for categorizing the late stages of evolution is the mass of the star when first born, when it begins to burn hydrogen in its center. This initial mass determines the size of the core, which in turn controls the late stages of evolution. In those late stages some portion of the total mass can be lost without significantly altering the outcome of the evolution.

As the central regions condense, being transmuted first to helium and then to carbon and oxygen, a new physical phenomenon enters, that of electron degeneracy. Particles like electrons, protons, and neutrons (those with half-integral intrinsic spin) cannot occupy the same position with the same energy and the same spin orientation. Thus, as the density rises, these particles crowd together but cannot overlap. Then the uncertainty principle comes into play. This principle dictates that as the volume occupied by a particle diminishes, its energy must increase. When the exclusion energy (the particle is "excluded" from occupying a larger volume) of a collection of particles becomes greater than the kinetic energy of thermal motions, degeneracy is said to occur. Electrons become degenerate at much lower densities than the more massive protons or neutrons.

The special characteristic of the energy and hence pressure due to degenerate electrons is that it is very insensitive to temperature. Sufficient degeneracy breaks the cycle whereby a star contracts and heats up as it loses energy. The thermal energy can be removed from a highly degenerate star, thus cooling it, while the degeneracy pressure that supports the star is unaltered. Degeneracy pressure supports the white dwarf stars, which theoretically will cool to near absolute zero with no significant change in structure.

In other contexts the degeneracy pressure, rather than causing extreme stability of the star, leads to its disruption. This takes place when a nuclear fuel ignites under degenerate conditions. Normally a star which is not burning is contracting, as described previously. As it heats up, an element that was previously ash becomes fuel. As the new fuel burns and gives off heat, the temperature and hence the pressure go up and the contraction is halted as the star settles into its new burning phase. But when the star is degenerate the temperature generated by the newly ignited reactions has little effect on

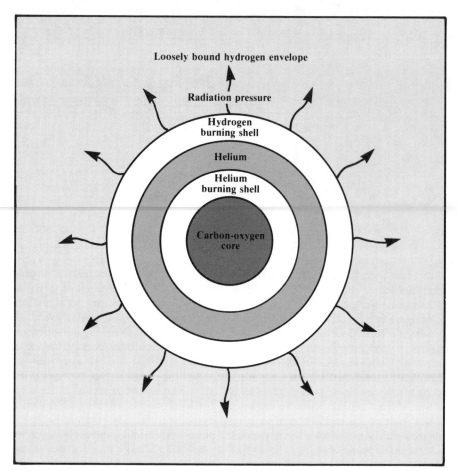

Figure 1. Schematic diagram of the structure of a several solar-mass star in the double-shell burning red-giant phase, when it is susceptible to loss of the loosely bound hydrogen envelope due to the effects of radiation pressure. The outer radius of the star is some 10^4 times the radius of the core and hence is off scale even in this schematic diagram. The degree of the shading denotes qualitatively the relative densities of the various regions.

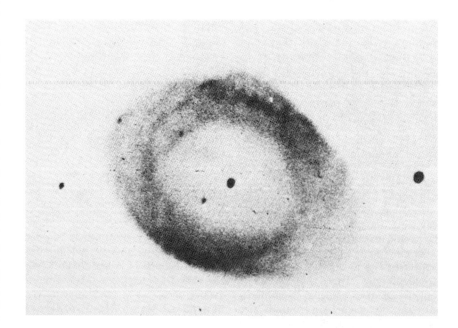

Figure 2. NGC 6720—the well-known Ring Nebula—is a prime example of the planetary nebula phenomenon. This photograph shows clearly the compact central star, which will directly become a white dwarf, and the expelled envelope. (Courtesy Hale Observatories.)

the pressure so the contraction does not halt and the temperature continues to rise. The reaction rates themselves are extremely temperature sensitive, and, by the time the pressure begins to react to the increasing temperature, great amounts of energy can be released, with catastrophic results for the star. Thus with no nuclear burning a degenerate star can be exceptionally stable, but with burning the same star can be equally unstable.

A natural way to discuss the fate of exploding stars is according to their mass. As mentioned previously, the relevant mass in this regard is that which they had when first born. The original mass determines the pattern for the more advanced stages of evolution. One other characteristic of stellar structure should be mentioned in this context. A general rule is that a given stage of central nuclear burning, i.e. hydrogen, carbon, etc., occurs at *lower* densities the *higher* the mass. This is important because it implies that whereas a lower-mass star might ignite a given fuel under degenerate, hence explosive, conditions, a higher-mass star will burn the same fuel in a gentle, unperturbed manner. We will take note of this rule in the ensuing discussion.

Fate of the lowest-mass stars

Starting from lower masses, perhaps the most important evolutionary calculations bearing on the present subject were those of Paczyński (1970). The problem with advanced stages of evolution is that they are subject to thermal instabilities wherein the structure of the star changes by large amounts in a short time as thermonuclear fuels flash on and off in the center and in various shells in the star. To follow these instabilities in any kind of detail requires an untoward amount of computer time.

Paczyński's bold stroke was to assume that while the structural changes during instability were significant they were also transient, and that the star would eventually relax back into its original state. He thus artificially suppressed the thermal instabilities in his calculations. There may be latent dangers in this artificial treatment, but the benefits are immediate. Paczyński was able to calculate evolution from the initial core hydrogen burning phase through hydrogen shell burning, helium core and shell burn-

ing, and finally to carbon and even oxygen ignition in the most massive stars he considered. The mass range investigated was from 0.8 $M_\odot$ to 15 $M_\odot$ ($M_\odot$ equals one solar mass; this is not mere anthropocentrism. The sun is the best studied star, and it is also a very average star, hence eminently suitable as a natural unit. The nightmarish thought, evoked by the solar neutrino problem, that the sun is not so average, is discussed by Fowler 1972).

Paczyński finds that for masses less than 3.5 $M_\odot$ no explosion takes place. A less violent phenomenon begins when the star has completed central helium burning and thus has a carbon-oxygen core surrounded by helium and hydrogen burning shells and the distended hydrogen envelope. In this phase the star is a highly luminous red giant. The pressure due to photons bouncing off electrons—radiation pressure—becomes so large that the loosely bound hydrogen envelope is driven off. The situation is illustrated schematically in Figure 1. The end result of this process is the formation of a planetary nebula by the expelled envelope, an example of which is shown in Figure 2. The central core presumably cools to become a stable white dwarf.

Other investigators, notably Rose (cf. Rose and Smith 1970), maintain that the thermal instabilities in the burning shells may cause the formation of the planetary nebula. The situation may thus be more complicated than that presented by Paczyński. The important point for the present, however, is that below some mass supernovae do not occur due to the formation of planetary nebulae and white dwarfs. Paczyński's cutoff mass of 3.5 $M_\odot$ is roughly in accord with the observed number of white dwarfs.

Weidemann (1971), however, points out that the white dwarf masses predicted by Paczyński show a greater spread than observations allow. Weidemann suggests that added effects, perhaps rotation, will cause stars with mass between about 1 $M_\odot$ and 3.5 $M_\odot$ to form neither white dwarfs nor, at least immediately, supernovae. He suggests a connection to X-ray sources whose number is roughly consistent with a progenitor mass in the range 1–3.5 $M_\odot$. Another suggestion is that, since most stars in

this mass range are members of multiple systems, a white dwarf could indeed form, whereas the envelope would be accreted by the companion rather than forming a planetary nebula. Such a system might very well generate an X-ray source later when the companion star evolves to the red-giant stage and loses mass to the white dwarf.

Carbon flash

For masses between about 3.5 $M_\odot$ and 10 $M_\odot$ Paczyński (1970) found a remarkable phenomenon. In the double-shell burning red-giant phase, the carbon-oxygen cores of all the stars throughout this mass range become essentially identical before carbon ignition. This is indicated by the convergence of the evolutionary paths for the 4 $M_\odot$ and 8 $M_\odot$ stars shown in Figure 3. Barkat (1971) has accounted for this phenomenon in terms of a playoff between neutrino energy losses from the core (the star is utterly transparent to neutrinos; any neutrino created immediately leaves the star carrying its energy with it) and the rate at which mass is accreted onto the carbon-oxygen core from the helium burning shell. An important point is that for this mass range radiation pressure is unable to blow off entirely the loosely bound envelope, thus halting the core-mass accretion; some mass loss is possible without radically altering the scenario. The convergence of the cores to nearly the same structure is significant because it provides a unique pre-supernova model applicable throughout this mass range. The structure of these stars is still qualitatively similar to that given in Figure 1.

Pulsars were discovered nearly concurrently with Paczyński's computations. In an empirical study, Gunn and Ostriker (1970) examined the number of pulsars and their distribution about the galactic plane. More massive stars tend to be found nearer the plane. From this Gunn and Ostriker estimated that the parent masses of the pulsars were between about 2 $M_\odot$ and about 10 $M_\odot$. Just from the number of pulsars they estimated that, in fact, the lower limit could not be much less than about 4 $M_\odot$. Proportionately more stars are found in the lower mass ranges; if the parent mass were much less than 4 $M_\odot$ there would be more pulsars than are observed.

Sofia (1972) has noted that none of the pulsars appears in a binary system, whereas about 75 percent of the stars in the range 4–10 $M_\odot$ do. Assuming that the pulsar-forming supernova disrupts the binary system and that the companions become the so-called "runaway O and B stars" (Blaau 1961), Sofia estimates the mass ejected in the supernova explosion must be greater than 5 $M_\odot$. This is roughly consistent with the conclusions of Gunn and Ostriker. A star with mass less than 5 $M_\odot$ might make a neutron star without disrupting a binary system. In this case the presence of the companion can affect the appearance of the neutron star, and it might be manifest not as an ordinary pulsar but as an X-ray source. This occurs when the companion reaches the red-giant stage and loses mass to the neutron star. The great impact velocities of material raining on the neutron star surface then generate X-rays. The pulsating X-ray source Centaurus X-3 may be the first object identified to be of this genre (Schreier et al. 1972).

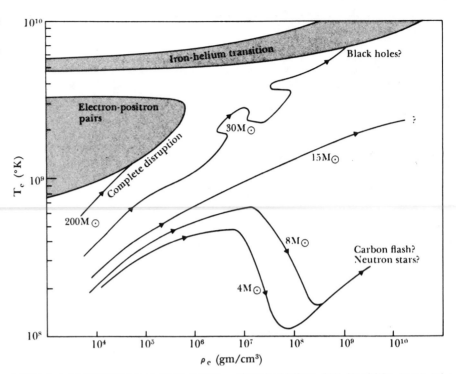

Figure 5. This evolutionary diagram depicts the central conditions of stars of various masses in the central density (ρ_c)–central temperature (T_c) plane. Arrows indicate the direction of the evolution. Also shown are the regions where the iron-helium transition and electron-positron pair formation cause collapse of the stars, and the various (speculated) outcomes for each type of evolution. Both the evolutionary tracks and the masses associated with them here are only qualitatively correct.

The similarity of the mass range where all the cores are identical at carbon ignition and the mass range from which pulsars arose immediately suggested that carbon ignition in these cores triggered the formation of pulsars. In fact, a relevant dynamical calculation had already been performed by Arnett (1969). Carbon ignites under degenerate conditions in these cores, and hence the process is potentially explosive, as discussed earlier. Arnett found that the carbon began to burn rapidly when the nuclear energy generation rate exceeded neutrino losses from the core. Energy transport by convection controlled the burning only temporarily, and an explosion eventually occurred. To the great disappointment of the pulsar-hunters this explosion left no remnant whatsoever!

There ensued a scramble to find reasonable ways of altering the parameters of Arnett's calculation in such a manner that a neutron star remnant would be formed. An idea hit upon by the author and his collaborators (Barkat, Buchler, and Wheeler 1971; Barkat, Wheeler, and Buchler 1972), and independently by Bruenn (1972), was that if the central density of carbon ignition were postponed to higher values, electron capture reactions (of the type proton + elec-

tron → neutron) on the burning products from the explosion would be enhanced. Since the core is supported by the pressure of degenerate electrons, removal of the electrons via electron capture reduces the pressure and raises the possibility that some of the core would succumb to this pressure decrease and collapse to a neutron star state while the rest exploded into space.

Paczyński (1970) found that the carbon ignites at a central density of $\rho_c = 3 \times 10^9$ gm/cm³. Barkat et al. (1971) estimated that if the central density at ignition were 6×10^9 gm/cm³ electron capture reactions would cause the formation of a neutron star. Barkat et al. (1972) calculated that physically reasonable conditions could be found which would yield that high an ignition density. Bruenn, however, did the actual dynamical calculations and found that the real case failed to conform to the expectation of Barkat et al. (1971). Bruenn found that a central density at ignition of greater than 2×10^{10} gm/cm³ was necessary before electron-capture reactions would induce the formation of a neutron star! This ignition density was far above that considered reasonable on evolutionary grounds by Barkat et al. (1972).

Before this dilemma could be fully appreciated the whole situation was conceptually altered by a new suggestion regarding neutrino losses by Paczyński (1972). Neutrino energy losses are very important in shaping the evolution in density and temperature of the carbon-oxygen cores. Barkat et al. (1972) considered a variety of neutrino processes including, in a nominal way, the so-called Urca process. The simplest example of this process is when a proton (p) captures an electron (e) to form a neutron (n) and the neutron subsequently decays back to a proton and an electron. Each of these reactions involves the production and loss of a neutrino, i.e. $p + e \rightarrow n + \nu$ followed by $n \rightarrow p + e + \bar{\nu}$, where ν and $\bar{\nu}$ represent a neutrino and an anti-neutrino, respectively. Thus, the nature of the Urca process is that after a cycle one has the same composition but has lost the energy associated with the two neutrinos. (The process was given its name by George Gamow, who saw the analogy between this nuclear transformation and the gambling operation at the Casino de Urca near Rio de Janeiro. There also, no matter how you played the game, you seemed to lose.)

There is a threshold density active in the Urca process above which the

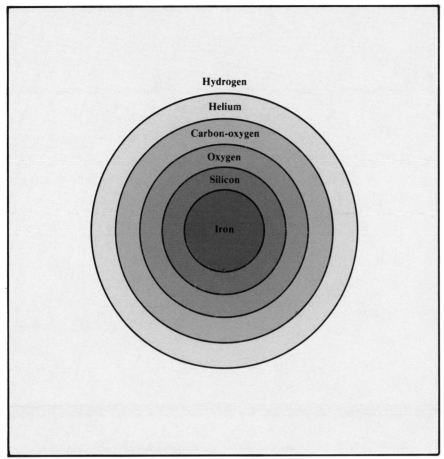

Figure 4. Schematic diagram of the structure of a star with mass in the range 20 $M_\odot$ to 60 $M_\odot$ prior to collapse of the core due to the iron-helium transition. As in Figure 1, the outer edge of the star is way off scale and the shadings denote relative densities.

system must be raised in order to drive the electron capture and then below which it must be lowered to allow the electron decays. The rates for Urca processes for various compositions driven by stellar pulsations have been studied by Tsuruta and Cameron (1970). Paczyński (1972) noted that for the same composition Urca losses would be drastically enhanced by the density variations which occur naturally due to convection. Convection takes place for some time after carbon ignites and stirs material through much larger density variations than are possible with stellar pulsation. The further the density is varied from the threshold for electron capture, the stronger are the Urca neutrino losses.

Paczyński estimated that with solar abundances (that is, presumably' abundances corresponding to those present when the star was born) of the seed nuclei that are susceptible to the Urca process electron captures, the Urca process will prevent carbon from burning rapidly enough to explode. Carbon burning cannot be damped completely since some burning is necessary to cause the convection which drives the Urca losses; however, if detailed calculations support Paczyński the evolution of these stars will be much different than that envisaged previously.

At this point one can only list the fascinating but serious speculations as to what future work will show. As long as carbon continues to burn nonviolently, the density of the core continues to grow even more rapidly as mass is added to the core. This condition leads toward dynamical collapse due to electron capture on seed nuclei or the products of carbon burning, or due to the destabilizing effects of general relativity. Collapse will lead to detonation of the fuel, as increases in density result in burning rate increases, which convective Urca losses eventually will be unable to control. One question is whether the detonation will begin while the star is still nearly in a static state, while it is collapsing full

bore, or somewhere in between. Bruenn's work implies that a nearly static detonation will leave no remnant, but perhaps large densities can be reached by collapse before detonation, with the concomitantly higher post-detonation electron-capture losses. Sufficiently high central density at detonation will lead to continuing collapse of the core.

The most likely possibility is that helium and, particularly, the carbon burning itself would lead to more seed nuclei and hence that the assumption of solar abundances represents a lower limit. On the other hand one might entertain the possibility that the seed nuclei used by Paczyński are consumed in previous nuclear reactions. Not all workers are convinced that such reactions would lead to more seed nuclei. Perhaps the convective Urca loss mechanism works, but not to the efficient degree predicted by Paczyński.

If the Paczyński mechanism functions to limit burning in the center, it may just serve to promote off-center ignition in a shell, a possibility that exists in even the simplest case. The complication of rotation tends to prevent contraction of the core and would also promote off-center ignition.

Off-center ignition does not necessarily lead to detonation and thence to explosion as some suppose. Off-center ignition might lead to some complicated, nonexplosive evolution. If detonation does occur, it is not clear whether it will propagate only outward, perhaps leaving no remnant, or whether it might propagate both ways, giving a chance to form both a neutron star and a supernova explosion. Much work needs to be done in this regard.

In a related work Ostriker and Gunn (1972) have pointed out that if the envelope is supported while a neutron star forms in the core, the protopulsar could pump the star full of magnetic dipole radiation and actually drive the supernova explosion. Only a moderate injection of energy is necessary to give the requisite support to the envelope. Thus it is possible that the various conceivable modes of carbon ignition, central or off-center, do not have to result in the full-blown supernova explosion. They only have to kick the envelope and trigger or accompany the formation of a spinning magnetized neutron star. The role of the pulsar may then prove to be crucial to the existence of this particular supernova mechanism, which was itself designed to account for the existence of pulsars!

Iron-core collapse

The other set of evolutionary calculations which have significantly advanced the quality of pre-supernova models are those of Rakavy, Shaviv, and Zinamon (1967). These calculations were of the evolution of fixed-mass cores, totally neglecting the envelopes on the assumption that they have no effect on the evolution. This treatment simplifies the calculations but leaves some uncertainty as to the original total mass to which a given core mass corresponds. For this reason the masses mentioned in the next two sections are educated guesses not corresponding to any direct calculations.

Between roughly 10 $M_\odot$ and 20 $M_\odot$ carbon ignites nonexplosively under nondegenerate conditions, sometimes off-center. Before explosive oxygen burning takes place, complications like rapid neutrino loss, electron capture, and the effects of general relativity occur. Models of these stars run into numerical difficulties which leave their fate, unfortunately, completely obscure. A schematic evolutionary path for a 15 $M_\odot$ star is illustrated in Figure 3.

Models for stars between about 20 $M_\odot$ and 60 $M_\odot$ are on better footing. Their evolution is typified by that of a 30 $M_\odot$ star in Figure 3. These stars burn all their fuel nonexplosively—carbon, oxygen, and finally silicon—to form a core of iron as shown in Figure 4. That this evolution would occur for some stars was pointed out by Burbidge, Burbidge, Fowler, and Hoyle in 1957. They also discussed the next step, which leads to catastrophe for the star. As the iron core is compressed, the iron decomposes endothermically into helium, protons, and neutrons. The energy to drive this transformation comes from the thermal energy supporting the star, and hence the process results in the collapse of the core.

The natural assumption was that the collapse of the iron core would trigger the collapse, compression, and ignition of overlying unburned nu-

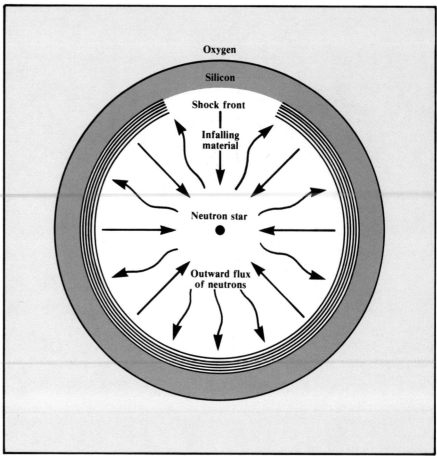

Figure 5. This schematic diagram depicts events after the collapse of the iron core to a neutron star state in stars in the mass range 20 $M_\odot$ to 60 $M_\odot$. As material continues to rain in on the surface of the neutron star, it is heated violently and gives rise to a great flux of neutrinos. A shock wave is generated which propagates into the overlying silicon and oxygen fuels before they begin to collapse as well.

clear fuels, silicon, oxygen, and carbon. Colgate and White (1966) performed the corresponding dynamical calculation and found that when the fuels ignited in free fall the ensuing explosion was swallowed by the continuing catastrophic collapse.

They then invented the ingenious neutrino-transport mechanism. Here it was supposed that a copious supply of neutrinos would be generated as material collapsed to the nuclear densities of neutron star matter. The infalling matter is very hot and dense and not totally transparent to the neutrinos. Under certain assumptions regarding the neutrino transport, Colgate and White found that enough neutrino energy could be redeposited in the infalling matter to create a suitable supernova explosion. The general situation is illustrated in Figure 5.

Wilson (1971) has recently redone this calculation with better treatment of the neutrino transport. He finds that the interaction of the neutrinos with the infalling matter is insufficient to cause any explosion. The collapsing iron core forms a neutron star properly enough, but then there is nothing to prevent the remaining mass of the star from collapsing as well. The total mass of the star is well above the amount that neutron matter can support even if the entire hydrogen envelope, about half the mass of the star, were lost. The star must suffer total gravitational collapse. Thus Wilson predicts that stars in the 20–60 $M_\odot$ range form nothing but black holes!

Wilson's calculations are idealized in certain respects. The inclusion of the effects of rotation or magnetic fields may alter the final outcome of these events. The potentially explosive nature of the layers immediately above the iron core was not taken into account by Wilson. According to Wilson's results these outer layers do not go into free fall immediately, but rather are subject to an outward-traveling shock administered by the core, as indicated in Figure 5. This

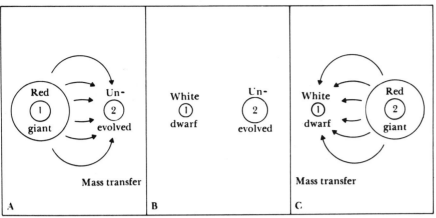

Figure 6. Possible scenario for the production of a Type I supernova. In A the originally more massive star (1) evolves to the red-giant phase and loses an appreciable amount of mass to its binary companion (2), which is still essentially unevolved. This results (B) in a white dwarf (1) composed of potentially explosive fuel orbiting the now more massive unevolved star (2). Finally (C), star (2) evolves to the red-giant phase and transfers mass back to the white dwarf (1). The sudden transfer of mass compresses the white dwarf and can cause explosive ignition of its fuel.

comes about because the collapsing core does halt temporarily in a neutron star configuration. The shock is dissipated in the noncombustible envelopes Wilson incorporated.

The author is presently engaged in a project to explore whether this shock can trigger an explosion in a combustible envelope which will allow the original iron core to be left as a neutron star. Such "conservative" possibilities must be explored thoroughly before we accept the ineluctability of the formation of black holes by these stars.

The Gunn-Ostriker mechanism for driving the supernova explosion by pulsar radiation from the newly formed neutron star is probably not applicable in this case. The layers surrounding the collapsing core are considerably denser and they collapse much more rapidly, the relevant time scale going inversely as the square root of the mean density. If they cannot be exploded from the star, the dense silicon-oxygen layers will rain down on the temporary neutron star yielding a black hole before the pressure of the pulsar radiation builds to full effect. For the carbon flash case, the surrounding helium and hydrogen layers are slower to fall, easier to support, and in general more lenient.

Pair-formation instability

Rakavy, Shaviv, and Zinamon (1967) also calculated the evolution of dense cores corresponding to masses in the range from about 60 $M_\odot$ to several hundred solar masses. Before these stars ignite oxygen, their centers reach the regime of high temperature and relatively low density where photons can be transformed into electron-positron pairs in copious amounts, as illustrated in Figure 3 for a 200 $M_\odot$ star. The photons contribute directly to the support of the star via radiation pressure, whereas most of the energy in the electron-positron pairs is tied up in rest mass which contributes no pressure.

Thus, while creating electron-positron pairs out of photons is energetically favorable, it is disastrous for the star which finds itself with insufficient pressure. The star collapses rapidly and heats up until its oxygen fuel ignites. In this case the result is the complete explosion of the star, leaving no remnant whatsoever. These results have been obtained independently by Barkat, Rakavy, and Sack (1967), and by Frayley (1968).

Some justification is necessary for discussing these very massive stars. These stars are pulsationally unstable in the hydrogen-burning phase and do not approach truly static evolution until they form significant helium cores. In addition, the largest star for which the mass is known is about 60 $M_\odot$. The natural conclusion was that stars more massive than 60 $M_\odot$ shook themselves apart, or, at least, lost mass until they became stable with less than 60 $M_\odot$. This situation held until only recently, when dynamical calculations of the pulsations were actually performed. Indepen-

dently Appenzeller (1970), Ziebarth (1970), and Talbot (1971) found that the pulsations grew to a large but finite size. No disruption at this stage was indicated. On these grounds there was no reason to think that these stars would not then form helium cores, become stable, and evolve to the point where they became pair-formation instability supernovae. The possibility exists that a similar evolution holds through masses up to about 10^5 $M_\odot$, where general relativistic instability is important even in the hydrogen-burning phase.

The existence of these massive stars has been questioned on other grounds by Larson and Starrfield (1971). They point out that radiation pressure acting during star formation may prevent stars with mass greater than 60 $M_\odot$ from ever reaching the hydrogen-burning phase. This mechanism depends on the opacity of the infalling protostar matter, which is sensitive to the abundance of elements heavier than helium. If the results of Larson and Starrfield hold up, there may still be a role for these very massive stars. At times previous to the formation of the galaxies, the abundance of the heavy elements was very low—zero, in fact—according to current cosmological theories. Perhaps then formation of very massive stars was possible or even favored. One can speculate that a pre-galactic phase of massive stars generated the small but finite amount of heavy elements observed in the very oldest stars in our galaxy.

Explosive nucleosynthesis

Marked advances in the theory of nucleosynthesis have been made in the last few years, as outlined by Arnett and Clayton (1970). Some of the most encouraging results come from the so-called "explosive nucleosynthesis" calculations. Using this technique the proper abundance ratios (to some standard, e.g. carbon) are calculated for many elements and their various isotopes ranging from carbon through iron.

The nuclear reaction networks incorporating hundreds of reactions among other hundreds of constituents are too cumbersome to allow complete coupling with detailed hydrodynamic calculations at this time. Instead, density and temperature histories are assumed which are a reasonable first approximation to the behavior in a portion of a star. For instance, the temperature could be assumed to jump to a peak corresponding to the passage of a detonation wave and then to drop as if the star were expanding adiabatically. A corresponding density history then enables the nuclear reaction networks to compute the abundances as a function of time and permits the results to be compared with observation (e.g. Cameron 1968).

Recently, temperature-density histories corresponding to the behavior of selected mass zones from actual dynamical calculations have been used in nucleosynthesis calculations (Arnett, Truran, and Woosley 1971). This is the first step toward a complete calculation coupling nucleosynthesis with the dynamics.

The result of these calculations is the conclusion that explosive carbon burning is able to produce the elements from carbon (atomic weight $A = 12$) through phosphorus ($A = 31$). Oxygen burning produces those from sulfur ($A = 32$) through titanium ($A = 46$), and silicon burning produces the heavier elements up to around iron ($A = 56$). Two aspects of these results should be emphasized, both relating to the complexity of the calculations. On the optimistic side the results are very dependent on the conditions of the explosion, so that when conditions are empirically discovered that give good agreement, intuition dictates that something near the truth is at hand. In counterbalance the calculations are so complex that anything like a uniqueness proof seems highly unlikely. Furthermore, the calculations are sufficiently delicate that it is not at all clear yet that detailed calculations of the structure and dynamics will yield just the conditions that seem to be required.

With the latter provisos in mind the initial model suggested by the nucleosynthesis calculations can be examined. Inasmuch as the observed abundances seem to be uniform over the galaxy and no way is known to mix the effluvia of many widely dispersed supernovae, the assumption seems reasonable that we are looking for one type of supernova that produces a nearly unique set of abundance ratios over the whole range from carbon to iron. On the other hand one should perhaps note that the most detailed observations by which

abundances are determined are those obtained in the solar system, from the sun itself or meteorites. The possibility exists that we are pasting together the model of a unique supernova in the solar neighborhood, say the one which triggered the formation of the planetary system, if there were such a thing.

Be that as it may, to obtain all the abundances from a unique type of event, explosive carbon, oxygen, and silicon burning would appear to be necessary in one explosion. This requires a multilayered or "onionskin" model with a silicon layer under an oxygen layer, in turn under a carbon layer. Fortunately, the natural course of stellar evolution gives this structure in many cases. We see that for masses greater than about 10 $M_\odot$ stars have a hydrogen envelope outermost, followed by a layer of helium. Next comes a layer composed of the burning products of helium, a mixture of mostly carbon and oxygen. Next is a layer of oxygen mixed with the burning products of carbon. Then, if the evolution reaches this far without catastrophe, will be a layer of predominantly silicon, the outcome of oxygen burning.

To a certain approximation a given fuel is found at the same temperature regardless of the star in which it exists. This is a result of the nature of nuclear burning. A certain element can only be found in abundance under conditions of density and temperature where it has been formed by a previous stage of burning but has not yet become the fuel for the next stage. These layers of fuels must have the proper pre-explosion conditions in order to satisfy the requirements of nucleosynthesis. Fortunately the calculations show that, at least roughly, these conditions are met.

What, then, do the nucleosynthesis calculations say about the various supernova models? The carbon flash model does not weigh favorably because at the high densities where it occurs the carbon itself is totally consumed. From the great relative abundance of carbon in the solar system we conclude that the explosive carbon burning must not be allowed to burn all the carbon. The same holds true for oxygen. The next category of stars, those between roughly 10 $M_\odot$ and 20 $M_\odot$, may have the requisite layering to be promising sites for

nucleosynthesis. The unfortunate dearth of knowledge about the late evolution of these stars prevents any stronger statement.

The iron-core collapse models have a very promising structure. As we have seen, however, at present these stars seem more likely to form black holes with no associated supernova explosion. In addition, the fact that other sites for nucleosynthesis are available precludes the argument that these stars are necessary to account for nucleosynthesis and therefore some explosion must have occurred. The pair-formation supernovae also have a structure reasonably compatible with nucleosynthesis. Silicon forms as oxygen burns during the collapse phase. The resulting explosion is not so violent as other cases, being more similar to a rapid combustion than to a detonation. Still, the conditions seem favorable for "explosive nucleosynthesis."

Type I versus Type II

The supernovae we have discussed so far fit roughly in the empirical category Type II (Zwicky 1965). These are thought to be relatively massive stars arising in the later generations of galactic star formation. Type I supernovae, which are distinguished by a characteristic peak and exponential decrease in their light output, are found everywhere Type II are, but they are also found in elliptical galaxies where no Type II has ever been observed. In elliptical galaxies all stars are thought to have been formed in one burst about 10^{10} years ago. Thus one needs to look for a star 10^{10} years old which can erupt as a supernova. The ages of the stars we have previously discussed are less than about 10^8 years at explosion and hence are unsuitable for models of Type I supernovae.

Two classes of models have been proposed that might account for the Type I phenomenon. Finzi and Wolf (1967) pointed out that a white dwarf, which could in principle be stable forever, might be rendered unstable after an arbitrarily long time by a suitable choice of composition that could undergo slow electron capture. This model, which awaits evolutionary verification of its requisite initial conditions, would probably create a neutron star. The other class of model seeks a long pre-supernova lifetime by

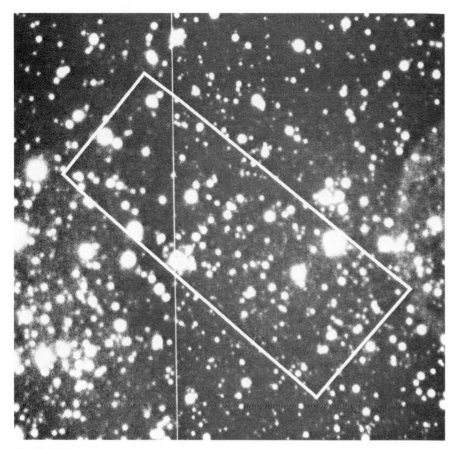

Figure 7. The X-ray pulsar Centaurus X-3 lies somewhere in the indicated error box. Cen X-3 may be a neutron star of very small mass, less than 0.1 M⊙. (Taken by William Liller and James L. Elliot with the 60-inch reflector at the Boyden Observatory, South Africa.)

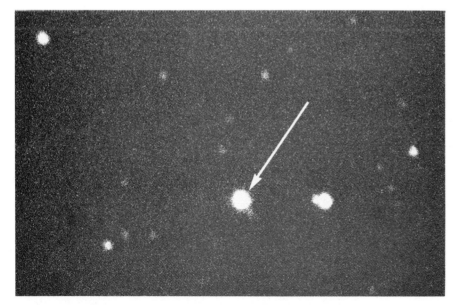

Figure 8. The bright star indicated in the center is the optical counterpart to the X-ray source Cygnus X-1. The mystique of this otherwise ordinary plate is that Cyg X-1 may be the first object to be identified as a black hole. (Taken by William Foreman with the 60-inch reflector at the G.R. Agassiz Station, Harvard, Mass.)

using low mass stars which naturally exist in their ordinary phases for 10^{10} years or more. To generate a supernova explosion it is necessary to put the star into a binary system and invoke mass transfer between the components to provide the necessary "kick" to ignite the explosion. This mass transfer naturally occurs in a binary system when one component evolves to the red-giant phase and creates the characteristic large, loosely bound envelope.

Truran and Cameron (1971) suggested that a proper system could be found in which the more massive star shed its hydrogen envelope to the secondary, leaving its degenerate helium core. As the secondary evolved it would then transfer mass back to the first star, compressing the helium core and triggering an explosion. The basic scheme is illustrated in Figure 6. The details of this process are being worked out by one of their students, T. Mazurek. A similar mechanism involving low mass hydrogen stars, one of which becomes degenerate and hence potentially explosive before mass transfer occurs, has been suggested by Hartwick (1972). This event would happen in the first stage shown in Figure 6.

Future directions

A scenario describing the outcome of the late stages of stellar evolution can be constructed. For original masses below about 3.5 M⊙, stars form white dwarfs. Stars with masses in the range 3.5 M⊙ to 10 M⊙ generate supernovae and leave pulsars as a remnant. An important site for nucleosynthesis may lie with stars in the 10 M⊙ to 20 M⊙ range. Black holes are produced by stars with mass between about 20 M⊙ and 60 M⊙ due to their inability to generate an explosion. The most massive stars, with mass greater than about 60 M⊙, may have played a role only in the very early history of our galaxy, or even at pregalactic times; but they appear to be promising sites for nucleosynthesis.

This brief scenario can now operate as a hypothesis about the outcome of stellar evolution, to be tested as more calculations are performed and more experimental data are collected.

At this writing two new pieces of information await confrontation with the theory. One is the estimate by Wilson (1972) that the X-ray pulsar Centaurus X-3 (see Fig. 7) is a neutron star with mass on the order of $0.1 \ M_{\odot}$. On simple grounds one might expect neutron stars of mass about $1.4 \ M_{\odot}$, since that is the mass of the exploding core in the carbon flash model. Do the dynamics actually conspire to leave only a small remnant? If the rest of the core is exploded away, is too much neutron-rich material produced in this event to square with observations? If so, is this neutron star a rarity, produced by some different mechanism than is the ordinary pulsar?

The other piece of information concerns the identification of the optical counterpart of the X-ray source Cygnus X-1 by Bolton (1972), shown in Figure 8, and the estimate that the mass of the secondary must be greater than $3 \ M_{\odot}$. This is greater than the maximum mass a neutron star can support, and hence Cyg X-1 is probably a black hole. Is such a black hole consistent with our theoretical expectations? Such questions are bound to arise faster than we can answer them in the near future.

References

Appenzeller, I. 1970. *Astr. and Ap.* 5:355.

Arnett, W. D. 1969. *Ap. and Space Sci.* 5:180.

Arnett, W. D., and D. D. Clayton. 1970. *Nature* 227:780.

Arnett, W. D., J. W. Truran, and S. E. Woosley. 1971. *Ap. J.* 165:87.

Barkat, Z. 1971. *Ap. J.* 163:433.

Barkat, Z., G. Rakavy, and N. Sack. 1967. *Phys. Rev. Lett.* 18:379.

Barkat, Z., J.-R. Buchler, and J. C. Wheeler. 1971. *Astrophys. Lett.* 8:21.

Barkat, Z., J. C. Wheeler, and J.-R. Buchler. 1972. *Ap. J.* 171:651.

Blaau, A. 1961. *B.A.N.* 15:265.

Bolton, C. T. 1972. *Nature* 235:271.

Bruenn, S. W. 1972. *Ap. J. Suppl.* 24:283.

Burbidge, E. M., G. R. Burbidge, W. A. Fowler, and F. Hoyle. 1957. *Rev. Mod. Phys.* 29:547.

Cameron, A. G. W. 1968. In *Origin and Distribution of the Elements*, L. H. Ahrens, Ed. New York: Pergamon Press, p. 125.

Colgate, S. A., and R. H. White. 1966. *Ap. J.* 143:626.

Finzi, A., and R. A. Wolf. 1967. *Ap. J.* 150:115.

Fowler, W. A. 1972. *Nature* 238:24.

Fowler, W. A., and F. Hoyle, 1964. *Ap. J. Suppl.* 9:201.

Frayley, G. S. 1968. *Ap. and Space Sci.* 2:96.

Gunn, J. E., and J. P. Ostriker. 1970. *Ap. J.* 160:979.

Hartwick, F. D. A. 1972. *Nature Phys. Sci.* 237:137.

Hoyle, F., and W. A. Fowler. 1960. *Ap. J.* 132:565.

Larson, R. B., and S. Starrfield. 1971. *Astron. and Ap.* 13:190.

Ostriker, J. P., and J. E. Gunn. 1972. *Ap. J. Lett.* 164:L95.

Paczyński, B. E. 1970. *Acta Astron.* 20:47.

Paczyński, B. E. 1972. *Astrophys. Lett.* 11:53.

Rakavy, G., G. Shaviv, and Z. Zinamon. 1967. *Ap. J.* 150:131.

Rose, W. K., and R. L. Smith. 1970. *Ap. J.* 159:903.

Schreier, E., R. Levinson, H. Gursky, E. Kellogg, H. Tananbaum, and R. Giacconi. 1972. *Ap. J. Lett.* 172:L79.

Sofia, S. 1972. *Ap. J.* 172:53.

Talbot, R. J. 1971. *Ap. J.* 165:121.

Truran, J. W., and A. G. W. Cameron. 1971. *Ap. and Space Sci.* 14:179.

Tsuruta, S., and A. G. W. Cameron. 1970. *Ap. and Space Sci.* 7:374.

Weidemann, V. 1971. *Astrophys. Lett.* 9:155.

Wilson, J. R. 1971. *Ap. J.* 163:209.

Wilson, R. E. 1972. *Ap. J. Lett.* 174:L27.

Ziebarth, K. 1970. *Ap. J.* 162:947.

Zwicky, F. 1965. In *Stellar Structure*. L. H. Aller and D. B. McLaughlin, Eds. Chicago: University of Chicago Press, p. 367.

The Origin of Galaxies

Richard B. Larson

How can we account for the observed diversity of characteristics of galaxies?

Most of the visible matter in the universe is contained in stars, and most of the stars are concentrated in galaxies (Fig. 1)—large independent stellar systems or "island universes" held together by gravity, with masses between about 10^5 and 10^{13} solar masses. Our own Milky Way galaxy is believed to be a typical representative of the most numerous class of galaxies, the spiral galaxies, which are relatively flat and disclike in structure, with a central bulge; most flat galaxies show a spiral pattern in the disc, often with a central straight section or bar. Galaxies of the second major type, the elliptical galaxies, are relatively round and featureless, with ellipsoidal shapes and centrally condensed mass distributions. A minority of galaxies are so irregular or distorted in appearance that they defy any simple geometric classification. Even the more regular galaxies, if examined closely, show so many individual differences and peculiarities that it is impossible for any simple summary of properties or any selection of textbook photographs to do full justice to their diversity. A magnificent collection of galaxy photographs, well worth the perusal of the interested reader, has been published in *The Hubble Atlas of Galaxies* (Sandage 1961).

After obtaining his Ph.D. at the California Institute of Technology in 1968, Richard B. Larson joined the faculty of the Astronomy Department at Yale University where he has been Professor of Astronomy since 1975. His research has included computational studies of stellar dynamics and of the dynamics of collapsing gas clouds as relevant to the formation of stars and galaxies. Address: Yale University Observatory, Box 2023 Yale Station, New Haven, CT 06520.

Figure 1. Part of the cluster of galaxies Abell 1367, photographed by Augustus Oemler, Jr., of Yale University, using the 4-m telescope of the Kitt Peak National Observatory. This cluster contains several hundred galaxies of various types; perhaps one-third of them are within the area shown here. Elliptical galaxies predominate in the dense central part of the cluster, while some spiral and irregular galaxies are seen in the outer parts. The distance of the cluster is roughly 80 Megaparsecs, and the angular field of view shown here is about 30 × 40 minutes of arc, corresponding to about 0.7 × 0.9 Megaparsecs at the distance of the cluster (1 parsec = 3.3 light-years).

The elliptical galaxies (Fig. 2) appear to constitute the most homogeneous class, having similar radial surface-brightness profiles, although they vary considerably in their apparent flattening and its variation with distance from the center. Their spectra indicate that they consist predominantly of old stars formed approximately 10^{10} years ago, and in most cases there is no evidence that they contain young stars or interstellar gas or dust. However, some elliptical galaxies show anomalies such as patches or filaments of dust and gas, concentrations of young stars, and even strong radio emission. A particularly remarkable and occasionally spectacular phenomenon sometimes observed in such galaxies is the emission of large amounts of energy from compact nonstellar sources of unknown nature (black holes? dense clusters of pulsars?) in the galactic nucleus; the nucleus can even outshine the rest of the galaxy, causing the system to appear as a quasi-stellar object, or quasar, when observed at a great distance.

The spiral galaxies are more complex in structure and correspondingly more varied in their properties. The central bulge component can vary in relative prominence from inconspicuous to dominant; the content of gas, dust, and young stars may vary from undetectable to dominant in the appearance of the galaxy; and the spiral pattern, if present, can take a seemingly limitless variety of forms, sometimes with a beautiful symmetry but usually with a more irregular appearance. Hubble (1936) noted some correlations among these properties and proposed a classification sequence beginning with galaxies having prominent bulges and smooth, tightly wound spiral arms, and progressing through systems with smaller bulges and more prominent, open spiral arms, increasingly marked by bright concentrations of young stars and ionized gas (Figs. 3–5). However, such correlations are far from perfect, and van den Bergh (1976) has recently proposed a two-dimensional classification scheme that treats the disc-to-bulge ratio and the relative content of gas and young stars (which dominate the spiral pattern) as independent parameters.

Anomalies or irregularities are also fairly common among spiral galaxies; some galaxies appear distorted or asymmetrical in structure, and the discs of many spiral galaxies, including our own, are warped at the edges, sometimes by very large amounts. Some spiral galaxies, again including our own, contain at their centers particularly massive and dense concentrations of gas and young stars. Finally, nuclear activity closely resembling that in the quasars is seen in some spiral galaxies called Seyfert galaxies, and there is strong evidence for a continuity in properties between Seyfert galaxies and quasars.

Figure 2. Hubble's classification sequence is illustrated by Figures 2–5. The giant elliptical galaxy M87 in this photograph is smooth in structure and nearly round in outline. The very sharp bright nucleus, which cannot be seen in the burned-out central part of this reproduction, resembles in some respects a "mini-quasar." The many apparently stellar objects in the outer part of this galaxy are globular star clusters, each containing at least a million stars. (Figs. 2–5 and 15 from *The Hubble Atlas of Galaxies,* courtesy Allan R. Sandage.)

Figure 3. The "Sombrero galaxy" M104 is dominated by the smooth elliptical-like bulge component, which contains globular clusters like those in M87 (Fig. 2). The thin flat disc component is evident mainly as a dark band caused by the absorption of bulge light by dust particles in the disc. The disc is relatively smooth, and the spiral arms are relatively tightly wound and inconspicuous.

Figure 4. The nearby large spiral galaxy M81 has a smaller central bulge and more prominent and open spiral arms than M104 (Fig. 3). The spiral arms, which are more symmetrical in M81 than in most galaxies, are marked not only by concentrations of bright stars and ionized gas but by dark dust filaments.

Figure 5. The giant spiral galaxy M101 has an inconspicuous central bulge and is dominated by the relatively open and patchy spiral pattern. The bright patches are regions of ionized gas produced by massive hot stars recently formed in the spiral arms.

Evolutionary processes

How can this great diversity of galaxy properties be accounted for? It is important to understand first which properties and which differences can be explained in terms of the normal evolutionary processes and dynamics of galaxies and which remain to be explained by the galaxy formation process. Several of the most conspicuous features of galaxies depend on the presence of gas and the continuing formation of stars from it; these include the appearance of the spiral arms and probably even their existence, since they are evident primarily as concentrations of gas and young stars, and the presence of nuclear activity, which is probably related to the condensation of gas at the center of a galaxy. The gas in a galaxy is steadily consumed by star formation, and unless it is somehow replenished the gas content will decay, and such related features as prominent spiral structure and nuclear activity will gradually die out as the galaxy ages. Thus those differences between galaxies that are related to the gas content might result from differences in age or in the rate of gas consumption.

However, processes other than star formation can also alter the gas content of galaxies. Under some conditions, supernova explosions may heat gas sufficiently to drive it out of a galaxy in a hot outflowing wind similar to the solar wind. A galaxy can also be swept clean of its gas by rapid motion through an external intergalactic medium. One or both of these processes probably accounts for the absence of detectable gas from most elliptical galaxies and from some disc galaxies; the fact that these gas-free galaxies tend to be concentrated in dense clusters containing hot intergalactic gas (as observed at X-ray wavelengths) suggests that the sweeping mechanism plays a role in removing their gas. On the other hand, it is also possible for the gas content of a galaxy to be maintained or even increased by the infall of uncondensed surrounding gas left over from the galaxy formation process, or even of gas previously lost from galaxies. Accretion of gas is favored over sweeping when the velocity of the galaxy relative to the surrounding gas is small and the gas is relatively cool. In some nearby groups of galaxies, extended envelopes or clouds of cool intergalactic gas are in fact observed around some particularly gas-rich galaxies, so it is possible that recent gas infall has been important in rejuvenating these galaxies.

Dynamics of stars

The random orbital motions of the stars in a galaxy tend to smooth out any initial irregularities in their distribution, producing after only a few orbital periods a system with a relatively regular and symmetrical structure. It is therefore not surprising that the galaxies with the oldest stellar content—i.e. the gas-free elliptical and disc galaxies—generally also have the smoothest and most regular structures, since this means that the "astrophysical age," measured by the average age of the stars, is correlated with the "dynamical age," indicated by the degree of regularity of the mass distribution.

The regularity and symmetry of a galaxy can, however, be temporarily destroyed by effects such as the accretion of new material, a close encounter with another galaxy, or even a direct collision and coalescence with another galaxy. If two galaxies pass near each other at low velocity, they are subjected to several tidal distortions, and material can even be thrown out in extended sheets or filaments. Some examples of tidally distorted or colliding galaxies are shown in Figures 6–9. Such tidal interactions have been studied by Toomre and Toomre (1972, 1973) and others who, using numerical simulations of galaxy encounters, have been strikingly successful in explaining a number of very peculiar or distorted galaxies, including some with long narrow arms or filaments that can be interpreted as tidally ejected material (Figs. 7 and 8). Also successfully explained are the rare and particularly interesting ring galaxies (Figs. 10 and 11), which are dynamically very unstable and hence must be of very recent origin; they can be produced by the passage of an intruding galaxy through the center of a disc galaxy, causing a strong compression wave or ripple to propagate outward in the disc (Lynds and Toomre 1976).

When galaxies undergo a close encounter or collision, tidal drag forces will dissipate their energy of relative motion and may cause them eventually to spiral together and merge. A related effect called dynamical friction acts on a small galaxy when it enters the outer envelope of a larger galaxy and can cause the smaller galaxy to spiral inward and merge with the larger one. The number of galaxies with a disrupted appearance suggestive of a recent tidal encounter or merger is significant, and Toomre has suggested that after the chaos has died out and stellar motions have reestablished a regular structure, they will resemble elliptical galaxies; many of the elliptical galaxies that we observe may even have originated in this way. As we shall see, models of galaxy formation that involve only stellar dynamics do not directly reproduce the observed dense cores of elliptical galaxies, but if an appreciable amount of gas remains in such a system, the gas can continue to condense at the center and form a dense nucleus.

A disturbance to a disc galaxy such as a tidal encounter may generate wavelike density fluctuations, or density waves, that propagate through the galaxy. A special type of density wave that is possible in some circumstances and has been extensively studied is a nearly stationary spiral wave pattern with a two-armed symmetry; such spiral density waves may help to explain many observed features of the spiral structure in galaxies (Shu 1973; Wielen 1974; Roberts, Roberts, and Shu 1975). However, such density wave patterns cannot last indefinitely—they decay in about 10^9 years—so the observed persistence of spiral structure for longer times requires a regeneration mechanism or continual input of energy into the wave pattern. Tidal encounters may sometimes serve to regenerate a spiral pattern, but it seems more likely that in most cases the presence of gas and the tendency of the gas to form condensations play an important role. A spiral pattern, perhaps with a central bar, can also develop spontaneously as a favored mode for the large-scale instability of a gas disc (Bardeen 1975). In any case, a conspicuous spiral pattern is probably a transient feature of the evolution of disc galaxies, being dependent on the presence of a sufficiently high gas content.

Gas dynamics and star formation

The primary distinguishing feature of the dynamics of gas in a galaxy as

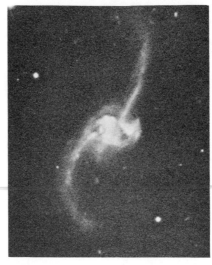

Figure 6. Figures 6–8 illustrate some effects of gravitational interactions between galaxies; in particular, they show a possible sequence of tidal interaction and coalescence of two spiral galaxies, leading perhaps to the formation of an elliptical galaxy as suggested by A. Toomre. NGC5257 and 5258, shown here, are a pair of strongly tidally interacting spiral galaxies. (Figs. 6–11 from the *Atlas of Peculiar Galaxies,* courtesy Halton Arp and Hale Observatories.)

Figure 7. NGC 2623 probably represents a later stage of tidal interaction when long tails have been ejected, as predicted by numerical simulations, and the main bodies of the interacting galaxies have partially merged.

Figure 8. NGC7252 has a number of looping filaments or tails suggestive of tidal interaction, yet only a single main body is evident; perhaps two colliding galaxies have here merged into one, which will eventually resemble an elliptical galaxy.

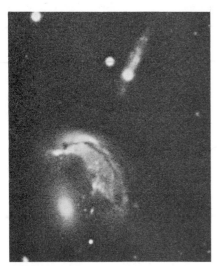

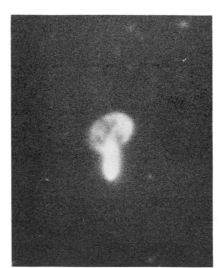

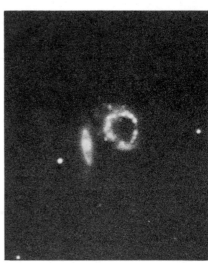

Figure 9. Some further possible effects of gravitational interactions are illustrated in Figures 9–11. NGC2936 and 2937 (Fig. 9) constitute a system in which one of the galaxies, containing dust and concentrations of gas and young stars, is distorted into a spiral shape suggesting that it is being gravitationally drawn toward and accreted by the other galaxy. In this case, because of its large orbital angular momentum, it may eventually form a disc or ring orbiting around the other galaxy.

Figure 10. This system, known as Mayall's object or Arp 148, may be a colliding system in which one galaxy is passing through another. Lynds and Toomre (1976) have shown that the passage of a second galaxy through the center of a disc galaxy can cause the disc to be temporarily transformed into a ring, so this object might represent an early stage in the formation of a ring galaxy like that illustrated in Figure 11 (Arp 147).

Figure 11. This system, known as Arp 147, is an example of a ring galaxy with a nearby companion that may have been responsible for producing the ring as a result of a recent collision.

compared with stellar dynamics is that moving gas elements or gas clouds collide with each other, usually supersonically, creating shock fronts that dissipate the kinetic energy of the gas motions. When the gas in a galaxy loses kinetic energy, the gravitational field of the system causes the gas to condense into a more compact spatial distribution: if it has little angular momentum or loses angular momentum because of viscosity or other effects, it concentrates at the center of the galaxy, whereas if it has a large angular momentum it condenses into a thin disc orbiting around the center.

As would therefore be expected, in most galaxies the largest concentrations of gas and young stars are observed to be either in the central regions or in an extended thin disc (or both, as seems to be the case in our galaxy). Also, there is a strong correlation between the flatness and the gas content of galaxies; the flattest galaxies generally have the largest gas contents, and nearly all the galaxies with large amounts of gas are flat disc systems (or systems that can be interpreted as discs that have recently been disrupted by encounters with other galaxies). We can therefore readily understand the predominance of disc galaxies in the universe if we suppose that galaxies were formed from gaseous material and that in most cases it remained in the form of gas long enough for it to settle into a disc before forming stars.

Gas-dynamical effects may also help to explain some apparently anomalous features of galaxies. Since the gas in the inner part of a spiral galaxy dissipates its random motion and settles into a disc more rapidly than the gas in the relatively tenuous outer parts, there will be a stage in the evolution of a spiral galaxy when the outermost material has not yet settled into the plane of the inner part; if this outer material initially has a somewhat different plane of rotation, the system may then appear warped. The frequent occurrence of warps in spiral galaxies may then be explained if these systems are dynamically young, or if their outer parts contain material that has been added to them only recently.

Another effect of gas dynamics in galaxies with a density wave pattern is a loss of angular momentum from the gas as it passes through shock fronts associated with the density wave, causing the gas to spiral inward; this effect is expected to be particularly severe in barred spirals, where the central bar may act as an efficient mechanism for funneling gas into the nucleus (Sørensen, Matsuda, and Fujimoto 1976). The presence of concentrations of gas and young stars in the nuclei of barred spirals is in fact a common feature, and perhaps similar mechanisms may help to concentrate gas into the nuclei of Seyfert galaxies and quasars.

If a part of the gas in a galaxy is somehow compressed to a density much higher than the average density of matter in the galaxy, it can cool by the emission of radiation and become gravitationally unstable; that is, its self-gravity can become dominant over both the internal gas pressure and the tidal dispersive forces due to the rest of the galaxy and thus can cause the compressed gas to collapse to form stars. Recent discussions of star formation have emphasized the probable role of shock fronts in compressing the gas to the densities required for stars to form. For example, in the density wave theory of spiral structure most of the observable manifestations of the wave pattern are caused by the compression of interstellar gas in a spiral shock front, triggering the formation of the young stars that most conspicuously delineate the spiral arms (Shu 1973).

A possible direct demonstration of the effectiveness of shock compression in producing rapid star formation is provided by the observation that some galaxies which appear to have suffered violent disturbances or collisions also appear to have experienced recent intense bursts of star formation, as judged from their very blue colors, which indicate a high content of young stars. However, because of many complex details of the star formation process that remain to be better understood, there is as yet no quantitative theory relating the rate of star formation to factors such as the strength of shock compression.

Galaxy formation

We have seen that many observed properties and differences of galaxies can be understood as resulting from normal evolutionary processes or dynamical phenomena occurring after galaxies have formed. However, there remain several basic structural properties of galaxies, such as their sizes, masses, angular momenta, and flattenings or disc-to-bulge ratios, that cannot be explained in terms of evolutionary processes and must therefore be determined by the formation process; except for the effects of occasional collisions mentioned above, no processes are known that could greatly alter these fundamental characteristics of galaxies after they are formed. A further understanding of the properties of galaxies therefore requires an understanding of the galaxy formation process (Larson 1976b).

At present we have little direct information about how galaxies form, so our account must now become more speculative. It is believed that the universe is expanding from a previously much denser and hotter state (the "big bang"), in which the matter was distributed much more uniformly and the large density fluctuations now observed as galaxies and clusters of galaxies were not present or were much smaller. Although the universe as a whole is probably not gravitationally bound and will continue to expand forever (Gott, Gunn, Schramm, and Tinsley 1974, 1976), regions of higher than average density may still be gravitationally bound and may therefore evolve independently of the rest of the universe, reaching a maximum radius and then recollapsing under gravity. The densest such regions will stop expanding at relatively small maximum radii and recollapse first, while regions of lower initial density will expand to larger radii and recollapse at later times, as illustrated in Figure 12, which shows the time dependence of the radius of regions with the same mass but with differing initial density. Because a region of excess density in the early universe will in general have a point of maximum density surrounded by a zone in which the density decreases outwards, the center of such a density enhancement will recollapse first and material farther and farther from the center will turn around and fall back at later and later times, producing a continuing infall of material into the central part of the recollapsing region.

How does this simple picture of the expansion and recollapse of primordial density fluctuations relate to the formation of galaxies? The simplest possibility is that the density fluctuations had masses comparable to those of the presently observed galaxies and expanded to radii comparable to or somewhat larger than the observed radii of galaxies; the collapse of these galaxy-sized condensations or "protogalaxies" may then have formed directly the galaxies that we observe. This simple hypothesis has formed the basis of most attempts to understand galaxy formation.

However, more complicated sequences of events may also have taken place: for example, the recollapsing density fluctuations may have been smaller than the present galaxies, forming small objects (such as gas clouds, star clusters, or small galaxies) that later accreted together to form larger systems. It is also possible that the recollapsing regions began with very large sizes comparable with those of clusters of galaxies rather than those of individual galaxies and that they later fragmented to form galaxies. However, this latter possibility seems less likely in view of the fact that many clusters of galaxies appear to have such long collapse times that they have only recently stopped expanding and begun to collapse; the clusters of galaxies thus appear to be younger than the galaxies themselves.

A still more complex (and speculative) possibility is that the early universe was highly turbulent, either as the result of a primordial state of chaos with which the universe was born or as a result of the early formation of condensed objects that then released large amounts of energy into the pregalactic universe. Galaxies may then have originated as turbulent eddies or as condensations generated by the primordial turbulence.

The various theories that have attempted to understand the origin of galaxies as a result of either the gravitational amplification of small initial density fluctuations or the development of primordial turbulence have been comprehensively reviewed by Jones (1976). Unfortunately, despite much work on the subject, these theories have as yet yielded few real predictions and have provided no convincing explanation of even such basic properties as the sizes and masses of galaxies. This may in part be because the early development of galaxies is a more complex process than is envisioned in any of the theories, involving gravitational interactions between protogalaxies, collisions, coalescence, disruption, expulsion of matter by winds, etc., so that it is difficult to identify the observed galaxies directly with density fluctuations or turbulent eddies in the early universe.

A possibly more fruitful approach to understanding the sizes and masses of galaxies has recently been considered

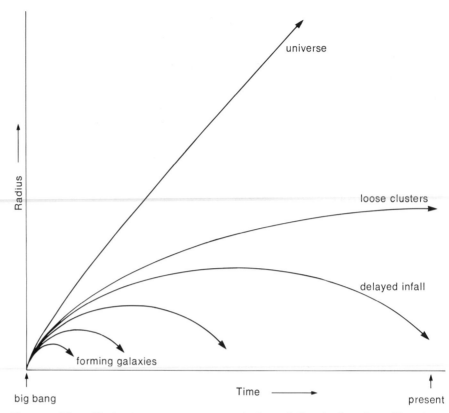

Figure 12. The radii of regions representative of different parts of space may vary with time as illustrated here; all regions contain the same mass (e.g. the mass of a typical galaxy), but they begin with slightly different initial densities. A typical part of the universe is believed to be unbound and will expand forever, while galaxies are believed to have formed in regions of higher density that were gravitationally bound and recollapsed. The less dense peripheral regions of protogalaxies fell in at later times, and loose clusters of galaxies may only recently have begun to collapse or may be still expanding.

by Ostriker and Rees (in prep.). They propose that the characteristic dimensions and masses of galaxies are largely determined by astrophysical processes occurring in collapsing protogalaxies rather than by the poorly understood properties of pregalactic density fluctuations. In particular, they suggest that collapsing protogalaxies are heated to high temperatures by shock fronts generated by supersonic collapse motions and that the formation of stars and a galaxy can only occur if the time required for the gas to cool again is shorter than the collapse time. This requirement is met if the mass does not exceed about 10^{12} solar masses and the radius does not exceed about 75 kiloparsecs, which may help to explain why the observed maximum masses and radii of galaxies are indeed of this order.

Collapse models

Most of the recent progress in understanding the basic structure of galaxies has come from the calcula-

tion of detailed models for the collapse of protogalaxies. Since the state of the matter in a protogalaxy when it begins to collapse is not known, different types of initial conditions have been considered. The simplest possibility is that all the pregalactic gas is quickly turned into stars before a protogalaxy begins to collapse; then only stellar-dynamical processes need be considered during the collapse. This might conceivably be the case during the formation of the elliptical galaxies, most of which appear to contain only old stars and no gas.

Calculations of the collapse of a system of stars (Gott 1973, 1975) show the expected tendency of stellar motions to establish a smooth and regular structure, and the resulting systems have moderate flattening and a density that decreases smoothly outward, qualitatively resembling the observed properties of elliptical galaxies. Quantitative agreement with the density distributions of the envelopes of elliptical galaxies can be obtained if an extended envelope is built up by the continuing infall of stars from a surrounding region of lower density. However, a shortcoming of such purely stellar models is that they do not directly reproduce the observed dense cores of elliptical galaxies, nor do they explain the observed radial variation of color, which is probably due to an inward increase in the relative abundance of heavy elements in the stars. These properties of elliptical galaxies suggest that gas condensation as well as stellar dynamics is important for the formation of their central regions and that the inflowing gas in a forming galaxy is progressively enriched in heavy elements ejected from evolving stars.

Other variants of the collapse of a system of stars have also been proposed. Binney (1976) has suggested that pregalactic gas clouds first collapse into flat sheets and form thin layers of stars that then undergo a further stage of collapse to form elliptical galaxies. In this model, the flattening of elliptical galaxies is a vestige of the original highly flattened structure rather than a result of rotation; if recent indications that elliptical galaxies do not rotate fast enough to explain their flattening are verified, this point of view may receive some support. Toomre (1974, in prep.) has suggested that many elliptical galaxies have formed as a result of collisions and mergers between smaller disc galaxies and has shown by a numerical simulation of a collision between two stellar discs that the resulting system is nearly round and has a density profile very similar to that observed for the envelopes of elliptical galaxies.

A second type of model proposed to explain the formation of elliptical galaxies is based on the assumption that protogalaxies are still gaseous when they begin to collapse, so that their development is dominated by gas dynamics and continuing star formation rather than by stellar dynamics (Larson 1974, 1975). It seems likely that the gas in most protogalaxies, even if initially very hot, would rapidly cool and form dense condensations, either because of gravitational instabilities or because of compression by any remaining hot gas; accordingly, it has been assumed in these calculations that the gas is distributed in discrete clouds. The random motions of the clouds are dissipated by collisions with other clouds, causing the gas distribution to become more and more condensed. The random cloud motions also provide a viscosity, just like the molecular viscosity of a gas, which acts to transfer angular momentum outward and allow low-angular-momentum material to collect at the center. Another very important assumption of these models is a prescription of the rate at which gas is turned into stars during the collapse; unfortunately, in the absence of a detailed theory of star formation, this prescription must be fairly arbitrary.

When the parameters are chosen in appropriate and plausible ways, the systems predicted by these calculations closely resemble elliptical galaxies in all their well-observed properties, including their density profiles, flattening profiles, and radial color gradients. One must be cautious in attributing too much significance to this agreement because of the number of adjustable parameters in the models, but the fact that they naturally predict highly condensed nuclei like those observed in galaxies does seem significant, since the purely stellar models need additional assumptions to yield highly condensed nuclei. Also, the gas-dynamical models almost inescapably predict a radial variation in the abundance of heavy elements, since inflowing gas is enriched in elements created and ejected by stars formed during earlier stages of the collapse. If current observational work confirms that composition gradients extending out to large radii are a common property of elliptical galaxies, this will provide good evidence that continuing star formation and element synthesis took place throughout the collapse of elliptical galaxies.

While several seemingly opposing hypotheses have thus been advanced for the formation of elliptical galaxies, it is entirely possible that in reality the formation process combines elements of all the proposed models. For example, small galaxies may have formed first and then merged into larger systems, and stellar-dynamical effects may have helped to produce a smooth mass distribution while still preserving some vestiges of the initial nonspherical shape. If the colliding galaxies still contained large amounts of gas, then gas-dynamical effects and continuing star formation may also have been important in establishing the observed highly centrally condensed structures of elliptical galaxies and the radial gradients of chemical composition.

Origin of galaxy types

Galaxies with prominent disc components, including the spiral galaxies, which are the most common type, cannot have formed by any process of stellar dynamics but can only have formed from gaseous material as a consequence of dissipation and settling of the gas into a flat layer with nearly circular motion. Calculations of the collapse of protogalaxies containing both gas and stars (Gott and Thuan 1976; Larson 1976a) show that if an appreciable amount of gas survives the initial collapse without being turned into stars, this residual gas can form a disc; the resulting galaxy then has two components, a spheroidal bulge formed during the initial stages of the collapse and a disc component formed later from the residual gas.

Regardless of the detailed model considered, it is clear that the essential differences between the formation of an elliptical galaxy (or the bulge component of a spiral) and the formation of a disc system is that in the former case the stars must be formed quite rapidly—i.e. within a time not exceeding the collapse time—whereas for a disc system the rate of star formation must be much slower so that the gas has time to settle into a disc before turning into stars. Thus the primary distinction between elliptical and spiral galaxies is probably not in their angular momenta, as has often been supposed, but in the efficiency of star formation during early stages of the collapse and in the amount of gas remaining at later times to form a disc. Detailed calculations with various assumptions about star formation (Larson 1976a) also show that the formation of distinct disc and bulge components in a galaxy requires two more or less distinct modes of star formation: one that rapidly turns part of the gas into stars during early stages of the collapse and a second that is much slower and does not become effective until the gas has settled into a thin disc. The disc-to-bulge ratio of a spiral galaxy then depends on the relative importance of these two modes of star formation.

Are there plausible ways to understand the very different rates of star formation required to explain the formation of spheroidal and disc systems? Detailed answers cannot yet be given, but a qualitative understanding may be possible if star formation is caused by the shock compression produced when gas clouds or streams collide supersonically. The collapse of a protogalaxy is quite likely to involve supersonic gas motions and collisions, and these may be important in producing rapid early star formation. The density to which gas is compressed in a radiating shock front increases with increasing preshock density and velocity, and therefore the effectiveness of early star formation in a protogalaxy may depend both on the density of the gas and on the random velocities with which gas clouds collide. Star formation may also be enhanced if the protogalactic gas is strongly clumped, so that the colliding clouds already have a density much higher than average.

If these speculations are valid, we might expect the formation of elliptical galaxies or large bulge components in spirals to be favored if either the protogalactic gas density or the velocity of collisions is high, whereas the formation of spiral galaxies in which the disc predominates may be favored if the protogalactic gas is more tenuous or quiescent. In agreement with these expectations, the most massive galaxies, which have the largest internal velocities, are predominantly elliptical galaxies and systems with large bulges. Smaller galaxies are predominantly spiral or irregular in structure, while small elliptical galaxies may not even exist except as satellites of much larger galaxies. In addition, elliptical galaxies are observed to be strongly concentrated in dense clusters, where both the mean density of matter and the velocities of motion are much higher than average, while spiral galaxies avoid the dense clusters and are mostly found in the low density regions between clusters or in small loose groups, where both the space density and the velocities of the galaxies are relatively small. These observed correlations between the types and the locations of galaxies thus contain important clues about the origin of galaxies and galaxy types.

If spiral galaxies form in regions of

Figure 13. NGC2685 appears to be a flattened disclike galaxy seen nearly edge-on and surrounded by helical filaments of gas, dust, and young stars orbiting around it in a plane perpendicular to the plane of the disc. These filaments almost certainly represent relatively recently accreted material. (Figs. 13–14 from the *Atlas of Peculiar Galaxies*, courtesy Halton Arp and Hale Observatories.)

Figure 14. M82, a small companion of the large spiral M81 (Fig. 4), is one of the most striking and extensively studied peculiar galaxies. Once believed to be an exploding galaxy, it now appears more likely to be a galaxy that has recently accreted or is still accreting large amounts of gas and dust from an extended gas envelope that surrounds M81, M82, and the similar peculiar galaxy NGC3077. The filaments that appear to radiate outward from the center of this galaxy reflect light from bright concentrations of young stars and ionized gas that have recently formed near the center of the galaxy, perhaps from newly accreted gas, but are largely obscured from direct view by dust.

Figure 15. NGC5128 is well known as the radio galaxy Centaurus A. It appears to be a giant elliptical galaxy to which has been added a broad, chaotic dark band of gas, dust, and young stars orbiting around the center like the disc of a spiral galaxy. The large angular momentum of the dark band makes it unlikely that it has been ejected from the interior of the galaxy with the radio-emitting plasma; it more likely represents material recently accreted from outside the galaxy, which is dissipating its turbulent motions and settling into a thin flat disc that may eventually resemble the one in the Sombrero galaxy M104 (Fig. 3).

relatively low density, they will have longer collapse times than elliptical galaxies, which form in regions of higher density. This will be particularly true of the outermost parts of collapsing spiral galaxies, where the density decreases to values not much greater than the background density of the universe, which is currently estimated to be quite low (Gott, Gunn, Schramm, and Tinsley 1974, 1976). Therefore the outer parts of proto-spiral galaxies will continue

expanding for a long time before finally turning around and falling back, and this will produce a continuing infall of low-density matter that may remain important up to the present time and beyond. Many spiral galaxies may in this sense still be in the process of formation, including our own galaxy, where there is evidence (albeit still inconclusive) that gas is still falling in from outside. The continuing infall of low-density gas into spiral galaxies could then explain why they still have a high gas content, despite the fact that their gas is steadily being consumed by star formation. Also, it would make it easy to understand why most spiral galaxies are warped at the edges, since their outermost parts may have been added to them only recently and may not yet have had time to settle into the same plane as the inner parts. Some possible examples of galaxies with newly accreted material that has not yet settled into a disc are shown in Figures 9 and 13–15.

Young galaxies

A question of particular interest is whether we can observe very young galaxies that have only recently formed or are still forming, since this would tell us much about the processes of galaxy formation. From all indications it appears that nearly all the nearby galaxies are old, with ages that are a large fraction of the age of the universe. Therefore, to find very young galaxies we must look at distances great enough for the light that we now observe to have been emitted at much earlier cosmological epochs; because of the expansion of the universe, the radiation received from such objects is also considerably redshifted in wavelength. Galaxies must be extremely luminous to be detected at great distances, but forming galaxies are expected to be very luminous because of the large numbers of recently formed massive stars present. Thus it is possible that some young galaxies could be detected as distant, faint, high-redshift objects.

Collapse models in which star formation occurs during the collapse of a protogalaxy predict that the peak luminosity of a forming galaxy is achieved when it develops a dense nucleus; observed at a great distance, it would then have an almost starlike appearance (Meier 1976b). Since such "primeval galaxies" are also expected to have high luminosities and redshifts, their predicted properties resemble in several ways those of quasars.

Could the quasars in fact be very young galaxies observed at a formative stage? The predicted concentrations of massive young stars near the centers of forming galaxies cannot account directly for the nonstellar spectra and the rapid variability of many quasars, which are probably due to extremely compact energy sources such as black holes or clusters of pulsars. However, the most plausible origins of such highly condensed objects involve the collapse of the cores of evolved massive stars, in which case the nuclear energy radiated by the massive stars is comparable with the gravitational energy available from the collapsed objects. One might then expect to see a stellar contribution to the light of at least some quasars. Some quasars do in fact have spectra that closely resemble those predicted for the massive young stars expected to exist in primeval galaxies (Meier 1976a). This is suggestive support for the hypothesis that quasars represent an early stage in the evolution of galaxies, when a dense nucleus of massive stars is formed by the infall of gas into the center of a galaxy. Many more observations of quasars and other high-redshift objects using powerful instruments, such as the planned Large Space Telescope, will be necessary to clarify the very early evolution of galaxies, but the possibility is strong that eventually we can study by direct observation the formative phases of galaxies and check the validity of the still rather speculative theories outlined in this article.

References

Bardeen, J. M. 1975. Global instabilities of discs. In *Dynamics of Stellar Systems; IAU Symposium No. 69*, ed. A. Hayli, pp. 297–320. Dordrecht/Boston: Reidel.

Binney, J. 1976. Is the flattening of elliptical galaxies necessarily due to rotation? *Monthly Notices Roy. Astron. Soc.* 177: 19–29.

Gott, J. R. 1973. Dynamics of rotating stellar systems: Collapse and violent relaxation. *Astrophys. J.* 186:481–500.

Gott, J. R. 1975. On the formation of elliptical galaxies. *Astrophys. J.* 201:296–310.

Gott, J. R., J. E. Gunn, D. N. Schramm, and B. M. Tinsley. 1974. An unbound Universe? *Astrophys. J.* 194:543–53.

Gott, J. R., J. E. Gunn, D. N. Schramm, and B. M. Tinsley. 1976. Will the Universe expand forever? *Sci. Am.* 234(3):62–79.

Gott, J. R., and T. X. Thuan. 1976. On the formation of spiral and elliptical galaxies. *Astrophys. J.* 204:649–67.

Hubble, E. 1936. *The Realm of the Nebulae.* Yale Univ. Press. (Repr. 1958, Dover.)

Jones, B. J. T. 1976. The origin of galaxies: A review of recent theoretical developments and their confrontation with observation. *Rev. Mod. Phys.* 48:107–49.

Larson, R. B. 1974. Dynamical models for the formation and evolution of spherical galaxies. *Monthly Notices Roy. Astron. Soc.* 166:585–616.

Larson, R. B. 1975. Models for the formation of elliptical galaxies. *Monthly Notices Roy. Astron. Soc.* 173:671–99.

Larson, R. B. 1976a. Models for the formation of disc galaxies. *Monthly Notices Roy. Astron. Soc.* 176:31–52.

Larson, R. B. 1976b. The formation of galaxies. In *Galaxies; Sixth Advanced Course of the Swiss Society of Astronomy and Astrophysics*, ed. L. Martinet and M. Mayor, pp. 67–154. Geneva Observatory.

Lynds, R., and A. Toomre. 1976. On the interpretation of ring galaxies: The binary ring system II Hz 4. *Astrophys. J.* 209:382–88.

Meier, D. L. 1976a. Have primeval galaxies been detected? *Astrophys. J. Letters* 203: L103–05.

Meier, D. L. 1976b. The optical appearance of model primeval galaxies. *Astrophys. J.* 207:343–50.

Ostriker, J. P., and M. J. Rees. In prep.

Roberts, W. W., M. S. Roberts, and F. H. Shu. 1975. Density wave theory and the classification of spiral galaxies. *Astrophys. J.* 196: 381–405.

Sandage, A. 1961. *The Hubble Atlas of Galaxies.* Carnegie Inst. of Washington.

Shu, F. H. 1973. Spiral structure, dust clouds, and star formation. *Am. Sci.* 61:524–36.

Sørensen, S. A., J. Matsuda, and M. Fujimoto. 1976. On the formation of large-scale shock waves in barred galaxies. *Astrophys. Space Sci.* 43:491–503.

Toomre, A. 1974. Gravitational interactions between galaxies. In *The Formation and Dynamics of Galaxies; IAU Symposium No. 58*, ed. J. R. Shakeshaft, pp. 347–65. Dordrecht/Boston: Reidel.

Toomre, A. In prep.

Toomre, A., and J. Toomre. 1972. Galactic bridges and tails. *Astrophys. J.* 178:623–66.

Toomre, A., and J. Toomre. 1973. Violent tides between galaxies. *Sci. Am.* 229(6):38–48.

van den Bergh, S. 1976. A new classification system for galaxies. *Astrophys. J.* 206: 883–87.

Wielen, R. 1974. Density-wave theory of the spiral structure of galaxies. *Publ. Astron. Soc. Pacific* 86:341–62.

Margaret J. Geller

Large-Scale Structure in the Universe

How do galaxies form clusters? What can galaxy clustering reveal about cosmology? How can the cluster environment affect the properties of individual galaxies?

Millions of light years separate us from galaxies which map the large-scale structure of the universe. It is now a matter of definition that galaxies are stellar systems external to our own Milky Way, but little more than fifty years ago today's commonplace was a subject of debate. In 1924 Edwin Hubble clinched the argument by using the new 100-inch telescope at Mount Wilson to study Cepheid variable stars in Andromeda (Fig. 1) and in other nearby galaxies. The absolute (intrinsic) luminosity of a Cepheid is a function of its period of variability. A distance estimate is derived from measurements of the period and the energy flux incident on the Earth from the star. Hubble used this technique to show that we are separated from Andromeda by a distance roughly ten times the diameter of our own galaxy.

To extend the map to distances at which Cepheids could not be picked out, Hubble sought for other objects that have a small spread in absolute luminosity. The brightest supergiant star in a galaxy and the fifth brightest galaxy in a cluster of galaxies are the "standard candles" Hubble used to bootstrap his way to a distance (modern calibration) of 800 megaparsecs (1 parsec [pc] = 3.26 light

Margaret Geller received her Ph.D. from Princeton University in 1975. Since then she has been a member of the staff of the Center for Astrophysics, where she is now a Research Fellow of the Harvard College Observatory. Her research has included studies of the nature of the galaxy distribution, and she has also investigated models for compact radio sources associated with quasars and galactic nuclei. Address: Center for Astrophysics, Harvard University, 60 Garden Street, Cambridge, MA 02138

years = 3.086×10^{18} cm) (Hubble 1936). This region contains some 2×10^7 average galaxies, and its extent is approximately 15 percent of the radius of the visible universe!

If the galaxies were randomly distributed, there would be one or two of them in each 100 Mpc^3. Figure 2 shows the distribution in the sky of galaxies that are nearer to us than about 100 megaparsecs. The central region of the Virgo Cluster (the region outlined in Fig. 2) is a striking example of the nonrandomness or clumpiness of the galaxy distribution on scales less than a few megaparsecs. Some galaxies form fairly isolated binaries; some lie in small aggregates like the Local Group, which is dominated by our galaxy and Andromeda; and some are members of rich (large, dense) clusters that may contain thousands of galaxies. Many clumps of galaxies with a variety of sizes and numbers of members are visible in Figure 2.

There is a continuous hierarchy of structures from galaxies through groups and clusters of galaxies to clusters of clusters of galaxies (Peebles 1974). The radius of the optically bright region of an average galaxy like our own is in the 20 to 30 kiloparsec range. The core region of a rich cluster has a typical radius of half a megaparsec (Dressler 1976), and recent studies show that the outer regions can extend for 10 to 20 megaparsecs (Seldner and Peebles 1977). Recent statistical investigations also reveal clusters of clusters of galaxies, which, on average, contain two or three rich clusters (Bogart and Wagoner 1973; Hauser and Peebles 1973). Over this wide range of sizes—30 kiloparsecs to several tens of megaparsecs—there are no apparent preferences for particular clustering scales. Any border lines between groups, groups of groups, clusters, and clusters of clusters of galaxies are merely arbitrary.

If we go to even larger scales and compare regions of the universe that have a volume of 10^6 megaparsecs or more, the count of galaxies in one sample is not very different from the count in another. When averaged over these hundred megaparsec scales, which are still small relative to the size of the visible universe, the galaxy distribution is remarkably uniform. It is a good approximation to say that on these large scales the universe is homogeneous—it has the same appearance when viewed from any point—and isotropic—it has the same appearance in all directions. Figure 2 gives a qualitative feeling for these characterizations of galaxy clustering.

The formation and development of large-scale structures from galaxies to clusters of clusters of galaxies is linked to cosmology. In the next section I shall review briefly the standard cosmological models. Within this context I shall then discuss models for the galaxy-clustering process. Observations of galaxy clustering may in turn contain clues to cosmology. (See R. B. Larson 1977 for a more detailed discussion of the formation of galaxies.) Finally I shall move to yet smaller scales and examine clustering as a process that can affect the morphology of galaxies. Studies of the distribution of galaxies as a function of their morphology can be used to measure the effectiveness of the environment in modifying the imprint of conditions during the epoch of galaxy formation. The underlying theme of the discussion is that structures on one physical scale can be

Figure 1. The Andromeda Galaxy (NGC 224), shown here, and the Milky Way dominate the Local Group of galaxies. Two elliptical companions of Andromeda, also members of the Local Group, are visible too. (Figs. 1, 9, and 12 from *The Hubble Atlas of Galaxies,* by Allan R. Sandage; courtesy Carnegie Institution.)

used to probe phenomena on other scales.

The standard cosmological models

The picture of the distribution of galaxies is impressive but fundamentally incomplete. The universe evolves: it is dynamic, not static. During the period 1929–31 Hubble provided the first observational evidence for an expanding universe (1929). Galaxies serve as markers for the expansion. The velocity of a galaxy is derived by interpreting the redshift of its spectral lines as a Doppler shift. Hubble used his own distance estimates and Slipher's redshift measurements to demonstrate that, on the average, the apparent recession velocity of a galaxy is proportional to its distance from us. This relation is the Hubble law. The constant of proportionality during the present epoch, H_0, has units of (time)$^{-1}$ and is called the Hubble constant. Galaxies also have so-called peculiar motions relative to the Hubble flow, which are a combination of residual random velocities relative to the mean flow and of velocities induced by galaxy clustering. The Hubble law is, of course, valid only when the peculiar motions make a negligible contribution to the observed redshift.

The Hubble law is a direct mathematical consequence of the assumption of homogeneity and isotropy in the Friedmann–Lemaître cosmological models. These models are based on Einstein's theory of general relativity. What can we learn about cosmology by simply extrapolating the Hubble flow to earlier epochs? If we assume that galaxies have been moving apart from one another at constant velocities, we conclude directly that at a time H_0^{-1} ago, all the matter now lumped in galaxies was densely packed together: the mean density of matter in the universe was greater in the past than it is now. The characteristic time H_0^{-1} is a measure

of the age of the universe. More precisely, H_0^{-1} is an upper limit to the age in the standard models. The rate of separation of pairs of galaxies on the Hubble flow has not been constant but has been decreasing under the influence of mutual gravitation (see Fig. 3).

A variety of observational determinations place the Hubble constant in the range of 50–100 km sec^{-1} Mpc^{-1} (Gott et al. 1976). That is, the recessional velocity of a galaxy 1 Mpc from us is 50–100 km sec^{-1}. The peculiar dimensional units obscure the fact that if H_0 is 50 km sec^{-1} Mpc^{-1}, H_0^{-1} is 2×10^{10} years. This age is consistent with the maximum star ages and with radioactive-decay ages of the elements. This consistency fortifies the impression that the expanding-universe postulate is a reasonable working hypothesis.

During earlier epochs the universe was hot as well as dense—we have even seen radiation from this "primeval fireball"! In 1964 Arno Penzias and Robert Wilson (1965; Peebles and Wilkinson 1967) at Bell Tele-

phone Laboratories first discovered the relic radiation as excess radio noise in an antenna designed for use in the first satellite communications experiments. Further observational studies have shown that this cosmic radiation field is highly isotropic and that its spectrum is consistent with that of a 3°K black body (Weinberg 1972; Peebles 1971). As we shall see in the next section, the existence of the 3°K relic radiation has important implications for the formation of mass concentrations such as galaxies.

In this sketch of the early stages of the hot big-bang model, we have seen that the universe is in a state of violent expansion as a result of some explosive phenomenon at birth, and we are prompted to ask what the future holds in store. The big-bang cosmological models offer a range of possibilities. One of the goals of astrophysics is to discover which, if any, of these possibilities is the appropriate one.

If the universe is, for example, "closed," it is of finite extent. Its

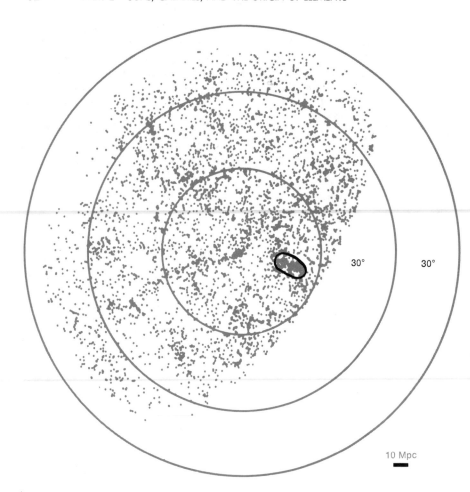

Figure 2. The distribution of galaxies with apparent magnitude m ≤ 15.0 is plotted here (from the Zwicky catalog). The galaxies in this sample are typically nearer to us than 100 Mpc. The plot is an equal-area polar projection with coordinates centered on the north Galactic pole. The central region of the Virgo Cluster (outlined in black) is about 20 Mpc from us; the cluster may extend to include our galaxy and the other members of the Local Group. The length of the black bar corresponds to 10 Mpc at a distance of 100 Mpc. The observed density of galaxies is lower nearer the Galactic equator (the outer edge of the plot) because of obscuration by dust along the plane of the Milky Way. The empty region at the right has not been surveyed. (Courtesy P. J. E. Peebles.)

spatial geometry is the three-dimensional analogue of the surface of a sphere. The average motions of galaxies are like the motions of points on the surface of an expanding spherical balloon. Each point moves away from any other point with a velocity proportional to their separation. In the closed models, expansion eventually ceases and the universe contracts back to indefinitely high density. There is some mind-tickling speculation that the universe oscillates through many cycles of expansion and contraction. Structures from atomic nuclei to clusters of clusters of galaxies are supposed to be formed during expansion and then destroyed during the succeeding contraction.

If the universe is "open," its spatial extent is infinite, and expansion continues forever. Stars eventually die. The temperature drops asymptotically to absolute zero. The end is cold and dark; the open universe goes from fire to ice. The borderline case between open and closed models is called the Einstein–de Sitter, critical, or flat case. The spatial geometry is Euclidean and, as in the open case, expansion continues indefinitely.

Einstein's theory of general relativity specifies the connection between geometry and gravity. The relative motion of a pair of galaxies separating with the Hubble flow is analogous to the motion of a rocket shot from the Earth. If the initial velocity of the rocket is less than the velocity required to escape from the gravitational field of the Earth, the rocket falls back to Earth, but if the initial velocity equals the escape velocity, the rocket can just coast to infinity. The escape velocity defines the border line between bound and unbound trajectories.

To pursue this analogy, consider any spherical region of the universe that is initially expanding with the Hubble flow. Provided that this region is small compared to the size of the universe, matter outside the sphere has no effect on the motion of galaxies within it, and Newtonian physics is perfectly applicable for treating this motion.

We compare the Hubble velocity with the velocity required to escape from the gravitational field of the matter contained within the sample sphere.

In a uniform-density, Einstein–de Sitter universe, the Hubble velocity at the surface of any sample region is just equal to the escape velocity. This condition implies that the critical density at epoch t satisfies the relation $\rho_c(t) = 3H(t)^2/8\pi G$, where $H(t)$ is the Hubble constant at time t, and G is the gravitational constant. At this density the region is marginally unbound, and expansion therefore continues forever. In a closed universe the density of matter within every sample region is greater than ρ_c, and the Hubble velocity is less than escape velocity. Because the galaxies in each region are bound to one another, expansion eventually ceases and the galaxies fall back on one another. The open case has mean density less than ρ_c and is unbound. The galaxies have excess kinetic energy; they are still moving apart from one another when the region has expanded to infinity.

Measurement of both H_0 and the present mean density is one route to distinguishing among open, critical, and closed models. With a Hubble constant $H_0 = 50$ km sec^{-1} Mpc^{-1}, the present critical density is 4.8×10^{-30} g cm^{-3}. Lower mean densities point to open models; higher ones, to closed models. Current best estimates suggest that the present mean mass density of the universe is greater than 5×10^{-31} g cm^{-3} and is at most equal to the critical density. Open models now seem favored by these and other observations, but there is lively debate over the interpretation of the clues (cf. Gott et al. 1976).

Figure 3 shows the evolution of the scale of the universe for open, critical, and closed models. In the early stages, the evolution is the same for all cases.

Later the expansion slows more rapidly for the critical and closed cases. In the open case, matter becomes less and less effective in decelerating the expansion, and the expansion rate reaches a constant value. These results can all be derived from the Newtonian arguments. I shall extend these arguments in the next section to discuss the nature of galaxy clustering and its relationship to cosmology.

Formation of galaxy clusters

Designing scenarios for the formation and evolution of galaxies and clusters of galaxies is rather like the task of a film director who is presented with a running sequence of scenery and is then asked to fit the sequence with characters whose story produces a number of specific effects. The precision of the requirements and the details of the scenery limit the director's options. Those who work on the problem of the formation and evolution of structure in the universe are faced with elegantly simple scenery—the hot big-bang picture—and with an almost bewildering array of reasonable possibilities for the development of galaxies and clusters of galaxies.

One of the central problems is the choice of appropriate initial conditions. If galaxies and clusters of galaxies are the product of some special set of initial conditions, the difficulties of producing these structures are only replaced by the difficulties of explaining the particular initial conditions. Alternatively, the structures we observe now may be a result of random initial conditions that have been appropriately modified by the physical processes in the primeval fireball. As we project back to earlier and earlier epochs, the energy density of radiation increases more rapidly than the density of matter. Thus the primeval fireball is important in determining the dynamic as well as the thermal history of the early universe.

Depending on the particular cosmological parameters, the energy density in radiation exceeds the matter energy density when the scale of the universe is less than one-thousandth of its present size. Also, during approximately this epoch the temperature of the radiation field is

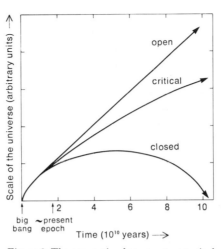

Figure 3. The separation between two typical points in the universe evolves as shown here in the standard closed, critical, and open models. The open and critical models expand forever; the closed models recollapse. Current estimates indicate that it has been less than 2×10^{10} years since the big bang.

greater than 3000°K, hot enough to decompose hydrogen. Photons cannot travel unimpeded: they are repeatedly scattered by the swarm of free electrons. In other words, matter and radiation are coupled. The mean time between scatterings is short compared to the time scale for the expansion of the universe. The system is in near thermal equilibrium, and the radiation has a blackbody spectrum which is preserved as the universe expands. As expansion continues and the temperature falls below 3000°K, matter and radiation decouple: by this time most of the electrons have combined with protons to form hydrogen atoms.

There are at least two major schools of thought on some other aspects of the initial conditions. One view is that the early universe was chaotic and that departures from homogeneity and isotropy were large. Galaxies developed as remnants of turbulent eddies. The largest scale on which turbulence is important is set at the time when the energy densities of radiation and matter are equal. It has been suggested that the mass selected is of order 10^{15} $M_\odot$ (1 solar mass = 2 $\times$ 10^{33} g), the mass of a typical rich cluster of galaxies. Clusters of galaxies are the condensations that form first; they later fragment into galaxies. The detailed sequence of events is complex because both gravity and fluid dynamics play important roles.

This model provides a natural explanation for the angular momentum of galaxies. However, a particular cluster mass can be selected out that may conflict with the observation that there are no preferred clustering scales. A more fundamental problem is that the gravitational collapse of overdense regions could occur too early, and some additional arguments are required to produce the global homogeneity and isotropy we observe now (for more detail, see the review by Jones 1976).

The second major school neglects the possible role of turbulence and focuses on the growth by gravitational instability of statistical density irregularities (see Weinberg 1972; Peebles 1971). It is argued that the structures we see have grown from overdense statistical fluctuations in the primordial soup; the statistical density fluctuations might be similar to those in a classical gas. In this picture, the universe is globally homogeneous and isotropic from the beginning. A major difficulty here is that the growth rate for the overdense regions is slow. Galaxies, which are density enhancements a factor of 10^6 to 10^7 over the mean density, have to be produced. To end up with galaxies as we observe them, we need inhomogeneities during the epoch of decoupling that are larger than the fluctuations we expect for a classical gas in thermal equilibrium. Perhaps the difficulty hinges on the rigorously incorrect assumption that equilibrium statistical mechanics applies. The universe is not truly in thermal equilibrium. For lack of a less problematic approach, the usual course is to assume that the primordial fluctuations are large enough.

Galaxies originate and grow from the density irregularities present during the epoch of decoupling. During the 10^6 years or so before decoupling, the soup of electrons, nuclei, and radiation in the universe is so hot and dense that matter cannot clump into stars and galaxies under the influence of self-gravitation. A clump can collapse only when the force of gravity compensates the pressure of matter and radiation.

Consider a medium of uniform density and pressure. The gravitational force at the surface of a spherical lump increases with the radius of the lump, but the pressure is independent of the lump size. There is, then, a minimum mass, known as the Jeans mass, above which the gravitational force exceeds the pressure. Before decoupling, the radiation pressure is so great that the Jeans mass is several orders of magnitude greater than the mass of a rich cluster of galaxies. At decoupling, radiation pressure ceases to play a role, and the Jeans mass drops to 10^5 $M_\odot$, the typical mass of globular star clusters like those seen around the nucleus of our Galaxy and other nearby galaxies (Peebles and Dicke 1968) (Fig. 4).

In this picture, globular clusters are the first gravitationally bound mass concentrations to form, and they are the basic building blocks for larger structures (cf. Tremaine et al. 1975). Galaxies form as the protoclusters accrete. When the lumps, which probably contain a mix of stars and gas, collide with each other, shock waves may be produced in the gas. The presence of gas implies that shock heating as well as dissipative processes can be important during the process of coalescence. In short, nongravitational effects are important in galactic formation and could determine the mass spectrum of galaxies.

In the gravitational instability picture, the problem of the clustering of galaxies is simpler than the problem of the formation of galaxies themselves. The galaxies form before clustering begins. To study clustering we need consider only the gravitational interactions among the galaxies in the evolving cosmological arena (Peebles 1974; Davis et al. 1977).

The simplest starting assumption is that the galaxy formation process leaves galaxies more or less randomly distributed in space. If we sample regions of fixed size, the number of galaxies we find will fluctuate statistically from one region to the next. Some regions will contain more than the average number of galaxies, some will contain fewer. Gravitational instability can cause the fractionally overdense regions to become increasingly overdense as the universe evolves. In other words, self-gravitation decelerates local expansion rates relative to the general flow, and the galaxy distribution becomes more clumpy with time. The rate at which the clumpiness increases depends on the cosmology. Eventually clusters

Figure 4. Globular clusters, such as NGC 5139 shown here, are thought to be the first gravitationally bound mass concentrations to form. This cluster is 5 kiloparsecs from the Earth, in a direction toward the Galactic center, and has a mass of about 10^5 $M_\odot$; its radius is about 20 parsecs. (Courtesy Cerro Tololo Inter-American Observatory.)

that are gravitationally bound stop expanding and fragment out of the general expansion. They collapse and relax to a quasi-equilibrium configuration.

In order to understand the clustering process qualitatively, recall the discussion of cosmological models in

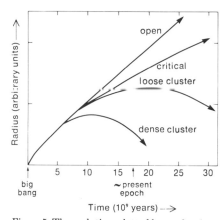

Figure 5. The evolution, plotted here, of regions of differing initial density in the universe shows how galaxy clusters may form. The curves represent the evolution of regions that contain a cluster mass of 10^{15} $M_\odot$. The initially most dense region reaches a maximum earliest, and the curve is a schematic for the evolution of a Coma-like cluster. Initially less dense regions reach maximum expansion later; like these regions, Virgo and Hercules may be still expanding or just starting to collapse. In an open universe regions with the critical density are overdense but, of course, unbound. The time scale shows that we are still very much in the era of cluster formation.

terms of escape velocity relative to the Hubble velocity. Initially all the galaxies are flying apart with the Hubble flow. If the Hubble velocity at the surface of the region is equal to or greater than the escape velocity from the region, expansion continues forever. In other words, if the density in a region is equal to or less than the critical density, galaxies in that region cannot form a gravitationally bound system. If, on the other hand, the density exceeds the critical density, galaxies in the region do form a gravitationally bound system.

Figure 5 shows the evolution of regions of varying density in the universe. Systems like the Coma Cluster, in which the core density is more than 100 times the mean density, appear to be relaxed. They may have reached maximum expansion and begun collapse perhaps 5×10^9 years ago. Loose irregular clusters like the Virgo Cluster (Fig. 2) and the Hercules Cluster (Fig. 7) have densities that are typically 5 or 6 times the mean, and they may have only recently begun to collapse.

During the early stages of the universe, the evolution of all models approximates the critical case. Thus the early evolution of density irregularities is not dependent on the particular cosmological model. Galaxies in any region where the density exceeds the mean form a clump. A critical universe is only marginally unbound. Therefore a density enhancement can cause formation of a bound system which extends over a region much larger than the size of the initial clump. In a critical universe a cluster can continue to grow forever. The density contrast grows over larger and larger regions as the universe expands.

In the open universe the situation is different. As the universe reaches the linear expansion phase (Fig. 3), matter condensations become ineffective in decelerating the expansion around them. The onset of the linear expansion phase sets a limit to cluster growth: the cluster can grow only over a region where the mean density is greater than critical. On any larger scale the density contrast remains constant. In an open universe not all density enhancements are bound: regions where the density is greater than the mean but less than critical will expand forever.

The clustering process perturbs the flow of galaxies relative to the average expansion. More precisely, the velocity dispersion about the Hubble flow is determined primarily by the clustering process. Measurement of the spread can be used as a cosmological probe; these measurements are among the challenges of modern cosmology.

Another link between the clustering process and cosmology is provided by the relationship between the cosmological epoch and the linear clustering scale as a function of density. If the universe is open, we should expect to see an indication of a maximum clustering scale in the data. This effect could be erased by appropriately tuned initial conditions, but this tuning seems unwarranted (Peebles 1974; Davis et al. 1977).

An important general feature of the observed clustering phenomenon is that galaxies do not appear to "prefer" to form clumps of any particular size. This sort of democracy of scale is perhaps not surprising. Gravity, probably the only important force in the problem, is an inverse square-law force. There is no way for this force law alone to select out scales of particular lengths. Statistical analyses of catalogs of galaxies like the one shown in Figure 2 indicate the absence of preferred scales in the three-decade range from 30 kpc to 30 Mpc (Peebles 1974 and references therein). Based on the prediction of a maximum linear clustering scale in the open case, the wide range of scales can be taken to favor a near-critical universe in apparent conflict with some other observations (Davis et al. 1977). There are many wrinkles that must be ironed out in order for us to judge the seriousness of the conflict. The agreement between the predictions of the gravitational instability model and the observations can be taken as an indication that galaxy formation took place prior to cluster formation.

The morphology of galaxy clusters

The broad statistical picture has provided some insights into the nature of clustering. The more classical and equally fruitful approach of studying individual clusters is also actively pursued (see, e.g., Oemler

Figure 6. In the very rich Coma Cluster the distribution of galaxies is smooth and regular. This photograph shows the central 15 × 20 arc minutes of the cluster, which is an extended X-ray source. At the cluster distance of 138 Mpc, this region is 0.6 × 0.8 Mpc. Nearly all the galaxies in the region are gas-free elliptical and S0 systems (see Figs. 8 and 12). (Figs. 6, 7, 8, 10, and 11 courtesy Kitt Peak National Observatory.)

1974). Rich clusters like the nearby Virgo Cluster (Fig. 2) and the Coma and Hercules Clusters (Figs. 6 and 7) are easily identifiable as large departures from the average density.

Clusters have a wide range of other characteristics which may indicate that they are in different stages of evolution. Coma is a very regular cluster, with its galaxies smoothly distributed around a well-defined core. The core region is dominated by a massive double galaxy. The smooth galaxy distribution is taken as evidence that the system is relaxed or dynamically evolved. The high density implies a short collapse time (Fig. 5), and the relaxed appearance is an indication that these clusters are indeed gravitationally bound systems. Near the center of other clusters with a similarly regular appearance we

Figure 7. The Hercules Cluster is irregular in appearance and has a mixed population of galaxies. Many spiral galaxies are visible in this 9.9 × 1.2 Mpc central region. The cluster is at a distance of 216 Mpc.

often find a giant cD galaxy. A cD galaxy is a bright elliptical galaxy (cf. Fig. 8) with a halo of stars that extends through a large fraction of the cluster (Oemler 1973). One fascinating suggestion is that the stars in these haloes are debris left from tidal interactions of galaxies (Richstone 1976). Giant galaxies may also evolve by accreting neighboring galaxies (Ostriker and Tremaine 1975).

Clusters like Virgo and Hercules have a rather ragged appearance and do not, in general, contain dominant cD galaxies. These irregular clusters are typically less dense than Coma and may still be in the process of collapse (Fig. 5); relaxation has not yet taken place. Irregular clusters are probably dynamically young.

Galaxies that are cluster members can be identified by measuring their redshifts. Galaxies that are background or foreground objects will have redshifts that differ substantially from the mean for the cluster. For rich clusters it is practical to measure redshifts only for a sample of the brightest members. Statistical arguments are used to estimate the number of faint members (Oemler 1974).

Once a number of redshifts have been measured, the virial theorem can be applied to the system. The virial theorem says that, for a gravitationally bound system in equilibrium, the absolute value of the potential energy is equal to twice the kinetic energy of the system. The application of this theorem to clusters of galaxies gives an estimate of galaxy masses. The result is roughly an order of magnitude greater than standard estimates of the mass we see in stars. If the clusters are bound, the mass must be present in a form we cannot see— but what is it? Low-mass stars, Jupiters, snowballs, black holes—we do not know.

The bright region of a large galaxy extends for 30 kpc. A dark surrounding massive halo may extend to 100 kpc or more (Ostriker et al. 1974). If this "missing mass" problem arose only from studies of individual rich clusters and not from studies of other systems, it would be reasonable to suppose that something had just gone awry in the application of the virial theorem to clusters. But the same problem arises from analyses of small groups of galaxies (Geller and Peebles 1973) and from studies of binary galaxies (Turner 1976). In these studies the statistical problems are tricky because the systems do not represent such large departures from the mean density. The typical galaxy mass obtained is 10^{12} $M_\odot$. Measures of galaxy masses are, of course, important benchmarks for theories of galaxy formation.

The nature of the galaxy population raises other interesting questions about the relationships between different levels of structure in the universe. Is there a correlation between the properties of a galaxy and the density of the region it inhabits? Which is more important, recent environment or conditions at the time of galaxy formation—nature or nurture?

Figures 8–12 are photographs of a representative selection of different types of galaxies. They have been chosen to be distinctive, but there is a continuous range of properties from one galaxy type to the next. On a sufficiently detailed level each galaxy is unique. However, in developing a general picture of the range of properties it is profitable to ignore details and discuss broad classifications (the major classifications are part of Hubble's legacy).

Elliptical galaxies (Fig. 8; the small companion galaxies of Andromeda in Fig. 1 are also ellipticals) are ellipsoidal stellar systems which usually contain no obvious traces of dust and show no evidence of recent star formation. The light of elliptical galaxies is dominated by an old, red stellar population.

Roughly 60 percent of the galaxies we see in a sample complete to some limiting apparent magnitude are flattened spiral systems similar to our own galaxy and Andromeda. Other examples are shown in Figures 9, 10, and 11. Spirals range continuously from galaxies with a dominant elliptical-like nucleus and a disc containing tightly wound, poorly defined arms (Fig. 9) to galaxies with a small nucleus and a disc with loosely wound, spectacularly prominent arms (Fig. 10). Spiral discs contain extensive dust and gas clouds. The discs are blue from the light of young supergiant stars. The arms of some spiral galaxies appear to originate at the

Figure 8. Elliptical galaxies, such as NGC 4472 shown here, have a greater range of size and brightness than spirals—from giant systems like the two visible in the central region of Coma (Fig. 6) to numerous but low-luminosity dwarfs.

Figure 9. The spiral galaxy NGC 2811 has thin, poorly defined arms that are tightly wound.

Figure 10. The spiral structure is prominent in galaxy NGC 5364. The arms are marked by concentrations of ionized gas and young stars.

Figure 11. In NGC 1530, a barred spiral galaxy, the arms extend from the ends of a luminous bar that crosses through the nuclear bulge. Dark filamentary dust lanes run through the bar and the arms.

Figure 12. NGC 7332, an S0 galaxy, has a disc and shows a smooth distribution of stars without spiral structure. The overall geometry is similar to the geometry of spirals, and the stellar population is similar to that of ellipticals.

ends of a luminous bar that extends symmetrically through the galactic nucleus (Fig. 11).

The geometry of S0 galaxies (Fig. 12) is similar to that of spirals. These S0 galaxies have a central bulge and a disc component, but the disc is generally gas-free and contains no young stars. There is some evidence that the light from the disc is bluer than the light from the central bulge. The color gradient could be evidence that star formation has taken place more recently in the disc than in the bulge. The interstellar medium could have been swept from the disc relatively recently. The gas-free systems—ellipticals and S0's—account for some 30 percent of the galaxy population.

The remaining galaxies are a mixture of irregular, peculiar, and dwarf systems. The classes of irregulars and peculiars are catch-alls for galaxies that lack symmetry or have other noticeably anomalous properties. Dwarf systems are small galaxies that may have masses as low as 10^8 $M_\odot$. The smallest known dwarf has a radius of only half a kiloparsec.

The interior regions of rich regular clusters like Coma are dominated by elliptical and S0 galaxies. Looser, more irregular systems like Virgo and Hercules have gas-rich galaxies in their central regions. In these clusters the mix of galaxy types is similar to the mix found in the general field. The outer, more diffuse regions of Coma-like clusters also have a similar mix (Oemler 1974).

These observations are the basis for a number of suggestions for relationships among different types of galaxies. Because of the similarity in their geometry, it is natural to postulate that spiral and S0 galaxies are related. One suggestion is motivated by the discovery that many clusters of galaxies are associated with extended sources of X rays. The X rays are the signature of a hot gas contained in the cluster. As spiral galaxies containing cool gas travel through this hot intergalactic medium at velocities of 1,500 km sec^{-1} (the characteristic internal relative velocity in a rich cluster), the gas is driven out of them (Gunn and Gott 1972).

This sweeping process works well in very dense regions but is not effective in low-density regions. The S0 galaxies found in low-density regions present a problem. Even if gas was driven from them during some early epoch, mass loss from stars should have replenished the gas by now. In elliptical galaxies, gas can be driven out by supernova-driven winds (Faber and Gallagher 1976). This mechanism is probably inadequate for S0 galaxies with extended discs. Perhaps the S0 galaxies in low-density regions differ in their properties and history from those in rich clusters. For example, it would be interesting to know whether S0 galaxies in rich clusters have more extended discs than those in lower-density regions. Perhaps initial conditions play an important role here.

The relationship between disc and elliptical galaxies is an even more speculative area of research. One possibility is that initial conditions play the dominant role. The formation of ellipticals might be favored in regions that are initially more dense (Gott and Thuan 1976; see also Sandage et al. 1970). If the rate of star formation were proportional to a power of the local density, stars would form earlier and more efficiently in more dense regions, and there would be little gas left to cool and collapse. In contrast, spirals might form in less dense regions, where a good deal of gas is left over in the star formation process. As the gas cools, it can no longer hold itself up against gravity, and it collapses to a disc, which is supported by rotation. The first stars formed end up in the bulge component. Stars are forming now in the discs of gas-rich galaxies.

Other proposals assume that initially all galaxies are systems with discs. The argument is that there are processes that can significantly change the morphology of a galaxy on a time scale that is less than the age of the universe. One proposal is that interacting disc systems can coalesce, leaving spheroidal systems of stars (Toomre 1977). Another model (Gold 1976; Shapiro and Marchant 1977) depends on tidal interactions among disc systems which gradually cause the systems to puff up into spheroidal systems. Because these mechanisms are environment-dependent, their effectiveness depends on the initial distribution of galaxies and its evolution. Interactions may be most important during or shortly after the epoch of galaxy formation. One ob-

servational check on the processes of interaction would be a study of ellipticity as a function of local density.

Statistical analysis of a nearby region of the universe which contains the Virgo Cluster but no Coma-like clusters provides other information on the relative distributions of different types of galaxies (Davis and Geller 1976). These studies indicate that elliptical and S0 galaxies are typically found in regions which are roughly twice as dense as regions around spirals. It would be helpful to compare these results with those from other samples containing rich clusters. If cores of rich clusters are the only regions heavily dominated by gas-free systems, we expect the result to be general. Only a small fraction of galaxies are in the cores of rich clusters, and these regions would not weigh heavily in the statistics. Because the mix of types seems to be affected only by large density variations, these results seem to point to the importance of initial conditions in determining morphology. We have seen that environment can certainly play a role and that, as is the case in biology, it may be difficult to distinguish its effects from the imprint of the formation process.

The broad-brush picture of large-scale structures leaves many questions unanswered. Structures on different scales are interrelated. More detailed studies of stellar populations and the gas content of individual galaxies might yield more clues about the effects and nature of the cluster environment. Studies of clustering may in turn give us new clues to cosmology. The next fifty years of discovery about large-scale structures in the universe promise to be as exciting and controversial as the first fifty years have been.

References

Bogart, R. S., and R. V. Wagoner. 1973. Clustering effects among clusters of galaxies and quasi-stellar sources. *Astrophys. J.* 181: 609–18.

Davis, M., and M. J. Geller. 1976. Galaxy correlations as a function of morphological type. *Astrophys. J.* 208:13–19.

Davis, M., E. J. Groth, and P. J. E. Peebles. 1977. Study of galaxy correlations: Evidence for the gravitational instability picture in a dense universe. *Astrophys. J.* 212:L107–11.

Dressler, A. A comprehensive study of twelve very rich clusters of galaxies. 1976 diss., Univ. of California, Santa Cruz.

Faber, S. M., and J. S. Gallagher. 1976. HI in early-type galaxies. II: Mass loss and galactic winds. *Astrophys. J.* 204:365–78.

Geller, M. J., and P. J. E. Peebles. 1973. Statistical application of the virial theorem to nearby groups of galaxies. *Astrophys. J.* 184:329–41.

Gold, T. 1976. IAU Symposium on Galaxy Formation, Grenoble.

Gott, J. R. 1977. Recent theories of galaxy formation. *Ann. Rev. Astr. and Astrophys.* 15:235–66.

Gott, J. R., J. E. Gunn, D. N. Schramm, and B. M. Tinsley. 1976. Will the universe expand forever? *Sci. Am.* 234:62–79.

Gott, J. R., and T. X. Thuan. 1976. On the formation of spiral and elliptical galaxies. *Astrophys. J.* 204:649–67.

Gunn, J. E., and J. R. Gott. 1972. On the infall of matter into clusters of galaxies and some effects on their evolution. *Astrophys. J.* 176:1–19.

Hauser, M., and P. J. E. Peebles. 1973. Statistical analysis of catalogs of extragalactic objects. II: The Abell Catalog of Rich Clusters. *Astrophys. J.* 185:757–85.

Hubble, E. 1929. Distance and radial velocity among extragalactic nebulae. *PNAS* 15: 168–73.

———. 1936. *Realm of the Nebulae.* Yale Univ. Press (Dover, 1958).

Jones, B. J. T. 1976. The origin of galaxies: A review of recent theoretical developments and their confrontation with observation. *Rev. Mod. Phys.* 48:107–49.

Larson, R. B. 1977. The origin of galaxies. *Am. Sci.* 65:188–96.

Oemler, A. 1973. The cluster of galaxies, A2670. *Astrophys. J.* 180:11–23.

———. 1974. The systematic properties of clusters of galaxies. I: Photometry of fifteen clusters. *Astrophys. J.* 194:1–19.

Ostriker, J. P., P. J. E. Peebles, and A. Yahil. 1974. The size and mass of galaxies, and the mass of the universe. *Astrophys. J.* 193: L1–4.

Ostriker, J. P., and S. D. Tremaine. 1975. Another evolutionary correction to the luminosity of giant galaxies. *Astrophys. J.* 202: L113–18.

Peebles, P. J. E. 1971. *Physical Cosmology.* Princeton Univ. Press.

———. 1974. The gravitational instability picture and the nature of the distribution of galaxies. *Astrophys. J.* 189:L51–53.

Peebles, P. J. E., and R. H. Dicke. 1968. Origin of the globular star clusters. *Astrophys. J.* 154:891–908.

Peebles, P. J. E., and D. T. Wilkinson. 1967. The primeval fireball. *Sci. Am.* 216:28–37.

Penzias, A. A., and R. W. Wilson. 1965. A measurement of excess antenna temperature at 4080 Mc/S. *Astrophys. J.* 142:419–21.

Richstone, D. O. 1976. Collisions of galaxies in dense clusters. II: Dynamical evolution of cluster galaxies. *Astrophys. J.* 204:642–48.

Sandage, A. 1961. *The Hubble Atlas of Galaxies.* Carnegie Institute of Washington.

Sandage, A., K. C. Freeman, and N. R. Stokes. 1970. The intrinsic flattening of E, S0, and spiral galaxies as related to galaxy formation and evolution. *Astrophys. J.* 160:831–44.

Seldner, M., and P. J. E. Peebles. 1977. A new way to estimate the mean mass density associated with galaxies. Submitted to *Astrophys. J. Letters.*

Shapiro, S. L., and A. B. Marchant. 1977. The formation of elliptical galaxies by tidal interaction. *Astrophys. J.* 215:1–10.

Toomre, A. In prep.

Tremaine, S. D., J. P. Ostriker, and L. Spitzer. 1975. The formation of the nuclei of galaxies. I:M31. *Astrophys. J.* 196:407–11.

Turner, E. L. 1976. Binary galaxies. II: Dynamics and mass-to-light ratios. *Astrophys. J.* 208:304–16.

Weinberg, S. 1972. *Gravitation and Cosmology.* Wiley.

"Well—the moons of Mars are even *smaller* than we thought!"

PART 3 *The Planets*

James W. Head
Charles A. Wood
Thomas A. Mutch

Geologic Evolution of the Terrestrial Planets

Observation and exploration have yielded fundamental knowledge of planetary evolution and have given rise to an exciting new view of Earth as a planet

Luna, Venera, Mariner, Ranger, Surveyor, Apollo, Lunokhod, Pioneer, Viking. These spacecraft with heroic names have taken us from the triumph of landing a pennant on another world to the reading of a weather report from Mars on the evening news. The accompanying avalanche of scientific information from 15 years of space exploration, combined with the simultaneous understanding of terrestrial plate tectonics, has led to an exciting new perspective on the nature and evolution of the planets of the inner solar system.

Physical, chemical, and dynamical studies of the planets have greatly contributed to the current renaissance of understanding, but simple observations of the types of terrain which form planetary surfaces and inferences of processes which pro-

James W. Head is an Associate Professor of Geology at Brown University, where his research centers on planetary geologic processes and history. He received his Sc.B. from Washington and Lee University and his Ph.D. from Brown and was involved with the Apollo Lunar Exploration Program from 1968 to 1972. Charles Wood is a Ph.D. student in the Department of Geological Sciences at Brown. His research and publications concern volcanology, lunar geology, and recent climatic change in Africa, where he taught and did research for four years. Thomas A. Mutch is a Professor of Geology at Brown and received his Ph.D. from Princeton. He is the author of books on the geology of the Moon and Mars and is presently the leader of the Mars Viking Lander Imaging Team. This work was performed under NASA Grants NGR-40-002-088 and NGR-40-002-116 from the Lunar and Planetary Programs Office. Thanks are extended to M. Cintala, R. Hawke, A. Gifford, R. Roth, S. Matarazza, and the National Space Science Data Center. Address: Department of Geological Sciences, Brown University, Providence, RI 02912.

duced the terrains have yielded fundamental knowledge of planetary evolution. In this article we review photogeologic and other evidence for each of the terrestrial planets—Mercury, Venus, Earth, the Moon, and Mars—but our discussion of each planet is grossly simplified, generally omitting descriptions of landforms such as craters and volcanoes (information readily available in the cited references); instead, we concentrate on the relationships between regional terrain types, ages, and planetary evolution.

The planets

Knowledge of the characteristics and processes operating on the Moon has been accumulating since the earliest visual observations centuries ago (summarized in Mutch 1972). Increasingly sophisticated Earth-based observational techniques have been complemented and extended by exploration by spacecraft and man. Soft-landing spacecraft and orbital vehicles provided abundant photographic evidence of the nature of the lunar surface and clues to the processes operating there. Manned exploration of the Moon during Apollo provided on-site investigations and sample return that opened an entirely new field—the petrologic and geochemical history of a planetary body other than Earth.

Galileo's early subdivision of the lunar surface into mare and terra is still significant in terms of lunar history and processes. The rough, cratered lunar highlands, or terrae, dominate the crust of the Moon, contrasting with the low-lying, dark mare units (Fig. 1). Geologic and petrologic evidence from Luna and Apollo samples (Taylor 1975) dem-

onstrates that the cratered highlands represent an early crust that resulted from global melting in the first few tens to hundreds of millions of years of lunar history. The global crust was continually modified by the impact of material from elsewhere in the solar system. The cratering record preserved in the early crustal units represents a distinct phase of intense cratering that began to decline rapidly about 3.8 billion years ago. Although other processes, such as volcanism, may have operated during this early period, the surface history of the Moon was written by craters of all sizes. Extremely large impacts excavated huge depressions, perhaps as large as 2,000 km in diameter, and spread ejecta over large areas of the planet, sometimes affecting a whole lunar hemisphere. The youngest of these large basins, such as Orientale and Imbrium, are surrounded by distinctive radially textured deposits that buried and gouged large regions of the lunar surface. Although seismometers placed on the Moon by Apollo astronauts prove that cratering continues to the present, relatively few large craters and no large basins have formed in the last 3.8 billion years.

The next stage of lunar history was dominated by the emplacement of the dark mare plains that cover approximately 17 percent of the lunar surface (Head 1976) and occur predominantly on the lunar Earth side. In general, maria are relatively thin layers of basaltic lava totaling less than one percent of the volume of the lunar crust. The time of initial mare emplacement is uncertain, but radiometric dates of returned Apollo mare rocks suggest that major outpouring of lava occurred from 3.9 to 3.2 billion years ago (Taylor 1975).

Moon

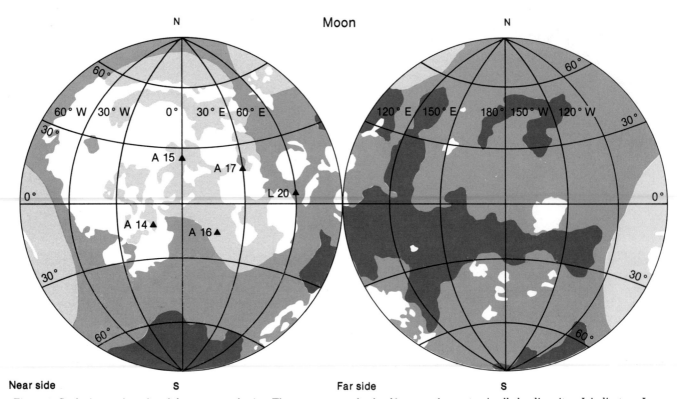

Near side   **Far side**

Figure 1. Geologic terrain units of the terrestrial planets have been synthesized, simplified, and modified from a variety of recent maps. The lunar surface is dominated by craters and relatively thin ponds of lava within frontside basins. There are no complex landforms such as the folded mountains on Earth or the giant canyons of Mars; the Moon is thus a primitive body whose surface was shaped by impact cratering and extrusion of lava flows. *A* indicates Apollo landing sites; *L* indicates a Luna site. (From Wilhelms and McCauley 1971, as simplified and extended by Howard et al. 1974.)

Although some mare deposits may be as young as two billion years (Boyce 1975), there has been no extensive igneous activity on the surface of the Moon for the last three billion years (Fig. 1).

Although major lineaments and fractures are observed in the lunar crust, there is no evidence of folded mountain belts or similar indications of compressional tectonic activity. Lunar mountain ranges were produced largely in conjunction with the formation of major impact basins.

The small size of Mercury (Fig. 2) and its proximity to the sun long hindered telescopic investigations, but in 1974 the Mariner 10 spacecraft photographed about 35 percent of the planet, revealing a cratered surface remarkably similar to the Moon's (Murray et al. 1975). Preliminary geologic mapping (Trask and Guest 1975) distinguished smooth and cratered plains and basin ejecta, as on the Moon, but there are a number of significant differences. Craters on the lunar highlands are densely packed, with the rims of the youngest superimposed on older craters. Lunar mare regions are sharply bounded and mostly contained in contiguous ba-

sins. On Mercury, by contrast, craters are often interspersed with relatively smooth plains, resulting in a speckled terrain map. At present there is a controversy concerning the relative ages and origins of the plains and the craters. Like most conflicts for which good evidence supports each side, probably each is partially correct: plains are probably older than crater units in some areas and younger in others, and have varied origins.

The most unusual terrain features on Mercury are lobate scarps that cut plains and craters alike. These scarps, which extend from tens to hundreds of kilometers in length, are interpreted as evidence of crustal shortening. Strom et al. (1975) calculated that the amount of shortening corresponds to a decrease in radius of 1–2 km. The same amount of contraction is estimated independently from theoretical modeling of lithospheric cooling (Solomon and Chaiken, in press).

Although Mercury has many impact basins (Wood and Head, in press), only a single large one, Caloris, is well shown in the area photographed. The interior of Caloris is surfaced by a lunar-mare-like plain wrinkled by

concentric ridges. Much of the radial ejecta from Caloris is apparently covered by a continuation of the basin fill material. Whether this is basaltic lava similar to lunar maria or a vast deposit of impact-emplaced material is currently under debate.

The pull of gravity on Mercury is more than twice as strong as on the Moon, resulting in a restricted distribution of ballistically emplaced ejecta from impact craters (Gault et al. 1975). However, central peaks and wall terraces of craters are approximately as abundant on the Moon as on Mercury and Mars, which have much stronger fields. This suggests that, in addition to gravitational effects, other factors such as impact velocity or target strength influence the morphology of craters (Cintala et al. 1976).

Crater and basin formation has thus dominated the early history of Mercury, fracturing and mixing an ancient crust. Volcanism may have produced plains deposits concurrently with the early intense cratering and may also be responsible for the Caloris plain, which formed later. On the basis of crater densities and flux estimates, virtually all the major ter-

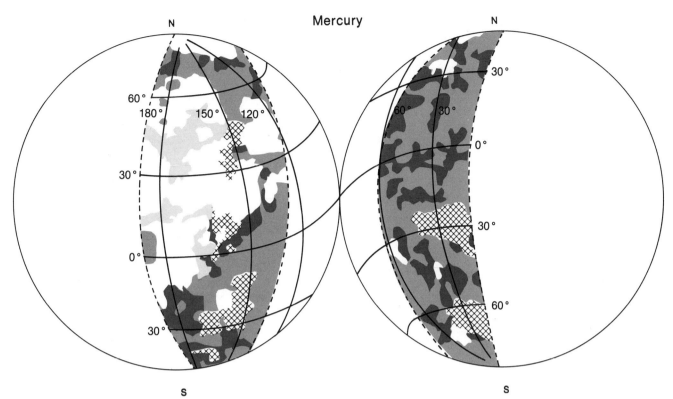

Figure 2. Only about 35 percent of Mercury has been photographed, but the surface we have been able to get a look at is remarkably similar to that of the Moon. Cratered terrain is common on the planet, and large tracts of smooth lava-like plains occur around the Caloris Basin (30°N, 190°). (From Trask and Guest 1975.)

Key to terrain units

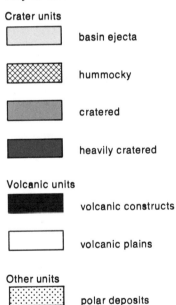

Crater units

basin ejecta

hummocky

cratered

heavily cratered

Volcanic units

volcanic constructs

volcanic plains

Other units

polar deposits

channel deposits

platform deposits

folded mountains

rain units (Fig. 2) formed in the first 1.5 billion years of Mercury's history.

Earth-based observers have long been intrigued by the red color, dusky markings, and the waxing and waning polar caps of Mars. Early Mariner

missions provided the first close-up views of the surface of Mars and portrayed a cratered terrain not unlike that seen on the Moon. However, these missions had photographed only a small portion of the surface of the planet, and the global view provided by Mariner 9 in 1971 quickly demonstrated that Mars was a geologically diverse and complex planet that had evolved to a stage considerably beyond the surface of the Moon (Masursky 1973).

The geologic diversity revealed by Mariner 9 is illustrated by the variations within the cratered terrain, which comprises about 50 percent of the surface area of Mars (Fig. 3). Within this cratered terrain are plains units with variable crater densities perhaps reflecting early volcanic resurfacing of ancient cratered crust. The cratered highlands contrast with much of the northern hemisphere of Mars, which consists of flat, low-lying, sparsely cratered plains that cover about 30 percent of the surface of the planet. The boundary between the cratered terrain and the plains to the north is marked by a chaotic and hummocky material indicative of collapse and erosion at the edge of the cratered unit.

Some geologists believe that melting of a buried permafrost layer initiated the formation of the chaotic terrain of Mars (Sharp 1973), and evidence for surface water at an earlier epoch in the history of the planet is abundantly displayed in Mariner 9 and Viking photography. Meandering channels, tributaries, braided streams: the full lexicon of terrestrial fluvial geology is represented on Mars. Additionally, permanent caps composed of both water ice and carbon dioxide ice are found at each of the poles. Exposed layers of dust and ice in these caps are evidence for eolian erosion, transport, and deposition—processes that appear to have dominated the planet Mars for a billion years or more.

Martian dust storms and polar caps were known from Earth-based telescopic observations, but the nature of even the largest landforms, such as the 27 km high shield volcano Olympus Mons, was not appreciated. Such mountains represent the most recent volcanic activity on Mars; however, earlier shields exist, and flow fronts and wrinkle edges demonstrate that many plains units, in the northern lowlands and elsewhere, are volcanic in origin.

Mars

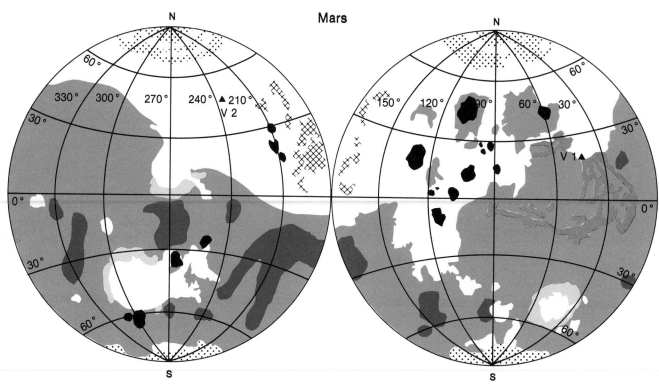

Key to terrain units

Crater units

▦	basin ejecta
▨	hummocky
▨	cratered
▨	heavily cratered

Volcanic units

■	volcanic constructs
□	volcanic plains

Other units

⠿	polar deposits
▨	channel deposits
▥	platform deposits
▨	folded mountains

Figure 3. Although large areas are covered by cratered terrain, major updoming and shield volcanism distinguish Mars from the less complex Mercury and the Moon. The concentration of volcanic smooth plains in the northern hemisphere is a fundamental observation which must be explained in any theory of the evolution of Mars. *V* indicates the Viking landing sites. (From Pollack 1975 and Carr et al. 1973.)

The ages of geologic units on Mars are somewhat uncertain due to a lack of detailed knowledge of the rate of impact crater formation. If martian cratering rates were similar to those of the Moon, the majority of the cratered terrain would have formed early in martian history, generally prior to about 3.5 billion years ago (Soderblom et al. 1974) and cratering has probably continued to the present, but at a much reduced rate. Volcanism has played a continuing role in the surface evolution of Mars, from the time of formation of the cratered terrain, through the surfacing of the northern lowlands, and, locally, by the building of large shield volcanoes in the last billion years (Carr 1973). The channels appear to be at least one to two billion years old. Tectonic features on Mars include radial structures associated with large impact basins; features similar to lunar mare ridges occurring in plains regions; and grabens and extensive fault valleys, such as Valles Marineris. The most prominent structural features on Mars are centered around the Tharsis region and formed approximately a billion years ago during the updoming of this region (Carr 1974).

The surface of Venus is obscured by a thick cloud cover, but the planet's similarities in size and density with Earth invite comparison of surface geology and processes. Photographs returned by Soviet unmanned landings in 1976 show the surface at the Venera 9 and 10 sites to be blocky, with indications of soil and bedrock. Recent analysis of radar data obtained by Earth-based observations (Rumsey et al. 1974) suggests that Venus may have a varied geologic terrain perhaps more similar to Mars than Earth.

Circular structures between 30 and 1,000 km in diameter have been detected and may represent impact craters and basins. Other large features include a 1,500 km linear trough similar in scale to the martian Valles Marineris and a large low circular dome with a central depression, similar in some aspects to shield volcanoes (Malin 1976). If the cratered terrain represents a surface modified by abundant impacts, then that portion of the Venus surface is relatively old and has not undergone the extensive tectonic recycling typical of Earth. Preliminary evidence of tectonic and volcanic processes suggests, however, that activity perhaps comparable to that of Mars has also taken place. More extensive high resolution radar coverage is required to determine if terrestrial-type tectonics have actually operated on the planet.

It is fortunate that mankind evolved and geology developed on our planet,

Earth

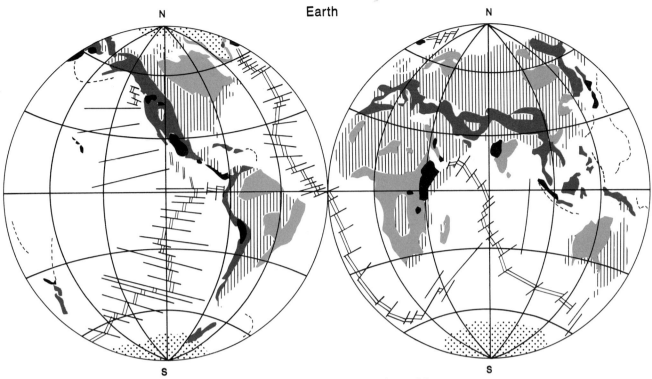

Figure 4. Earth is essentially a tectonically active and volcanic planet—young lavas erupt from spreading ridges (*double lines*), forming the ocean floors, which are sheared along transform faults (*single lines*) and are ultimately consumed at subduction zones (*dashed lines*). (From Wyllie 1971.)

since delineation of some unique terrestrial terrain units would have been difficult from space. Unlike other terrestrial planets, Earth has the majority of its crust concealed by liquid water and much of its exposed land mantled by a thin but dense veneer of vegetation. Atmospheric effects often add to surface obscuration. While the hydrosphere, atmosphere, and biosphere are not terrain units, they are dynamic geologic agents, which, with plate tectonics, make Earth a unique planet. Water, liquid and frozen, erodes and erases the land, recycling rocks as sediments that fill basins, form plains, and create new land. Such continental flatlands, perhaps, are the terrain equivalent of the smooth plains on other planets.

In the last ten years it has been recognized that plate tectonic movements of Earth's lithosphere are responsible for most of the large scale terrain features. Plate collisions produce folded mountains (the Alps, Atlas, and Appalachians), while continental rift valleys and vast plateau-forming deposits of basalts appear to be related to plate breakup. Where plates are consumed, lines of andesite volcanoes appear, and at

ocean ridges, lavas ooze out, creating new ocean floor.

Despite the dynamic crustal activity and resulting complexity of the Earth, some of the same terrain units occur as on the other planets (Fig. 4). The distribution of meteorite craters

Figure 5. The percent of the surface area of each terrestrial planet occupied by major terrain units is illustrated by the color code used in Figures 1–4.

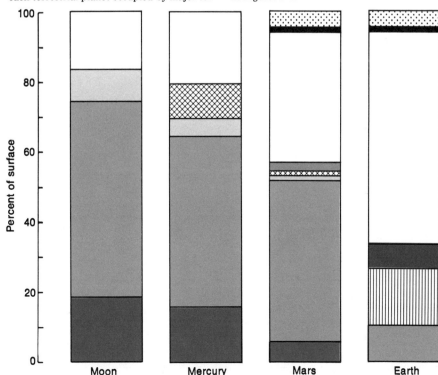

(especially large ones) suggests that the 10 percent of Earth surfaced by ancient rocks—the Precambrian shields—are the nearest terrestrial analog to cratered terrain, even though the shields' crater density is much less than that of cratered terrains on other planets. The most pervasive terrain unit on Earth is the basaltic plains of the ocean floor. These volcanic plains are among the youngest of Earth's rocks, ranging in age from 0 to 200 million years. Within the last 65 million years, separate phases of volcanism have built broad plateaus of basalts (Ethiopian and Indian Traps, etc.) as well as conical mountains and craters (Andean volcanoes, Hawaii, etc.).

Polar deposits similar to, but more massive than, those on Mars also occur on Earth. As on Mars, terrestrial polar terrain seasonally increases and decreases in area. The instability of Earth's polar terrain emphasizes that the map of terrain units is a snapshot of a particular geologic instant. Unlike the Moon, which has been largely unchanged during the last 3.5 billion years, Earth has a surface constantly in motion, with the positions and abundances of terrain units continually varying. Indeed, according to one hypothesis of the evolution of Earth's crust (Hargraves 1976), there was only a single terrain unit—water-covered plains—for the first billion years of Earth's history.

Comparative planetology

Our brief examination of the terrestrial planets reveals some important similarities and differences in the distributions of terrain types. Two terrain units occur on all planets. Cratered terrain and volcanic plains are ubiquitous, but their abundances vary from planet to planet. On Earth, volcanic units dominate 62 percent of the total crustal area (Fig. 5). Precambrian shields have an area of only about 10 percent. By contrast, the Moon has 70 percent cratered terrain and only 17 percent lava plains. Mars is somewhat intermediate, with abundant cratered terrain and significant volcanic plains. On the one-third of Mercury photographed, craters dominate, but smooth plains, largely associated with the Caloris Basin, occupy 25 percent of the known surface. Although there is disagreement on the interpretation of Mercury's smooth plains, their large

area, which is comparable to that of volcanic plains on Mars and the Moon, provides support for a volcanic origin.

Cratered terrain is generally believed to be the oldest unit on a planet, recording the impacts—and mass additions—of the terminal stages of planetary accretion and heavy bombardment. Preservation of cratered terrain implies a relatively stable crustal history. Volcanic material can only reach a planet's surface to form plains and build cones if the crust is fragmented. The ratio of volcanic to cratered terrain is a crude Planetary Evolution Index (PEI): high values represent highly evolved surfaces, and low values, primitive surfaces. The PEI for Earth is 6.2, an order of magnitude higher than that of any other terrestrial planet, and the PEI decreases from Mars (0.7) to Mercury (0.3) to the Moon (0.2). The limited extent of radar imagery of Venus apparently shows volcanoes and giant troughs, but craters appear to be the most prevalent landform (Malin 1976). Thus Venus appears more Mars-like than Earth-like, and Mars itself is more like the Moon and Mercury than like Earth.

Igneous processes

Evidence that both Earth and the Moon have undergone early global melting suggests that extreme igneous events accompanied the formation of all terrestrial planets (Lowman 1976). Subsequently, regional and local provinces of volcanic rocks were formed as a consequence of planetary melting associated with radioactive heating. The volumes and production rates of these postformation igneous processes varied from planet to planet and with time on individual planets, but their effects were widespread.

Igneous processes dominate Earth, considering that the thin veneer of water which occupies about 70 percent of the planet's surface is underlain largely by oceanic basalts and that perhaps one-third of the remaining continental rocks are granitic plutons, plateau basalts, and arc andesites. The northern lowlands of Mars (nearly one-third of the total surface area) are interpreted as volcanic flows (Carr et al. 1973), and various plains units on Mercury are probably of volcanic origin (Strom et al. 1975). Volcanoes have also been

interpreted from radar imagery of Venus (Malin 1976), 17 percent of the lunar surface is covered by mare basalts (Head 1976), and some meteorites have volcanic compositions (Bogard and Husain 1976). Thus all the terrestrial planets, at least one meteorite parent body, and perhaps some large satellites were host to volcanic processes.

The mode of occurrence of volcanic rocks on Earth varies according to the viscosity of the melt, which, in turn, is largely controlled by silica content. Silica-rich magmas tend to fragment explosively on eruption, producing plains and cones of ash, whereas silica-poor magmas usually extrude quietly, building shield volcanoes and basaltic plains. Geochemical evidence suggests that silica-poor rocks come directly from the mantle, whereas silica-rich rocks are petrologically complicated by interactions with continental crust (Carmichael et al. 1974). The distribution, chemistry, and form of terrestrial eruptive rocks are tectonically controlled, with lithospheric plate-margin volcanism more important than intraplate, hot-spot volcanism. Acidic volcanism is generally restricted to plate boundaries, but basaltic volcanism occurs within plates as well as at their margins.

The most widespread style of volcanism on the inner planets is basaltic plains. Terrestrial ocean floors, lunar mare fill, martian northern lowlands, and perhaps the smooth plains of Mercury are probably all basaltic plains. It seems safe to predict that basaltic plains exist on Venus. Mare ridges occur on all these units except the terrestrial examples, which instead are characterized by spreading ridges and transform faults. This emphasizes a major difference in the role of volcanism on Earth compared to other planets. Ocean-floor basalts are created as a major product of plate movement, whereas generation of basaltic magmas on the Moon and perhaps the other planets is unrelated to the tectonics of the depressions they fill. However, the common occurrence of flow fronts and sinuous rills/lava channels on various planets illustrates that the surface forms and processes of volcanic activity were similar. In fact, the variation in characteristic lengths of sinuous rills/lava channels—hundreds of kilometers on the Moon, tens of kilo-

Table 1. Properties of terrestrial planets. The symbol ⊕ indicates Earth; R, retrograde.

	Moon	Mercury	Mars	Venus	Earth
Diameter (km)	3,476	4,880	6,787	12,104	12,756
Diameter (⊕ = 1)	.27	.38	.53	.95	1
Mass (⊕ = 1)	.01	.05	.11	.81	1
Volume (⊕ = 1)	.02	.06	.15	.88	1
Density (gm/cc)	3.3	5.4	3.9	5.2	5.5
Surface gravity (⊕ = 1)	.16	.37	.38	.88	1
Rotation period (day)	27	59	1	243R	1
Distance from sun (10^6 km)	150	58	228	108	150
Main atmospheric constituent	—	—	CO_2	CO_2	N

SOURCE: Hartmann 1972 and Sagan 1975.

meters on Mars, and a few kilometers on Earth—demonstrates how a particular volcanic process is modified by differing environmental conditions.

A second volcanic landform that occurs on more than a single planet is the basaltic shield. Whitford-Stark (1975) has shown that heights of terrestrial shield volcanoes, lunar domes, and martian volcanoes are linearly proportional to their basal diameters. Hence, the shapes of the shields cannot be influenced strongly by gravity or atmospheric pressure, for these quantities vary widely from planet to planet whereas the proportions of the volcanoes do not. Since shield height is directly proportional to the volume of erupted magmas, the extreme heights of martian volcanoes may not be a consequence of a stable crust remaining over a hot spot, but rather may simply mean that a tremendous amount of lava was erupted from a single vent. In any case, Vogt (1974) has shown that the heights of terrestrial shield volcanoes are limited by plate thickness rather than by the movement of a plate away from a hot spot. Thus the giant martian volcanoes imply a lithospheric thickness on Mars much greater than on Earth. This observation, along with the lack of extraterrestrial stratovolcanoes or their calderas, such as occur above downgoing slabs, suggests that plate tectonic crustal motions occur only on Earth.

Tectonism

The Moon is a relatively small planetary body (Table 1) with a thick solid crust that formed early in its history. Tectonic features include grabens and arcuate mountains associated with major impact basins, lineaments located predominantly in the ancient cratered terrain, and compressional ridges associated with the lunar maria. The Moon exhibits no evidence of major lateral tectonic movement or major vertical displacements of crustal material. The Moon may represent one end of the spectrum of tectonic evolution of planets: a small, dry planetary body with a thick crust that precluded extensive vertical and horizontal tectonic movement.

Mercury is slightly larger and contains a large dense core. Although impact-basin tectonic features are seen on Mercury, extensive lobate scarps dominate the surface tectonic environment. These features appear to indicate a regional and perhaps global environment of compression. Little evidence exists for extensive tensional features. These observations strongly suggest that the surface tectonism of Mercury was dominated by volumetric contraction on a global scale, apparently related to systematic cooling of the planet (Solomon and Chaiken, in press).

Mars, on the other hand, is larger and, although it contains many of the tectonic features seen on the Moon, it is dominated by extensional features such as Valles Marineris and other lineaments related to the Tharsis uplift. This evidence suggests that, unlike the Moon and Mercury, Mars may have undergone planetary expansion during its history (Mutch et al. 1976; Solomon and Chaiken, in press). There is no compelling evidence for regional lateral crustal movement that might indicate plate tectonic activity.

In its present form, Earth is dominated by major lithospheric plates and compressional and tensional activity associated with their lateral movement. There is no need to invoke, or strong evidence for, planetary expansion or shrinkage.

The presence of abundant impact craters on Venus implies that plate tectonic processes have not recycled the crust. The evidence of tension provided by the large trough suggests that tensional features occur on large planets but are lacking on small ones.

Planetary elevations

The three terrestrial planets whose surfaces have been adequately documented exhibit differences in their distributions of elevations as well as in terrain types. The bimodal distribution of elevations of ocean floors and continents on Earth is dramatically illustrated by the well-known hypsographic chart (Fig. 6). There are also profound differences of age, petrology, and morphology in the two major terrain types. Ocean floors are homogeneous plains of young basaltic rocks (neglecting the volumetrically insignificant sediments, ridges, and islands), whereas continents are diverse in age, composition, and form.

A naked-eye view of the Moon shows a similar dichotomy of light and dark areas, which corresponds to cratered highlands and mare lowlands. The lunar hypsographic chart does not, however, indicate a bimodal distribution of elevations; rather, the distribution is unimodal and skewed. Although the lunar chart is preliminary (being based only on topographic data within 45° of the equator) one possible interpretation is that lunar surface heights have a Gaussian distribution except for excess low areas resulting from basin formation. It follows from this interpretation that the Moon has only a single crustal type, highlands—an observation consistent with geophysical and petrologic data.

On Mars, a fundamental dichotomy exists between the cratered terrain common in the southern hemisphere and the smooth northern lowlands. A preliminary hypsographic curve for Mars (Mutch et al. 1976) is dominated by a peak at 2–3 km elevation due to the cratered terrain, but the northern lowlands, unlike the terrestrial ocean floor, are not concentrated at a uniform elevation. Plains occur at two different elevations (Saunders 1976). Thus the distribution of topography on Mars appears to be intermediate between the unimodal Moon and bimodal Earth.

Ages of planetary surfaces

Ages of various terrain units on Earth and the Moon have been determined by field and laboratory investigation, but ages for surface units of Mars and Mercury depend upon models of impact cratering rates, since rocks from these planets have not been dated radiometrically. Figure 7 compares the estimated distributions of surface ages as a function of surface area for the terrestrial planets. Evolution of the surfaces of the Moon and Mercury was essentially complete 2.5 billion years ago or earlier. In contrast, 98 percent of Earth's surface is less than 2.5 billion years old, and 90 percent is less than 600 million years old. Fifty percent of Mars is less than 2.5 billion years old, although only 15 percent is as young as 600 million years; thus age distributions on Mars appear to be intermediate between the Moon and Mercury and the highly evolved Earth. The extreme youthfulness of most of the surface of Earth is due to the destruction of old terrain and the creation of new by plate tectonic motion that may have begun about 2.5 billion years ago

(Siever 1975). We have previously noted that major tensional structures occur on Mars and perhaps on Venus; yet there is no evidence of plate tectonics on those planets, suggesting that the grabens may represent an aborted attempt to establish plate tectonics.

Planetary evolution

Several basic themes and questions emerge from a review of the geology of the terrestrial planets and provide a new perspective with which to view Earth and its history. What are the fundamental processes that formed the terrestrial planets? Because of the youthful nature of Earth's surface, we would never have listed impact cratering as a fundamental process. Examination of other planetary surfaces, however, shows its great importance, particularly in the early history of the solar system. Prior to exploration of the ocean floors, we also would have underestimated the significance of volcanic processes. In terms of areal coverage, volume, and time duration, impact cratering and volcanism are the two processes dominating the surface histories of the terrestrial planets. Atmospheric and hydrospheric processes often are important agents of terrain modification, as on Earth and Mars.

Do terrestrial planets share a com-

mon early history? Analyses of returned lunar samples strongly suggest that the outer several hundred kilometers of the Moon underwent extensive melting, the heating being provided by the terminal stages of planetary accretion (summarized in Taylor 1975). Were all the terrestrial planets characterized by similar "magma oceans," or has accretion been slow enough in some cases to preclude initial melting (see Weidenschilling 1974)? Does cratered terrain represent remnants of this solidified crust, as it appears to do on the Moon? Do volatile-rich planets undergo evolutionary paths different from those for volatile-poor ones (e.g. evolution of thick atmosphere/hydrosphere)?

What are the significant factors in planetary evolution? Lewis (1974) and others have argued that differences in planet density are largely related to position and temperatures within the cooling nebula, with the more refractory elements common toward the sun and the more volatile elements abundant toward the outer planets. Although these factors may be significant in terms of planetary bulk chemical composition, there is no obvious correlation between stage or style of planetary surface evolution and distance from the sun. Correlations do appear, however, between stages of planetary evolution (PEI)

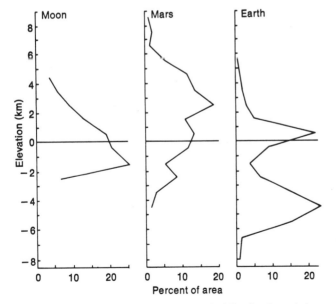

Figure 6. Hypsographic charts depict statistical distributions of planetary elevations. The terrestrial and martian charts are from Mutch et al. (1976). The hypsographic data were compiled from the lunar topography chart of Bills and Ferrari (1975). Data are available only for the zone between 45° N and 45° S.

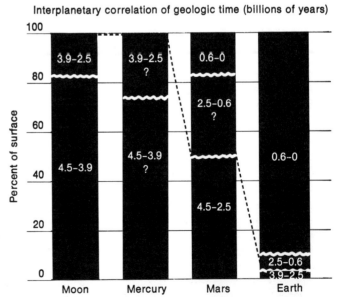

Figure 7. The areal distribution of age units on the terrestrial planets is based on radiometric dating of lunar and terrestrial rocks and on modeling of crater densities on Mars and Mercury. Here Mars appears to have age distributions intermediate between the Moon and Mercury on the one hand and Earth on the other.

and planetary mass and volume (see Table 1). The Moon and Mercury, with small volumes and masses, have a high proportion of primitive cratered terrain, while Mars shows a factor of two increase in both evolution index and mass and volume. Earth, on the other hand, shows close to an order of magnitude difference from the other planets in these values. Thus the abundance of cratered terrain decreases, being replaced by volcanic plains as planetary mass increases. Apparently, planetary size greatly influences thermal evolution (Kaula 1975). What factors relate these aspects to each other? Is there a minimum planetary mass below which mantle convection will not cause lithospheric rifting?

Do the present surfaces of planets represent stages in evolution experienced (or to be experienced) by each planet, or does each planet proceed along a separate path depending on its physical properties and position within the solar system? At present, the answer appears to be yes to both questions. Initial accretional heating seems to have produced an initial differentiation, and radioactive heating seems to have produced a second, but still early, differentiation in most terrestrial bodies. Cratering has dominated the early history throughout the inner solar system. However, on the basis of size and internal characteristics, it is unlikely that the Moon and Mercury will evolve past their present primitive state. Mars has a relatively high proportion of ancient cratered terrain, appearing closer to the Moon and Mercury in many respects than to Earth. But do the major uplifts and hemispheric asymmetry of Mars represent an incipient plate tectonic phase? Indeed, plate tectonics may represent a relatively recent stage in terrestrial evolution.

Exploration of the solar system over the past fifteen years has provided an exciting view of Earth as a planet. Man has traveled far enough into the solar system to marvel at the crescent Earth rising over the mountains of another planetary body. Planetary probes have unveiled surface features at scales far beyond our earthbound experience: craters the size of Europe, canyons for which our "Grand" Canyon would be but a puny tributary, and volcanoes nearly three times as high as Mt. Everest. The record of the early history of the solar system, which undergoes systematic destruction on the dynamic Earth, has been laid before our eyes. We will never again be able to view Earth history in isolation.

References

Bills, B. G., and A. J. Ferrari. 1975. *Proc. Lunar Sci. Conf., 6th Geochim. Cosmochim Acta.* Supp. 6, frontispiece.

Bogard, D. C., and L. Husain. 1976. A new 1.25 billion year old Nakhlite-Achondrite (abs.). *Div. Planet. Sci., 7th Ann. Meet. Am. Astr. Soc.,* p. 5.

Boyce, J. M. 1975. Chronology of the major flow units. *Conference on Origin of Mare Basalts and Their Implications for Lunar Evolution,* pp. 11–14. Houston: Lunar Science Institute.

Carmichael, I. S. E., F. J. Turner, and J. Verhoogen. 1974. *Igneous Petrology.* NY: McGraw-Hill.

Carr, M. H. 1973. Volcanism on Mars. *J. Geophys. Res.* 78:4049–62.

Carr, M. H. 1974. Tectonism and volcanism of the Tharsis Region of Mars. *J. Geophys. Res.* 79:3943–49.

Carr, M. H., H. Masursky, and R. S. Saunders. 1973. A generalized geologic map of Mars. *J. Geophys. Res.* 78:4031–36.

Cintala, M. J., J. W. Head, and T. A. Mutch. 1976. Characteristics of fresh Martian craters as a function of diameter: Comparison with the Moon and Mercury. *Geophys. Res. Let.* 3:117–20.

Gault, D. E., J. E. Guest, J. B. Murray, D. Dzurisin, and M. C. Malin. 1975. Some comparisons of impact craters on Mercury and the Moon. *J. Geophys. Res.* 80:2444–60.

Hargraves, R. B. 1976. Precambrian geologic history. *Science* 193:363–71.

Hartmann, W. K. 1972. *Moons and Planets.* NY: Bogen and Quigley.

Head, J. W. 1976. Lunar volcanism in space and time. *Rev. Geophys. Space Phys.* 14:265–300.

Howard, K. A., D. E. Wilhelms, and D. H. Scott. 1974. Lunar basin formation and highland stratigraphy. *Rev. Geophys. Space Phys.* 12:309–27.

Kaula, W. M. 1975. The seven ages of a planet. *Icarus* 26:1–15.

Lewis, J. S. 1974. The chemistry of the solar system. *Sci. Am.* 230:51–65.

Lowman, P. D., Jr. 1976. Crustal evolution in silicate planets: Implications for the origin of continents. *J. Geol.* 84:1–26.

Malin, M. C. 1976. Observations of the surface of Venus (abs.). *Div. Planet. Sci., 7th Ann. Meet. Am. Astr. Soc.,* p. 38.

Masursky, H. 1973. An overview of geological results from Mariner 9. *J. Geophys. Res.* 78:4009–30.

Murray, B. C., R. G. Strom, N. J. Trask, and D. E. Gault. 1975. Surface history of Mercury: Implications for terrestrial planets. *J. Geophys. Res.* 80:2508–14.

Mutch, T. A. 1972. *Geology of the Moon.* Princeton Univ. Press.

Mutch, T. A., R. E. Arvidson, J. W. Head, K. L. Jones, and R. S. Saunders. 1976. *The Geology of Mars.* Princeton Univ. Press.

Pollack, J. B. 1975. Mars. *Sci. Am.* 233:107–12.

Rumsey, H. C., G. Morris, R. Green, and R. Goldstein. 1974. A radar brightness and altitude image of a portion of Venus. *Icarus* 23:1–7.

Sagan, C. 1975. The solar system. *Sci. Am.* 233:22–31.

Saunders, R. S. 1976. Analysis of compensated regions of Mars. *Reports of Accomplishments of Planetology Programs.* NASA TM X3364, p. 53–56.

Sharp, R. P. 1973. Mars: Fretted and chaotic terrains. *J. Geophys. Res.* 78:4073–83.

Siever, R. S. 1975. The Earth. *Sci. Am.* 233:83–90.

Soderblom, L. A., C. D. Condit., R. A. West, B. M. Herman, and T. J. Kreidler. 1974. Martian planetwide crater distributions: Implications for geologic history and surface processes. *Icarus* 22:239–63.

Solomon, S. C., and J. Chaiken. In press. Thermal expansion and thermal stress in the Moon and terrestrial planets: Clues to early thermal history. *Proc. Lunar. Sci. Conf.* 7.

Strom, R. G., N. J. Trask, and J. E. Guest. 1975. Tectonism and volcanism on Mercury. *J. Geophys. Res.* 80:2379–507.

Taylor, S. R. 1975. *Lunar Science: A Post-Apollo View.* NY: Pergamon.

Trask, N. J., and J. E. Guest. 1975. Preliminary geologic terrain map of Mercury. *J. Geophys. Res.* 80:246–47.

Vogt, P. R. 1974. Volcano height and plate thickness. *Earth Planet. Sci. Lett.* 23:337–48.

Weidenshilling, S. J. 1974. A model for accretion of the terrestrial planets. *Icarus* 22:426–35.

Whitford-Stark, J. L. 1975. Shield volcanoes. In *Volcanoes of the Earth, Moon, and Mars,* ed. G. Fielder and L. Wilson. NY: St. Martin's.

Wilhelms, D. E., and J. F. McCauley. 1971. Geologic map of the near side of the Moon (Map I-703). Washington, DC: USGS.

Wood, C. A., and J. W. Head. In press. Comparison of impact basins on Mercury, Mars and the Moon. *Proc. Lunar Sci. Conf.* 7.

Wyllie, P. J. 1971. *The Dynamic Earth.* NY: Wiley.

William R. Muehlberger
Edward W. Wolfe

The Challenge of Apollo 17

The discoveries of the last manned lunar landing add much to our knowledge of the moon's geological history

The Apollo 17 mission was an outstanding success. Nearly flawless, it accomplished every goal, providing a spectacularly successful end to the beginning of manned planetary exploration. The challenge of unraveling the data of this one mission and, in turn, relating them to the results of earlier explorations should lead to a giant step forward in our understanding of the moon.

Our scientific capabilities grew rapidly in the few short years between Apollo 11, when in less than three hours on the lunar surface, Mission Commander Neil A. Armstrong and

Dr. Muehlberger, Principal Investigator for Apollo Field Geology Investigations for Apollos 16 and 17, has been on the faculty at the University of Texas at Austin since he received his Ph.D., and prior degrees, from the California Institute of Technology in 1954. His principal interests are in structural, volcanic, and lunar geology. He has studied the geology of New England, the Gulf Coast, the Southwestern United States, Mexico, Guatemala, and Honduras as well as the subsurface basement rocks of the continental United States. His lunar work is sponsored by NASA Contract T-5874A to the U.S. Geological Survey.
Dr. Wolfe, a geologist with the U.S. Geological Survey since 1961, received his B.A. at the College of Wooster and later taught there while completing his Ph.D., which he received in 1961 from Ohio State University. He has conducted field investigations in western Kentucky, southwestern Washington, California, and the San Francisco volcanic field of northern Arizona. Since 1968 he has been at the Center of Astrogeology in Flagstaff, Arizona, involved in Apollo mission planning, crew training, and postmission analysis. With V. L. Freeman, he was responsible for the detailed pre-mission mapping of the Apollo 17 site.
This publication was authorized by the Director, U.S. Geological Survey. Addresses: Dr. Muehlberger, Department of Geological Sciences, The University of Texas at Austin, Texas 78712; Dr. Wolfe, U.S. Geological Survey, Center of Astrogeology, 601 East Cedar Avenue, Flagstaff, Arizona 86001.

Colonel Edwin A. Aldrin deployed a geophysical station, took photographs, and collected loose rocks and soil from the immediate vicinity of the lunar module (LM), and Apollo 17, when, during 22 hours of surface exploration and experimentation, Captain Eugene A. Cernan and Dr. Harrison H. Schmitt deployed a sophisticated geophysical station, investigated all major geological formations in the valley of Taurus-Littrow, visited eleven major sampling locations (Fig. 1), traversed about 39 km of the valley floor, obtained more than 2,200 photographs, and collected 110.5 kg of rock and soil samples, including a 3 m core. Simultaneously, Commander Ronald A. Evans, in lunar orbit, made visual observations, took hand-held photographs of lunar features, and also operated a variety of scientific instruments.

Planning for Apollo 17

Prior to the Apollo program it was recognized that the near side of the moon consisted primarily of smooth, dark, topographically low maria (Fig. 2) with intervening, bright, heavily cratered highlands. Superposition, embayment relations, and crater development—all interpreted from telescopic views and from orbital photographs—indicated that the mare basin fillings were younger than the nearby highland materials. Apollo missions 11, 12, and 15 sampled mare fill and showed that it consists of basaltic lava flows. Radiometric dates on basalts from these three sites indicate that the enormous floods of mare basalt, representing a major volcanic episode in the moon's history, were produced within a relatively short interval of time, from

3.9 to 3.0 billion years (b.y.). The record of events that occurred on earth during this ancient period has been largely obliterated. The moon, however, has provided a window into the early stages of planet evolution.

Materials of the highlands or of the rocks that underlie the mare basalts have been sampled at every landing site. At the Apollo 11 site a small percentage of plagioclase-rich rock fragments in the otherwise basaltic soils was interpreted by Wood et al. (1970) to represent highlands material transported to the site by distant cratering events; an ancient lunar crust of anorthositic composition was inferred. Soil from the Apollo 12 site was found to include a component chemically like basalt but richer in aluminum than the local mare basalt (Baedecker et al. 1971). Hubbard et al. (1971) determined that this component, which they characterized as KREEP because it is distinctively enriched in potassium (K), rare-earth elements (REE), and phosphorous (P), occurs as numerous shocked glass particles and some crystalline plagioclase-pyroxene rock fragments. A ray from the crater Copernicus, 370 km to the north, crosses the landing site (Shoemaker et al. 1970). Hence the more aluminous component may be pre-mare rock excavated and transported by the Copernicus event.

It was hoped that ancient crustal material might be found in lunar highlands areas or in the walls or ejecta of large lunar basins. Figure 3 shows Orientale, the youngest of the giant impact basins; its form and deposits are used as models for interpreting the older basins. The debris ejected from a basin should contain a large percentage of crustal material, and thus samples of it should yield crustal material from depths to which we could not otherwise penetrate. Further, the high mountains ringing the basins should contain uplifted local crustal material beneath blankets of older basin ejecta.

Imbrium (the right eye of the Man-in-the-Moon) is the largest and second youngest of about 30 giant basins so far recognized on the moon (Stuart-Alexander and Howard 1970). Its position and its readily distinguished blanket of ejected debris have made it a prime target for study, beginning with early telescopic observations. The ejecta blanket probably covers nearly half the moon's near side, although much of its original extent is assumed to be buried beneath younger basalt lava flows of the maria. Apollo 14 was chosen to sample the Fra Mauro formation, which is interpreted as material excavated by the impact that formed the Imbrium basin (Swann et al. 1971). The samples are predominantly thermally metamorphosed feldspar-rich breccias composed of glass, mineral grains, basalts that are more feldspathic than mare basalts, deeper-seated igneous fragments that include gabbro, norite, and anorthosite, and feldspar-rich meta-breccias that resemble their host breccias (Wilshire and Jackson 1972). The variety of materials in the samples demonstrates that the pre-Imbrium lunar crust was already complex. In overall chemistry the Apollo 14 samples largely resemble the aluminous basalt composition of the KREEP component

Figure 1. Astronaut Schmitt and two of the five pieces that comprise Station 6 boulder (Fig. 7), which rolled down the North Massif and broke apart as it stopped. The total length of the boulder (when reassembled) is about 20 m. Scoop marks show where a sample was removed from its dirt cover. The photograph was taken from the uphill side of the boulder and shows the valley panorama. The lunar module is centered in the light streak on the valley floor to the right of the upper edge of the boulder. The streak was caused by the disturbance of dust during landing. The East Massif forms the left skyline. (NASA photograph AS17-140-21496.)

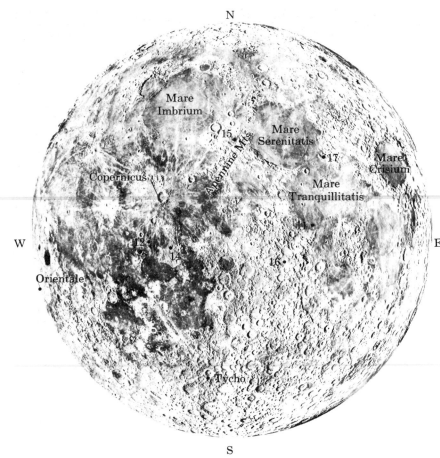

Figure 2. Composite photograph of the moon showing locations of Apollo landing sites and features discussed in the text.

at the Apollo 12 site (Hubbard et al. 1972; Rose et al. 1972; Philpotts et al. 1972).

From the mountainous margin of the Imbrium basin Apollo 15 astronauts collected feldspathic breccias and a fist-sized piece of rock identified as anorthosite. This was thought by some to be a piece of the original crust and was nicknamed the "Genesis Rock" by newsmen. Unfortunately this rock has been so intensely modified by impact events since it originally crystallized that the label is not wholly appropriate. Because a ray from one of the large craters Aristillus or Autolycus may cross the landing area (ALGIT 1972), it is uncertain whether the breccias and the anorthosite represent the materials of the mountainous basin margin or the materials underlying the Imbrium mare basalt fill.

The Apollo 16 crew landed in the central lunar highlands in order to study and sample the highlands plains-forming materials (Cayley formation) that filled and leveled

old craters and other topographically low areas. This landing site was also adjacent to lineated mountainous terrain (Descartes formation) that extends northward from the 50 km-diameter crater Descartes, which it fills. It was widely believed that these units were volcanic in origin, but instead they turned out to be thick sheets of breccia of anorthositic composition (AFGIT 1973a; LSPET 1973a). It now seems likely that the plains-forming material was derived from either the Imbrium or the younger and even more distant Orientale basin; the mountainous material might be from the Nectaris basin.

If these interpretations are true, then more material was transported farther by the giant impacts than previous estimates had indicated. Crater degradation studies (Soderblom and Boyce 1972) suggest that the ages of these highland plains surfaces are similar nearly everywhere on the moon. This conclusion, combined with the fact that all of the non-mare basalt rocks so far studied (including one sample

from Apollo 17) give nearly the same age, has led to the hypothesis (Tera et al. 1973) that the impacts that formed the major lunar basins and distributed the impact breccias sampled in the Apollo missions occurred in a relatively short period of time nearly four billion years ago.

Because of the many months needed for mission planning and crew training, it was necessary to select the Apollo 17 landing site long before the launch of Apollo 16. Highest priority for the final Apollo mission was given to investigating crustal material from a region of the moon unrelated to the Imbrium basin, (which was the goal for Apollos 14 and 15 and, as it turned out, perhaps for Apollo 16 as well). Second priority was given to sampling some of the moon's youngest volcanic materials, which were thought to be represented by dark, smooth-surfaced mantling deposits of local extent. The age and composition of the youngest lunar volcanic products are fundamental to an understanding of the thermal and dynamic history of the moon. Because the Taurus-Littrow area presented unique opportunities for sampling both very old crustal materials as well as a veneer of dark, apparently young volcanic material, it was selected as the landing site for Apollo 17.

Apollo 17 landed in the deep, narrow Taurus-Littrow valley, located within the rugged mountains that form the southeastern rim of the Serenitatis basin, which is older than the Imbrium basin. The potential for fulfilling both major geologic objectives was high because highland materials were accessible both at the mountain fronts and on an avalanche deposit that spread across the valley and because the dark material could be sampled on the valley floor.

Figure 4 shows the site as viewed from the LM in lunar orbit one revolution before landing. Mountains, thought to be uplifted relative to the valley along linear faults, rise sharply from the valley floor for 2,000 m in the case of the North Massif and 2,300 m in the case of the South Massif. The major movement on these faults, and hence the formation of the valley as a physiographic feature, is inferred to have taken place as the lunar crust adjusted to the stress of the Serenitatis impact. The mountains, both the massifs and the Sculptured Hills, were thought before the mission to consist most probably of brecciated lunar crustal material (Lucchitta 1972; Scott et al. 1972; Wolfe et al. 1972). Other major geologic units recognized before the mission include (1) the subfloor unit, a basin-filling unit that flooded and leveled the Taurus-Littrow valley floor; (2) the dark mantle, a fine-grained, relatively smooth-surfaced unit that was interpreted to veneer some of the uplands and to blanket the valley floor; and (3) the light mantle, which was interpreted as an avalanche deposit of debris from the face of the South Massif.

Traverse planning for Apollo 17 was greatly facilitated by the availability of orbital panoramic camera photographs from Apollo 15. Resolution is about 2 m, approximately an order of magnitude better than that of the pre-mission photographs of the Apollo 15 and 16 sites. Hence it was possible to select traverse routes based on details such as individual boulders that are visible in the photos. The traverses were to be the longest and most ambitious of any Apollo mission; the challenge was to make them a reality.

Ancient highlands materials

In our mission plan the base of the South Massif was the prime sampling site for old highlands crustal materials. While driving toward the South Massif, the astronauts recognized layering and color differences high on the mountainside: a blue-gray unit over a tan-gray unit (Fig. 5). Because of the steepness and height of the mountain slope, only its base could be reached, and samples were collected from three feldspathic breccia boulders (Fig. 6), one of which the astronauts recognized from a distance as having the appearance of the blue-gray unit near the mountain top and the other having the appearance of the tan-gray unit. The source of the third boulder is not known, but it too most likely originated from high on the mountainside. Additional rocks and also soil samples were collected.

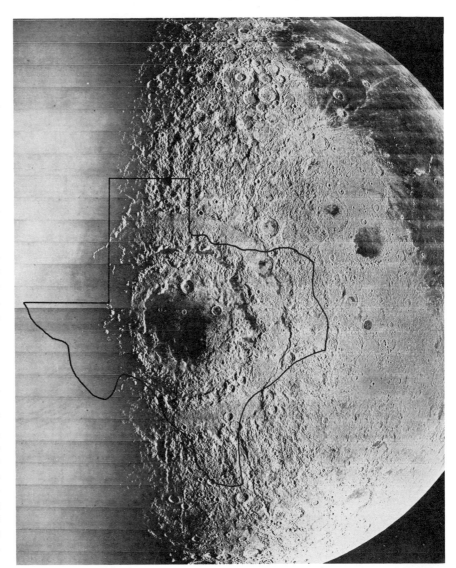

Figure 3. Orientale impact basin showing multiple mountain rings and widespread ridged ejecta deposits generated by the impact. The Cordilleran Mts. (outer ring) correspond to the Apennine Mts. (Apollo 15 site) of the Imbrium basin. The Rook Mts. (inner ring) mark the maximum possible size of the original impact crater. The mountains at the Apollo 17 landing site are comparable in position to the Cordilleran Mts. but are older. Outline of Texas shows scale of the features. Orientale is the youngest of the large basins; Imbrium is second youngest but has been flooded by basalt. Of the basins on this side of the moon, Serenitatis is sixth youngest and about the size of Orientale (Stuart-Alexander and Howard 1970). (NASA photograph LO IV, M-187.)

Extending about 6 km from the base of the South Massif is a large avalanche deposit of debris from the mountain face which may have been dislodged by the impact on the massif of a group of secondary missiles from a large distant crater, perhaps Tycho (Howard 1973; AFGIT 1973b). Samples from the avalanche deposit are all feldspathic breccias similar to those collected at the base of the South Massif.

The North Massif does not have a recognizable avalanche deposit at its base, but it does have a number of large boulders with prominent tracks that extend up the mountainside to their sources (Fig. 7). Most of the tracks originate in a light-colored zone one-third of the way up the slope, which is in sharp contrast to the South Massif samples, whose apparent source areas are from near the top of the mountain. One dark boulder, however, had rolled from high on the mountain, where a zone of dark boulders occurs. This light and dark zonation suggests that the North Massif is layered, even though the astronauts did not ·distinguish color banding. Using the telephoto camera system, they obtained photographs of the mountainside (Fig. 8) which allow us to examine the

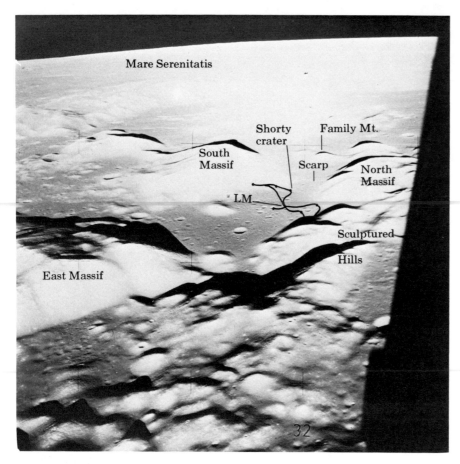

Mare Serenitatis

Shorty
crater
Family Mt.

South
Massif

Scarp

North
Massif

LM

Sculptured

East Massif

Hills

32

Figure 4. Apollo 17 landing site seen from the lunar module one revolution before landing. The light-colored material extending into the valley from the base of the South Massif is an avalanche deposit. The command module is visible near the base of the South Massif. The elevation above the valley floor of the South Massif is 2,300 m; the North Massif, 2,000 m. The distance across the valley between the South and North Massifs is 7 km. Traverse routes (lines) and sampling stations (dots) radiate from the lunar module landing point. (NASA photograph AS17-147-22465.)

source regions in detail, to search for possible causes of the downhill movement of the boulders, and to establish possible differences between the boulders on the mountainside and those sampled at the base.

Part of the giant boulder sampled at station 6 at the base of the North Massif is illustrated in Figure 1. Although in pre-mission photographs it appeared to be in two pieces, the boulder was found to be broken into five pieces that could be fitted back together with relatively little rotation of any of the blocks. These blocks turned out to be especially instructive. There is a contact through one of them across which an age sequence between an older blue-gray feldspathic breccia and a younger greenish-gray feldspathic breccia ("tan-gray" to the astronauts on the lunar surface) has been interpreted; the younger breccia has prominent elongate vesicles, an indication that it was hot enough to produce gas and flow, even though it consists mainly of rock fragments.

The older breccia includes large fragments of even earlier light-gray rocks, which, in turn, were found to be battered feldspathic rocks. At another point along the base of the North Massif (station 7, Fig. 7) the astronauts sampled a boulder which showed the same sequence and also contained dikes of blue-

Figure 5. View of the South Massif from the lunar rover, with the Scarp (Fig. 4) in the middleground. A large field of boulders is centered near the skyline in a dark area that the astronauts described as blue-gray in color. A sharp boundary sloping to the right (west) separates this dark area from light-colored areas below, which are tan-gray. The low hills and plains are covered with avalanche debris whose source area includes the boulder field. (NASA photograph AS17-135-20660.)

Figure 6. This part of the South Massif shows the area sampled at its base. Boulders are numbered in order of sampling. Boulder 1 is blue-gray; boulder 2 is tan-gray. Tracks associated with nearby boulders lead to the boulder field on the center skyline (same boulder field as in Fig. 5) and thus support the astronauts' correlations of boulder 1 with the top blue-gray layer and boulder 2 with the tan-gray layer. (NASA photograph AS17-138-21072.)

gray breccia penetrating the large light-gray breccia inclusions. In addition to the boulder samples collected at the North Massif, many small loose rocks were picked up which show a variety of feldspathic breccia types as well as some basalt typical of the valley floor.

If the massif rocks are in nearly horizontal layers, and if the stratigraphic sequences in the two massifs are similar so that we can correlate strata between them, then we can say that the samples from the South Massif represent younger material than do the samples from the North Massif because they are higher in the stratigraphic sequence. Older crustal materials might include Serenitatis or pre-Serenitatis basin ejecta such as that from the Tranquillitatis basin (North Massif samples?). Younger crustal materials might include Serenitatis ejecta or ejecta from younger basins such as Imbrium (South Massif samples?). Thus, one of the present challenges of Apollo 17 is to understand the boulder samples from the South and North Massifs, because of their great importance to understanding the history of the lunar crust. These samples probably represent a sequence of basin ejecta deposits that have furnished samples of lunar crust from many locations. In order to optimize our chances of success, teams of scientists representing many specialties have been assembled to make coordinated, comprehensive studies of these samples.

Figure 7. View of the North Massif from the lunar rover. Turning rock (TR) is composed of highlands material similar to that collected from sampling stations 6 and 7. The point labeled *source* is the beginning of the track of boulder 6. Most other tracks extend from the light-colored band to the right of *source*. The large dark boulder, *D*, is an exception; its track, 2,060 m in length, leads to a point high on the mountain. (NASA photograph AS17-141-21550.)

The materials of the Sculptured Hills offered another opportunity to sample ancient crust, because they are believed to be derived mostly from the Serenitatis basin (Scott et al. 1972). This assumption is supported by the similarity in topographic form and position of materials that ring the Orientale and Imbrium basins and are interpreted as ejecta. The sampling location at the foot of the Sculptured Hills, which was selected because it could be reached within the constraints of time and range, is in a band that appears mantled by dark smooth material. No large young, fresh craters have excavated Sculptured Hills material from beneath the veneer of debris that covers the surface, nor are there boulders with recognizable tracks. Thus it was

Figure 8. Telephoto mosaic of part of the North Massif from the lunar module showing most of the boulder track of dark boulder *D* in Figure 7 and the source of the boulder track leading to station 6 boulder (lower right of upper photo). The largest boulder at left center is about 8 m high. The dark boulder is 13 m long. (NASA photographs AS17-144-21991, 22128-9.)

necessary for the astronauts to collect materials from the surface debris in the hope that Sculptured Hills material could be separated and distinguished in the laboratory.

The Taurus-Littrow valley

We were unsure what type of material had filled and leveled the valley of Taurus-Littrow, especially after the plains surfaces at the Apollo 16 site were found to be underlain by impact-generated breccias rather than by sheets of solidified lava. In pre-mission photographs of the Apollo 17 landing site, large rocks, measuring up to several meters in diameter, could be seen in the vicinity of the large craters from

which they were apparently excavated. The astronauts' observations and the returned samples showed that these blocks are basalt similar to that from the Apollo 11 area. Even the largest craters, such as Camelot crater immediately west of the LM (Fig. 4), exposed only basalt. Geophysical data from the mission (Talwani et al. 1973; Kovach and Watkins 1973) suggest that the valley is filled by basalt more than a kilometer thick. Radiometric dates on subfloor basalt fragments range from about 3.71 (Husain and Schaeffer 1973) to 3.83 b.y. (Tatsumoto et al. 1973), but most ages cluster near 3.8 b.y. (Evensen et al. 1973; Tera et al. 1973). Hence these may be the oldest mare basalts visited on the moon.

As seen in orbital photographs, the surface of the valley floor is strikingly planar. It tilts one degree to the east for its entire length, most probably as a result of upward arching of the rim of the Serenitatis basin. This arching occurred after the Taurus-Littrow valley had been inundated and leveled by flows of subfloor basalt and prior to the main filling of Mare Serenitatis.

Many of the craters on the valley floor appear subdued in the pre-mission photographs and seem veneered by the dark mantling unit that was interpreted as a young, possibly pyroclastic blanket covering both the valley floor and portions of the nearby highlands. The contrast in tone of this dark material with the brighter surface of central Mare Serenitatis and with the even brighter steep slopes of the highlands is apparent in Figure 4. The dark mantle seemed to be thin—no more than a few meters in the traverse area, where large blocks appeared to project through it. In the extensive, smooth, apparently block-free regions beyond reach of the traverse, a somewhat greater thickness—perhaps tens of meters—was postulated.

Scattered fresh, small craters with dark ejecta were considered to be possible source vents for the dark mantling material, and the astronauts visited one of the most prominent—Shorty crater (Fig. 4). Because they had spent extra time at the foot of the South Massif and had made an additional brief stop on the avalanche earlier in the traverse, life support-system constraints reduced the time originally allocated for this crater's exploration. Upon reaching the rim of the crater, the astronauts discovered an orange "soil" (Fig. 9)—a color that is commonly present in steam alteration zones of terrestrial volcanoes. Because of the possibility, later disproved, that the orange "soil" was a product of young volcanism and hydrous alteration, it became the focus of attention at Shorty and was trenched (Fig. 10), sampled, and cored in the limited time available; all preplanned tasks other than those scheduled for the crater rim were cancelled.

At the same time, the command module pilot, who thought he could

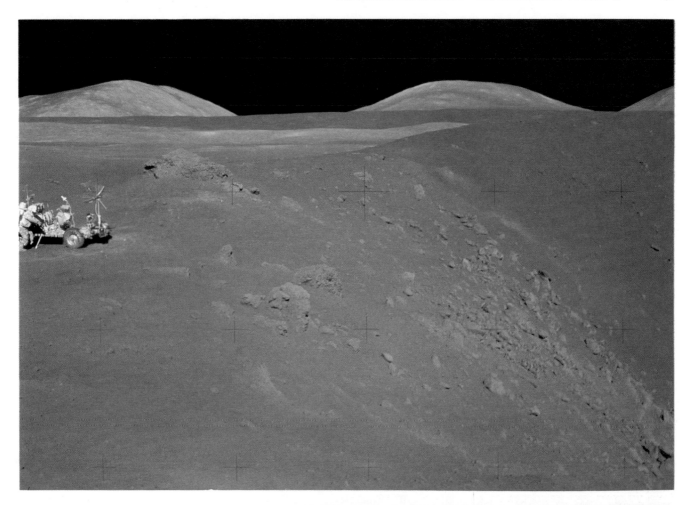

Figure 9. The rim of Shorty crater showing the sampling area of orange "soil" (beyond the lunar rover) and an orange streak (beyond block field) on the inner wall of the crater. A third orange-colored area is out of view on the crater rim, at left. The light band across the plains in the background is part of an avalanche deposit. Family Mt. is on the skyline to the right of center (see Fig. 4). The mountain mass on the left skyline is another massif located west of the South Massif. (NASA photograph AS17-137-21009.)

Figure 10. Closeup view of the trench in the orange "soil." The orange material is flanked on both sides by light gray material, and both are covered by a thin darker gray layer. A core taken from just beyond the trench shows that the orange material grades downward into the black at a depth of about 25 cm. Samples of fractured sub-floor basalt were collected from the big boulder in the background as well as from part of one of the small rocks behind the trench. (NASA photograph AS17-137-20990.)

barely see from orbit the orange zone at Shorty, was asked to look for orange-colored zones in similar settings elsewhere on the moon. He recognized and photographed two other such areas. An additional two had been noticed previously on Apollo 14 orbital photographs by Scott et al. (1971). Thus, observations on a small 100 m crater in the landing area acquired moon-wide significance.

On laboratory inspection of the samples, it was found that the orange "soil" consists of tiny orange glass beads and fragments. Also, black beads that are largely crystalline but are similar in composition to the orange glass fill the lower part of the core that was started in the orange "soil" at Shorty crater. Radiometric dates (Husain and Schaeffer 1973; Kirsten et al. 1973) show that although the orange glass is about 3.71 b.y. old it has been exposed at the lunar surface for only about 32 m.y.

Because orange glass beads have been found as a minor component in the soil of virtually every Apollo landing site (Mao et al. 1973), the process of their formation is not a feature unique to the Apollo 17 location. Two possible mechanisms of formation have been proposed: one volcanic, one impact-related. The prominent fire fountains of terrestrial volcanoes produce volcanic glass particles that are carried far and wide by the wind. On the moon, with its low gravity and no atmosphere, beads formed in such fountains could be thrown into trajectories that would cover a considerable area (Cadenhead 1973; Carter et al. 1973; McKay and Heiken 1973; Prinz et al. 1973; Reid et al. 1973). An alternative hypothesis is that a meteorite striking a large molten or only partially frozen lava lake splashed glass and crystals over a wide area (Roedder and Weiblen 1973; Prinz et al. 1973). Whatever the mechanism of origin and initial transport may have been, a layer of glass spheres and fragments was deposited on the subfloor basalt about 3.7 b.y. ago and was rapidly buried. It was excavated much more recently by the Shorty crater impact event (no more than about 30 m.y. ago).

The nature of the dark mantling

material is not yet resolved. Black beads are common in the soil (AF-GIT 1973b; LSPET 1973b) and may contribute to the dark color. Association of similar beads with orange glass in the core from Shorty crater suggests that black beads in the soil are also 3.7 b.y. old. If so, and if these are part of the dark mantle, the dark mantle is very old and has been thoroughly mixed into the soil—a hypothesis that is difficult to reconcile with the apparent youth of the dark mantle surface. However, this apparent youth may be a misconception based on the possibly erroneous observation that the older, more subdued craters are preferentially mantled. Abundant dark brown, highly vesicular particles in the soil sample taken from the large Camelot crater, west of the LM, are probably volcanic and might also be a component of the dark mantling material (Carter et al. 1973), although they could be fragments of the vesicular tops of subfloor basalt flows.

In conclusion, the available chemical and isotopic data indicate that the basalt of the valley floor is chemically different from and seemingly older than the overlying orange and black beads. This finding, combined with the basalt stratigraphy now recognized around the margin of Mare Serenitatis (Howard et al. 1973), suggests that the dark mantle is associated with the second of three distinct periods of basalt eruption. The earliest period is represented by the titanium-rich basalt of the Apollo 17 valley floor. The second period, which took place after the arching of the rim of the Serenitatis basin and included or immediately followed the main subsidence of the Serenitatis basin, may have been the period of generation of the dark mantle that covers this region. One of these first two periods may be represented by the basalts collected from the Apollo 11 landing site. The third episode of volcanism in this region was restricted to Mare Serenitatis and is characterized by a variety of basalt that is lower in titanium content (Soderblom 1970).

The Apollo 17 landing site provided important new data needed to decipher the early crustal history of the moon. Our ability to place some of our samples into their source posi-

tions on the mountain slopes gives us a unique opportunity to work out a geologic sequence that may form the basis for increased understanding of the geology of the previous highlands landing sites. In addition, we landed on what may be the oldest still-exposed mare basalt lavas on the moon, and we may have fundamental data, not yet fully understood, about the nature of its dark mantling deposits.

In the past decade we have made enormous strides in understanding our nearest neighbor in space. We expect that during the next decade an even better understanding of both the earth and the moon will evolve as these data are studied in depth.

References

Many of the Apollo 17 results referred to in this article were presented orally at the Fourth Lunar Science Conference, 4–8 March 1973. The abstracts published in *EOS* represent revised and updated studies presented at the annual American Geophysical Union meeting, 16–20 April 1973. A wealth of additional data on the samples distributed early in the preliminary examination of the returned samples can be found in the cited *EOS* issue. These very early allocations included orange soil and the adjacent gray soil from Shorty crater, soil and basalt of typical valley floor materials, and a loose breccia sample from station 6 that resembles the youngest breccia type found in the giant boulders at that locality.

AFGIT (Apollo Field Geology Investigation Team). 1973a. Apollo 16 exploration of Descartes: A geologic summary. *Science* 179:62–69.

AFGIT (Apollo Field Geology Investigation Team). 1973b. Geologic exploration of Taurus-Littrow: Apollo 17 landing site. *Science*, in press.

ALGIT (Apollo Lunar Geology Investigation Team). 1972. Geologic setting of the Apollo 15 samples. *Science* 175:407–15.

Baedecker, P. A., F. Cuttitta, and H. J. Rose, Jr. 1971. On the origin of lunar soil 12033. *Earth and Planetary Science Letters* 10:361–64.

Cadenhead, D. A. 1973. Lunar volcanic glass and cinder formation. *EOS, Transactions, Amer. Geophys. Union* 54:582.

Carter, J. L., H. C. J. Taylor, and E. Padovani. 1973. Morphology and chemistry of particles from Apollo 17 soils 74220, 74241, and 75081. *EOS, Transactions, Amer. Geophys. Union* 54:582–84.

Evensen, N. M., V. R. Murthy, and M. R. Coscio, Jr. 1973. Taurus-Littrow: Age of mare volcanism: Chemical and Rb-Sr isotopic systematics of the dark mantle soil. *EOS, Transactions, Amer. Geophys. Union* 54:587–88.

Howard, K. A. 1973. Avalanche mode of motion: Implications from lunar examples. *Science* 180:1052–55.

Howard, K. A., M. H. Carr, and W. R. Muehlberger. 1973. Basalt stratigraphy of southern Mare Serenitatis. In *Apollo 17 Preliminary Science Report.* NASA SP, in press.

Hubbard, N. J., P. W. Gast, J. M. Rhodes, B. Bansal, and H. Wieswann. 1972. Non-mare basalts. Part II, Proc. Third Lunar Sci. Conf., Suppl. 3. *Geochim. et Cosmochim. Acta,* 2:1161–79.

Hubbard, N. J., C. Meyer, Jr., and P. W. Gast. 1971. The composition and derivation of Apollo 12 soils. *Earth and Planetary Science Letters* 10:341–50.

Husain, L., and O. A. Schaeffer. 1973. Lunar volcanism: Age of the glass in the Apollo 17 orange soil. *Science* 174:1358–60.

Kirsten, T., P. Horn, D. Heymann, W. Hubner, and D. Storzer. 1973. Apollo 17 crystalline rocks and soils: Rare gases, ion tracks, and ages. *EOS, Transactions, Amer. Geophys. Union* 54:595–97.

Kovach, R. L., and J. S. Watkins. 1973. Apollo 17 seismic profiling: Probing the lunar crust. *Science* 180:1063–64.

LSPET (Lunar Sample Preliminary Examination Team). 1973a. The Apollo 16 lunar samples: Petrographic and chemical description. *Science* 179:23–34.

LSPET (Lunar Sample Preliminary Examination Team). 1973b. The Apollo 17 lunar samples: Petrographic and chemical description. *Science,* in press.

Lucchitta, B. K. 1972. The Apollo 17 landing site. *Nature* 240:259–60.

Mao, H. K., D. Virgo, and P. M. Bell. 1973. Sample 74220: Analysis of the Apollo 17 orange soil from Shorty crater. *EOS, Transactions, Amer. Geophys. Union* 54:598–99.

McKay, D. S., and G. H. Heiken. 1973. Petrography and scanning electron microscope study of Apollo 17 orange and black glass. *EOS, Transactions, Amer. Geophys. Union* 54:599–600.

Philpotts, J. A., C. C. Schretzler, D. F. Nava, M. L. Bottino, P. D. Fullagar, H. H. Thomas, S. Schuhmann, and C. W. Kouns. 1972. Apollo 14: Some geochemical aspects. Proc. Third Lunar Sci. Conf., Suppl. 3. *Geochim. et Cosmochim. Acta* 2:1293–1305.

Prinz, M., E. Dowty, and K. Keil. 1973. A model for the origin of orange and green glasses and the filling of mare basins. *EOS, Transactions, Amer. Geophys. Union* 54:605–06.

Reid, A. M., G. E. Lofgren, G. H. Heiken, R. W. Brown, and G. Moreland. 1973. Apollo 17 orange glass, Apollo 15 green glass and Hawaiian lava fountain glass. *EOS, Transactions, Amer. Geophys. Union* 54:606–07.

Roedder, E., and P. W. Weiblen. 1973. Origin of orange glass spherules in Apollo 17 sample 74220. *EOS, Transactions, Amer. Geophys. Union* 54:612–14.

Rose, H. J., Jr., F. Cuttitta, M. K. Annell, M. K. Carron, R. P. Christian, E. J. Dwornik, L. P. Greenland, and D. T. Ligon, Jr. 1972. Compositional data for twenty-one Fra Mauro lunar materials. Proc. Third Lunar Sci. Conf., Suppl. 3. *Geochim. et Cosmochim. Acta* 2:1215–29.

Scott, D. H., M. N. West, B. K. Lucchitta, and J. F. McCauley. 1971. Preliminary geologic results from orbital photography. *Apollo 14 Preliminary Science Report,* NASA SP-272, Chap. 18, Pt. B.

Scott, D. H., B. K. Lucchitta, and M. H. Carr. 1972. Geologic maps of the Taurus-Littrow region of the Moon. *U.S. Geol. Survey Misc. Geol. Inv.* Map I-800.

Shoemaker, E. M., R. M. Batson, A. L. Bean, D. H. Dahlem, E. N. Goddard, M. H. Hait, K. B. Larson, G. G. Schaber, D. L. Schleicher, R. L. Sutton, G. A. Swann, and A. C. Waters. 1970. Preliminary geologic investigation of the Apollo 12 landing site. *Apollo 12 Preliminary Science Report,* NASA SP-235, Chap. 10, Pt. A.

Soderblom, L. A. 1970. The distribution and ages of regional lithologies in the lunar maria. Ph.D. dissertation, Calif. Inst. of Tech., Pasadena.

Soderblom, L. A., and J. M. Boyce. 1972. Relative ages of some near-side and far-side terra plains based on Apollo 16 metric photography. *Apollo 16 Preliminary Science Report,* NASA SP-315, Chap. 29, Pt. A.

Stuart-Alexander, D. E., and K. A. Howard. 1970. Lunar maria and circular basins—a review. *Icarus* 12:440–56.

Swann, G. A., N. G. Bailey, R. M. Batson, R. E. Eggleton, M. H. Hait, H. E. Holt, K. B. Larson, M. C. McEwen, E. D. Mitchell, G. G. Schaber, J. P. Schafer, A. B. Shepard, R. L. Sutton, N. J. Trask, G. E. Ulrich, H. G. Wilshire, and E. Wolfe. 1971. Preliminary geologic investigations of the Apollo 14 site. *Apollo 14 Preliminary Science Report,* NASA SP-272, Chap. 3.

Talwani, M., G. Thompson, B. Dent, H. Kahle. 1973. Traverse gravimeter results on Apollo 17. *Lunar Science-IV,* J. W. Chamberlain and C. Watkins, eds., Houston, Tx.: Lunar Science Institute, pp. 723–25.

Tatsumoto, M., P. D. Nunes, R. J. Knight, C. E. Hedge, and D. M. Unruh. 1973. U-Th-Pb, Rb-Sr, and K measurements of two Apollo 17 samples. *EOS, Transactions, Amer. Geophys. Union* 54:614–15.

Tera, F., D. A. Papanastassiou, and G. J. Wasserburg. 1973. A lunar cataclysm at ~3.95 AE and the structure of the lunar crust. *Lunar Science-IV,* J. W. Chamberlain and C. Watkins, eds., Houston, Tx.: Lunar Science Institute, pp. 723–25.

Wilshire, H. G., and E. D. Jackson. 1972. Petrology and stratigraphy of the Fra Mauro Formation at the Apollo 14 site. *U.S. Geol. Survey Prof. Paper 785.* 26p.

Wolfe, E. W., V. L. Freeman, W. R. Muehlberger, J. W. Head, H. H. Schmitt, and J. R. Sevier. 1972. Apollo 17 exploration at Taurus-Littrow. *Geotimes* 17:11, 14–18.

Wood, J. A., J. S. Dickey, Jr., U. B. Marvin, and B. N. Powell. 1970. Lunar anorthosites. *Science* 167:602–04.

"From all indications, ours is one of the biggest universes there is."

John S. Lewis

The Atmosphere, Clouds, and Surface of Venus

Some chemical, geochemical, and meteorological perspectives on Earth's twin

Venus is Earth's twin in many respects. In mass, radius, density, and distance from the sun, Venus is far more Earth-like than the other planets in the solar system. In recent years, however, several unsettling disparities between Venus and Earth have come to light. Spurred by great advances in radio and radar astronomy, in earth-based infrared spectroscopy, and in spacecraft exploration during the past decade, our theoretical picture of Venus has changed repeatedly and unpredictably over a wide range of possibilities. Frequently it has been impossible for the workers in these fields to reach a consensus on working models for Venus's atmosphere and surface. There has frequently been a tendency toward an undisciplined proliferation of theories to explain a limited (and often conflicting) body of knowledge.

Partly as a result of the success of the Mariner II, Venera IV, V, VI, and VII, and Mariner V Venus probes, a number of theories about the planet have met their demise. Although unanimity is still lacking in numerous important questions, at least it is clear that essentially everyone is now talking about the same planet.

It is amusing but instructive to recall that, just a few years ago, when our data on Venus showed only that its atmosphere was CO_2-rich and that the planet

John S. Lewis is assistant professor of geochemistry and chemistry in the Departments of Chemistry and of Earth and Planetary Sciences at M.I.T. Since completing his doctoral studies under Harold C. Urey at the University of California, San Diego, he has been engaged in research on the composition and structure of planetary atmospheres, the thermodynamic and isotopic geochemistry of rare minerals in meteorites, models for the origins of the planets, and the origin and evolution of the atmospheres of the terrestrial planets. His teaching interests include courses on planetary atmospheres, meteorites, and planetology. He has served as a member of the National Academy of Sciences/Space Science Board's panels on the exploration of Venus and on the outer solar system. He is a member of the Science Advisory Group on the Outer Planets, which advises NASA on science objectives and implementation for the Grand Tour and related programs. Address: Department of Earth and Planetary Sciences, M.I.T., Cambridge, MA 02139.

emitted a substantial radio wavelength flux, it was argued by various parties that Venus was: (1) a moist, swampy planet teeming with life, or (2) a warm planet enveloped by a global ocean of carbonic acid, or (3) a cool, Earth-like planet, with water on its surface and a dense ionosphere, or (4) a warm planet covered by massive precipitating clouds of water droplets, producing intense lightning discharge activity, or (5) a planet with cold polar regions covered by ten-kilometer-thick icecaps, with hot equatorial regions far above the boiling point of water, or (6) a hot, dusty, dry, very windy planet covered by a global desert, or (7) an extremely hot, cloudy planet, with molten lead and zinc puddles at the equator and seas of bromine, butyric acid, and phenol at the poles. The emerging (but still very imperfect) view of Venus is genetically related to models (6) and (7), but the violent winds, the metallic puddles, and the chemical warfare in the polar regions have all fallen into disfavor.

In addition to the general acceptance of high ($>600°$ K) surface temperatures on Venus, two other marked differences between Venus and Earth have emerged. The overwhelming weight of evidence from infrared spectroscopy is that the abundance of water vapor above the visible clouds of Venus (at a temperature of near $240°K$) is less than one part in 10^4 of the total pressure, while searches for evidence of an H_2O absorption band in the thermal radio emission from Venus fail to disclose as much as $\sim0.8\%$ H_2O in the lower atmosphere. It is very likely that the amount of water vapor in the atmosphere of Venus is 10^{-3} or 10^{-4} times the amount of water in Earth's biosphere. It is of considerable importance to determine whether Venus differs from Earth in this respect because of different conditions of origin or because of divergent planetary evolution brought about perhaps by the closer proximity of Venus to the sun.

Active radar observations have disclosed that Venus

rotates extremely slowly on its axis and, for un-known reasons, Venus presents exactly the same side toward Earth every time it passes between Earth and the sun. Strangely, the rotation of Venus is retro-grade; that is, it rotates in the direction opposite to the rotation of Earth and the orbital motions of the planets. Although ad hoc explanations of this phe-nomenon can be offered that explain it as the natural result of unobservable interactions with objects which unfortunately no longer exist, I feel it is safe to say that we do not understand it. How Venus could find its rotation locked on to Earth despite the fact that the sun's tidal force on Venus is some 10^4 times larger than Earth's is far from clear.

Given the extremely slow rotation rate, the very high surface temperature, and the extreme aridity of Earth's twin, we are confronted by the unpleasant necessity of understanding these phenomena. But the exploration of Venus for its own sake is not advocated; rather, we should study Venus *in relation* to Earth and the other planets. This infant science is called planetology.

Earth in relation to other planets

Earth—with its atmosphere and oceans, its complex biosphere, its crust of relatively oxidized silica-rich sedimentary, igneous, and metamorphic rocks over-lying a reduced interior in which metallic iron is stable, with its ice caps, deserts, forests, tundra, jungle, grasslands, fresh-water lakes, coal beds, oil deposits, volcanoes, fumaroles, factories, automobiles, plants, animals, magnetic field, ionosphere, mid-ocean ridges, convecting mantle, and large axial inclination —is a system of stunning complexity. Let us look at Earth for a moment as an outside observer would and ask the questions which would arise in his mind as he tries to understand his discoveries.

Our postulated observer would probably discover water vapor and oxygen first by standard spectro-scopic means. He would have difficulty understand-ing the presence of a large amount of free oxygen on a planet with a metallic core. He could postulate that all the oxygen is derived from photolysis of H_2O by solar ultraviolet light and that the persistence of so much O_2 in the atmosphere is evidence for the com-plete oxidation of all ferrous iron in the crust to ferric oxides. He would be on sound theoretical grounds, but wholly wrong. Let him improve his observa-tional techniques somewhat, and he would find that CO_2 was about ten times more abundant than equilibrium with carbonate minerals would permit. He might explain this as due to the complete con-version of all calcium silicates in the crust to calcium

carbonate, or postulate that gases escaping rapidly from the interior of the earth were responsible. Again his arguments would be plausible but wrong.

Let him detect nitrogen oxides in the atmosphere; would he anticipate the existence of lightning dis-charges? Would the presence of a part per million of methane lead him to postulate anaerobic decay? Would his observations of carbon monoxide lead him to suspect that it was arising in large part from Earth's oceans? Let us attempt to imagine his chagrin when he discovers that the very nitrogen in the atmo-sphere ought to be severely depleted by chemical equilibration with atmospheric oxygen and liquid water to produce a dilute solution of nitric acid in the oceans! Would he be successful in postulating organisms living in the ocean and protecting them-selves from noxious nitric acid by destroying it enzymatically as fast as it is made? Would he attribute formaldehyde over Los Angeles to cars? Would SO_2 over Hawaii suggest volcanoes? How would he ex-plain terpenes in the air over the Great Smokies?

The unsettling truth about Earth's atmosphere is that virtually every component is produced, or de-stroyed, or both, by unique mechanisms. The number of different effects which must be postulated to describe the growth, decay, or regulation of the abundances of the atmospheric gases is probably greater than the number of gases present. This is true even of the rare gases, whose relative abundances and isotopic compositions are not by any means uni-versally constant. Clearly the main difficulty in the way of understanding these processes and of gaining the least appreciation of the origin of Earth's atmosphere and of life on Earth is that of separating or isolating these many competing processes.

I am not overstating my case. Such fundamental questions as, What is the composition of volcanic gas? How does it depend on the chemistry of its parent magma? and What was the composition of the earliest atmosphere of Earth? defy our best experi-mental and theoretical efforts. Virtually every volcanic gas analysis ever made betrays evidence of gross contamination by air.

One of the most crucial questions is the state of oxidation of the parent magma, but so facile are oxidation reactions involving CO, CH_4, H_2S, COS, etc. that the least contamination by air utterly destroys the fragile evidence. Also very common is contamination of volcanic gases by groundwater, organic matter, dissolved gases, and volatile consti-tuents of the host rock through which the magmatic gases escape.

A major reason for my interest in Venus is that the lower atmosphere presents us with an essentially unperturbed example of chemical equilibrium between atmosphere and lithosphere. In addition, the photochemistry of the upper atmosphere of Venus holds promise of being sufficiently different from our experience on Earth to assist greatly in understanding the general photolysis-recombination process. Isolating and understanding gas-rock reactions is an absolute necessity in any attempt to understand the origin and early compositional evolution of Earth's atmosphere, since only this can give us the boundary conditions under which life must have originated. The fundamental question of the ubiquity of organic material and of life will be much more tractable than it is now after such an understanding is attained.

Further, it is probable that there is an intimate connection between the origin and growth of continental blocks and the origin and evolution of the atmosphere. There is even a possible strong connection between the origin of the core of a planet by a catastrophic differentiation process and the sudden production of a substantial atmosphere by the heat pulse from core formation. Because of the great change in the rotational moment of inertia of a homogeneous planet as it differentiates to produce a silicaceous mantle and a metallic core, the rotational angular velocity of the planet may be considerably augmented, in effect, by a fundamentally geochemical occurrence.

In turn, the presence of a conductive core is essential for the existence of a large planetary magnetic field, which in turn determines whether the upper atmosphere of the planet is shielded by its magnetosphere or whether in fact the solar wind is able to "sweep" the planet's ionosphere. Because the rates of escape of hydrogen and helium are sensitive functions of the thermospheric composition and structure, secular changes in the chemistry of the atmosphere may be strongly dependent on the strength of the magnetic field (that is, on the rotation rate and the presence of a metallic core, which in turn depends on the path followed during geochemical differentiation). Current volcanic gases are frequently recycled volatiles from sedimentary ocean-floor deposits which are being forced down under the edges of continental plates by the forces driving sea-floor spreading and continental drift. All these effects working upon Earth's atmosphere can be seen to be very complexly interrelated.

Up to this point I have dealt exclusively with chemical problems of interpretation: but this is merely the one facet of Earth's atmosphere that I wish to discuss in detail. Let me also indicate to non-chemists another respect in which Earth's atmosphere is so complex that it defies comprehension.

If we attempt to understand planetary-scale atmospheric (or oceanic) circulation in a general way, we have a number of major effects to consider. First, there is the penetration of sunlight into the atmosphere. Second, there is absorption of sunlight at surfaces with attendant reemission of infrared radiation. Third, there is absorption of this infrared radiation by the atmosphere. Fourth, there is convection of strongly heated parcels of atmosphere. Fifth, there is radiative cooling of the atmosphere on the night side of the planet and at high latitudes. Sixth, there is mass motion of the atmosphere conveying heat from the equatorial region to the poles. Seventh, there is severe distortion of flow on rotating planets due to the Coriolis effects. Eighth, there is evaporation, condensation, and transport of latent heat. Ninth, there is modification of all radiative processes by the formation of cloud cover. Tenth, there is steering of winds by topography. Eleventh, there are local meteorological effects which can, by various feedback mechanisms, alter features of the planetary circulation. Twelfth, there is the most general effect of all: that our succinct formulation of effects is invariably incomplete, inaccurate, or simply misleading.

As we attempt to study the interactions of these effects in a reasonably general way, we soon find our attention drawn to the fact that the dynamic, radiative, and Coriolis forces are all of great importance on Earth. Mars is dominated by radiative effects, Venus and Jupiter by the dynamics. But Venus and Jupiter have a difference which is of prime importance: Jupiter is a rapidly rotating planet, while Venus is virtually free of Coriolis forces. We thus are in the remarkable position of being able to learn what we need to know about Earth's meteorology most readily by conducting a simultaneous study of Venus, Mars, Jupiter, and Earth. Venus is of particular interest in that the area of Earth's surface in which the circulation is most poorly understood is the tropics; near the equator the Coriolis forces are unimportant and Hadley cell circulation is found. Venus appears to present a nearly ideal opportunity to study such circulation with minimal disturbance from other forces.

Venus: background

In terms of the comparison between Venus and Earth which I suggested earlier, I have mentioned that the extremely high surface temperature of Venus came as a great surprise to many scientists. Why is the surface of Venus two to three times as hot as Earth? Exactly how hot is the surface of Venus, and how does the surface temperature vary with latitude and solar phase angle? Radio interferometer measurements of temperature differences on the surface of Venus imply that the coldest region on Venus may be within $10-20°K$ of the hottest region and that the coldest region lies on the equator near the sunrise terminator. Recent estimates of the surface temperature all fall in the range $700 \pm 100°K$.

It is not yet known whether the familiar greenhouse effect can provide such high surface temperatures in a rigorously self-consistent model. And if an evolutionary model can be found for Venus that gives the desired results, would it predict the same catastrophe for Earth? The most crucial data for the elucidation of this problem include precise absolute temperature profiles through the atmosphere, a measurement of the penetration of sunlight into the atmosphere, and reliable temperature-difference maps of the surface. An indirect technique that places useful limits on the surface temperature and pressure is surely worth pursuing.

The present data concerning the nature of the surface and crust of the planet are quite sparse. Radar astronomers have found that the dielectric constant of the surface is typical of silicates, which comes as a surprise to hardly anyone. Furthermore, it is found that the surface of Venus is remarkably free of high vertical relief near the equator. The planet appears generally flat to ± 1.5 km elevation, and a lone feature as high as 2.5 km above the mean has been detected. There is not the least shred of evidence for continental blocks and ocean basins on Venus, quite contrary to our knowledge of Earth and Mars. Plainly an indirect way to determine the chemical and physical properties of the surface would be of great value.

Our knowledge of the atmospheric composition is largely derived from Earth-based spectroscopic observations of Venus and thus refers to the atmosphere above the main cloud layer, near the $240°K$ level. The atmosphere is this region is nearly pure CO_2, with about one part in 10^4 each of H_2O and CO. In addition there is about one part in 10^6 of HCl and one part in 10^8 of HF. The detection of HF, a parts-per-billion constituent of the atmosphere of Venus, by Earth-based infrared interferometry is surely one of the most impressive achievements of planetary astronomy. Upper limits on dozens of other gases have been set by spectroscopists, and to date only these five constituents of the troposphere have been confirmed. It is interesting that none

of the sulfur-bearing gases H_2S, COS, SO_2, and SO_3 have been detected on Venus despite their abundance in terrestrial volcanic gases.

The clouds of Venus have been a favorite topic for controversy, and here the matter is still in a very uncertain state. Half a dozen species are currently favored by different individuals as making up the visible clouds. Among the most widely advertised are water or ice, silicate and carbonate dusts, ammonium chloride, compounds of the volatile elements mercury, arsenic, etc., carbon suboxide and its polymers, hydrochloric acid solution or solid hydrates of HCl, ferrous chloride dihydrate, etc. Each species has more detractors than supporters.

Several criteria may be used in judging the plausibility of these suggestions. First, we require compatibility with the observed atmospheric gases. Second, we require compatibility with the infrared reflection spectrum of the clouds. Third, we must match the refractive index for the cloud-top particles as derived from polarization of reflected sunlight. Fourth, the space-probe data on cloud structure should not be contradicted. Of course the question of the cloud composition and structure is intimately tied to the previously discussed question of atmospheric composition and structure. We therefore combine these problems to formulate the general question, Why does the atmosphere have its observed composition, and what are the implications for the clouds?

The upper atmosphere of Venus differs markedly from that of Earth in several important ways. The temperature of the exosphere of Venus is far lower than that of Earth's; the solar wind impinges upon the upper atmosphere of Venus but is held off at great distance from Earth by its large magnetic dipole moment; the processes governing the ionospheric structure on Venus are due almost exclusively to the photochemistry of CO_2, not O_2 as on Earth. Photolysis of HCl and traces of gases such as H_2S, HBr, and HI could dominate the photochemical production of free hydrogen. Since the rate of hydrogen loss is an important parameter in any discussion of chemical evolution of the atmosphere, and since the rate of photochemical production of free hydrogen may depend sensitively upon the abundance of trace gases such as those mentioned above, it seems that rather detailed chemical models of the atmosphere are needed before any sweeping generalizations can be made concerning the hydrogen escape rate averaged over geological time.

The topic of evolution of the atmosphere brings us

back to another point of tangency between Venus and Earth studies. Current models for the origin of the solar system suggest that modest differences between the conditions of temperature and pressure at the points in the solar nebula where Venus and Earth accreted may have resulted in profound differences in the degree of retention of volatile elements. It is quite plausible to suggest that Venus accreted at higher temperature than Earth, that it never contained as much water as Earth, and that even identical conditions of degassing of the interiors might have resulted in completely different early atmospheric composition. Because of the exponential temperature dependence of the dissociation pressures of solids containing volatiles, it is extremely difficult to assume constancy in the *relative* abundances of two volatiles in planets which may have accreted at very different temperatures.

All of these problems are difficult; none of them can at present be solved adequately. It is a recurrent misfortune that our understanding of many of these problems requires a detailed knowledge of the properties of the lower atmosphere and surface, while the great preponderance of the observational data at our disposal refers to the atmosphere above the clouds. Even the Venera IV–VI deep-entry probes left the lower 60 to 80% of the atmosphere unplumbed.

The Venera VII deep-entry probe, which reached Venus on December 15, 1970, was especially designed to withstand the high temperatures and pressures of the lower atmosphere. The scientific instrumentation of this probe was quite modest, with only four thermocouple gauges and four barometers aboard. One temperature gauge and one pressure gauge reported only very coarse measurements over a very wide range, while much more sensitive measurements over limited ranges of pressure and temperature were made by the other instruments. Unfortunately the commutator switch, which was supposed to select the outputs of instruments for transmission back to Earth, failed to function. As a result the *only* data returned by the probe were coarse temperature measurements, which were digitized in $\sim20°$ increments. Because the velocity of fall of the probe through the atmosphere can be measured from the Doppler effect on its radio signal, it is possible to reconstruct much of the lower atmospheric structure from aerodynamic calculations on the entry probe and its parachute.

It should not be necessary to stress that even the region probed by the Venera series to date is very poorly characterized. Even the temperature and pressure profiles currently available cannot be interpreted reliably without detailed chemical analyses of a sort not amenable to the extremely simplistic chemical analytical procedures used by the Venera probes. What is required is a detailed mass spectrometric analysis of the lower atmosphere, detailed temperature and pressure profiles, and some basic data on wind velocities and directions.

Geochemical modeling of Venus

But given our present knowledge of the atmospheric composition and clouds of Venus, what conclusions can be made regarding the surface conditions, the state of the lower atmosphere, the variability of surface temperature, and the origin and history of the atmosphere of Venus? How can we use the available data most effectively?

The basic working hypothesis in attempting to answer these questions is that chemical equilibrium may be attained in the atmosphere-lithosphere system on Venus. This simple assumption is remarkably powerful as a means of placing limits on the surface temperature because of the exponential dependence of the abundances of gases on the factor $-1/T$. Basically we shall attempt to consider every reaction between possible atmospheric constituents and surface minerals which can control the abundance of the gas. To illustrate the concept consider the equilibrium

$$MgCO_3 \rightleftharpoons MgO + CO_2(g). \qquad (1)$$

Here the equilibrium constant, K, for the reaction is given by

$$\log K_1 = \frac{-\Delta G_1°}{2.303\,RT} = \log P_{CO_2} \frac{a_{MgO}}{a_{MgCO_3}}. \qquad (2)$$

$\Delta G_1°$ is a function of T alone, while the expression on the right can often be simplified. For a system in which pure MgO and pure $MgCO_3$ are present, $a_{MgO} = a_{MgCO_3} = 1$, and we find

$$\log p_{CO_2} = f(T) = \frac{\Delta S°}{2.303R} - \frac{\Delta H°}{2.303RT}. \qquad (3)$$

This reaction has a unique CO_2 pressure corresponding to any temperature. Such a reaction is referred to as a *carbon dioxide buffer*. Because $\Delta S°$ and $\Delta H°$ are generally only very weak functions of temperature, we can write

$$\frac{\partial \log p_{CO_2}}{\partial(1/T)} = \frac{-\Delta H°}{2.303R} \qquad (4)$$

which is effectively a constant over a temperature range of a factor of two or more.

An illustration of a water buffer would be

$$Mg(OH)_2 \rightleftharpoons MgO + H_2O(g), \qquad (5)$$

and similarly for other hydrous minerals of interest.

In general the halogen acids are formed by reactions involving atmospheric water vapor: a simple example would be

$$MgF_2 + H_2O(g) \rightleftharpoons MgO + 2HF(g). \qquad (6)$$

However since we postulate atmosphere-lithosphere interactions for all gases, we are in fact solving equation (6) *simultaneously* with water buffer reactions such as (5) above. These systems of two equations in two unknowns are in principle soluble, but frequently give answers which are plainly irrelevant to Venus. The crucial points we must consider are:

1. An acceptable atmospheric model must be capable of explaining simultaneously the abundances of all observed gases within a narrow range of temperatures and pressures.

2. Because of our a priori ignorance of the mineralogy of the surface of Venus, we must consider *every* buffer reaction for which the thermodynamic data are known.

3. We will regard as unsatisfactory any atmospheric model in which blatantly incompatible mineral assemblages are predicted by buffers of different gases. Should a certain CO_2 buffer require the presence of Fe_2O_3 under conditions in which an independently derived CO buffer requires the presence of elemental Fe, we would be compelled to reject the entire suite of "agreeing" reactions as unacceptable.

This is not to say that we absolutely require that all the minerals participating in a set of simultaneous buffer reactions must be capable of coexistence in intimate contact. This would be equivalent to requiring complete geochemical uniformity of the entire surface of Venus, an assumption which virtually all geochemists would regard as completely untenable. However, should such a completely compatible buffer system with the ability to regulate the CO_2, CO, H_2O, HCl, and HF abundances at the observed levels be found, it would be most warmly received.

4. Once we think we have found a set of reactions which provide compatible buffers for all the observed gases in a narrow temperature and pressure range, then we must use the derived surface temperature and pressure to calculate the abundances of

literally hundreds of gases which have *not* been observed on Venus. We then compare these predicted gas abundances with the observational upper limits on these gases as determined by careful searches for absorption features in infrared and ultraviolet spectra of Venus. Should we find an otherwise satisfactory model which, for example, predicts 10^3 times more methane than the spectroscopists will allow, we must discard this model.

5. Once we know the temperature and pressure on the surface from the discovery of a wholly consistent model, we then may look at the possible melting or volatilization of surface materials, paying particular attention to the most abundant elements and the most volatile and fusible minerals of rarer elements. Because of the very high surface temperatures, we may find that traces of gases such as $FeCl_2$, HgI_2, arsenic and antimony sulfides, etc. would lend a peculiar pungence to the lower atmosphere even more memorable than that due to the traces of HCl and HF found above the clouds.

6. It is clear that at high surface temperatures there will be evaporation of some material that is much too involatile to be present as gases at observable ($T < 240°K$) levels. In other words, clouds of what may be described as "volcanic sublimates" must form. We hope to be able to say what materials are plausible cloud constituents.

7. Finally, once we have discovered a consistent set of buffer reactions, we find we also have begun to assemble a list of minerals that may be present on the surface. And with sufficient relevant data on the mineralogy of the surface, we may begin to guess at the petrology. I claim that the best data currently available on the geochemistry and petrology of Venus are our infrared observations of the atmosphere above the clouds.

Results and predictions: surface conditions

The most important single conclusion of present geochemical models of Venus is that there are two rather well-defined regions of temperature and pressure within which the observed atmospheric composition can be explained in terms of known chemical reactions with surface rocks. One such pressure-temperature region is on a narrow band connecting the points (190 bars, 630°K) and (31 bars, 595°K) on a plot of log P vs $1/T$. In this region Mg-bearing carbonates participate in the CO_2 buffer reactions and the CO abundance is regulated by $CO_2 + C(gr) \rightleftharpoons$ 2CO, the graphite precipitation equilibrium.

Several reactions are available to serve as H_2O, HCl,

and HF buffers. The only difficulty occurs when one attempts to account for the observational failure to detect any sulfur compounds in studies of the Venus infrared spectrum. The most stable sulfur-bearing gas, COS (carbonyl sulfide), can be minimized in abundance if the reaction regulating its pressure is

$$3FeS + 4CO_2 = Fe_3O_4 + 3COS + CO. \quad (7)$$

Even the amount of COS predicted by this reaction, $COS/CO_2 \, 3 \times 10^{-6}$, is already some ten times higher than the spectroscopic upper limit on COS above the clouds. We have three possible choices for explanation of this curious result: either there are errors in the thermodynamic and spectroscopic data, or there exists a mechanism for removing sulfur from the atmosphere below the visible clouds, or the crust of Venus is devoid of sulfur. Perhaps the least unpalatable of these alternatives is the second, but none can be conclusively dismissed at present.

The second possible pressure-temperature point is at $748 \pm 20\,°K$ and 120 ± 20 bars. Here the CO_2 buffer is the calcite decarbonation reaction

$$CaCO_3 + SiO_2 \rightleftarrows CaSiO_3 + CO_2, \quad (8)$$

and the CO abundance is controlled by

$$3FeMgSiO_4 + CO_2 \rightleftarrows 3MgSiO_3 + Fe_3O_4 + CO \quad (9)$$

or

$$3FeSiO_3 + CO_2 \rightleftarrows 3SiO_2 + Fe_3O_4 + CO. \quad (10)$$

It is interesting that the oxidation state of the lithosphere of Venus is found to be the same as that of Earth's upper mantle. This second pressure-temperature point is of extraordinary interest because of the fact that two completely independent estimates of the surface temperature and pressure, made by analyses of the wavelength dependence of the apparent temperature of Venus at radio wavelengths, find $770 \pm 25\,°K$ at 95 ± 20 bars and $790 \pm 20\,°K$ at 110 ± 15 bars, respectively. Very recently the Soviet Venera VII probe has successfully penetrated the atmosphere of the planet and provided a direct (albeit crude) measurement of the surface temperature: $747 \pm 20\,°$ K at a pressure calculated to be 90 ± 15 bars. The consensus of recent work is that the surface temperature is $\sim750\,°K$ at a total pressure of ~100 bars.

There is again one uncomfortable fact about this model for Venus: it is a near certainty that the predicted COS abundance is fully one hundred times larger than the spectroscopic upper limit! No matter what model we prefer for Venus we must account for the absence of detectable amounts of gaseous sulfur compounds. It is extremely unlikely that the data are

in error by this large a factor: a factor of ten error is barely credible. How then can we either remove sulfur from the atmosphere or from the entire crust of the planet?

If a chemical mechanism for precipitating sulfur from the atmosphere is required, then there must be in the lower atmosphere a volatile chemical agent capable of forming an involatile sulfur compound at some temperature intermediate between the surface temperature ($\sim750\,°K$) and the visible clouds ($\sim240\,°K$). Nearly as satisfactory would be a demonstration that photolytic decomposition of relevant sulfur compounds proceeds so rapidly near the visible clouds that COS, H_2S, etc. are irreversibly converted to unobservable materials such as solid sulfur. If the former is true, then a thermodynamic study of possible cloud-forming condensates would be necessary. If the latter is to be believed, then it must be shown that all the compounds actually observed in the Venus atmosphere are either more resistant to ultraviolet photolysis than these sulfur compounds or are regenerated by recombination reactions as rapidly as they are photolyzed.

Alternatively we may propose an origin or evolution for Venus which results in complete loss of sulfur from the crust of the planet. Plainly such an alternative is much less attractive and would be employable only as a last resort.

Surface composition

One result of these geochemical models for atmosphere-lithosphere reactions is the identification of certain buffer reactions that are compatible with the observed atmospheric composition. These buffer reactions, in turn, provide us with a list of plausible minerals which may be present on the Venus surface. The most important single feature of these models has been the apparently essential role of SiO_2 as a pure phase. The existence of free quartz strongly suggests that the crust of Venus, like that of Earth, is a geochemical differentiate rather than a "primordial" material in which, as in meteorites, SiO_2 is virtually impossible to find in uncombined form, and is largely tied up in ferromagnesian silicates.

Simultaneously, we find evidence that the oxidation state of the crust of Venus is closely similar to that in the upper mantle of Earth, in that ferromagnesian silicates with $Fe/(Fe+Mg) \simeq 0.2$ coexisting with magnetite, Fe_3O_4, determine the oxygen fugacity via the schematic equilibrium

$$3FeSiO_3 + 1/2\,O_2 \quad Fe_3O_4 + 3SiO_2. \quad (11)$$

Here $FeSiO_3$ is not present as a pure phase, but as a

solid solution with $CaSiO_3$, $MgSiO_3$, etc., and the actual oxygen pressure in this equilibrium at reasonable temperatures is $\sim 10^{-30}$ atm, a completely negligible proportion of the total atmospheric pressure. The observational parameter most closely related to the oxygen fugacity is the CO_2:CO ratio:

$$CO + 1/2\, O_2 \rightleftarrows CO_2, \qquad (12)$$

whence

$$\frac{p_{CO_2}}{p_{CO}} = K p_{O_2}^{\frac{1}{2}}. \qquad (13)$$

If the H_2 abundance in the lower atmosphere were known, the H_2O:H_2 ratio would provide an independent check on the parameter. Both the CO_2:CO ratio ($\sim 10^4$) and the H_2O:H_2 ratio ($\sim 10^2$) are within a reasonable range, while the CO_2:O_2 ratio of $\sim 10^{32}$ makes it clear that techniques aimed at direct detection of the equilibrium amount of O_2 are doomed to failure.

At presently accepted surface temperatures the most probable carbon-bearing mineral is calcite, $CaCO_3$. The presence of elemental carbon as graphite, a common feature of all low-temperature ($\leqslant 630\,°K$) equilibrium models for the surface, is not possible near $750\,°K$.

Atmospheric composition

In addition to the geochemically important parameter, the oxygen fugacity, one can also calculate the equilibrium abundances of countless other gases at the surface of Venus. Several general features of these calculations are of interest. First, there is the high predicted abundance of COS remarked upon earlier. Second, there is the complete absence of detectable amounts of O_2, Cl_2, and F_2.

Third, there is the strong possibility that volatilization of surface materials may greatly complicate the chemistry of the lower atmosphere. The chemical elements which may be involved of course include H, C, N, O, S, Cl, and F, but also very probably include Br, I, As, Hg, Sb, and other "volcanic sublimates" familiar in terrestrial settings. The crucial point here is that, however slow the degassing process is on Venus, the surface temperature is so high that no compound of these elements would ever condense and fall to the surface! Thus these materials, once degassed, will cease to participate in geochemical cycling and will reside permanently in the atmosphere. They will react with other atmospheric constituents to form gaseous products in the lower atmosphere and to form cloud condensates somewhere below the cold visible cloud layer.

Clouds. Infrared spectroscopy provides us with estimates of the temperature and pressure in the "visible" cloud layer, where optically thick, highly reflective clouds of some unknown substance are seen. The temperature within the upper portions of these clouds is near $235\,°K$, and the atmospheric pressure near this level is about 150 mb. In addition, observations of the polarization of sunlight reflected from the cloud tops give evidence for a haze layer near the 50 mb level, containing transparent liquid droplets of refractive index 1.45 ± 0.02. The temperature at the 50 mb level is probably near $200\,°K$, but this value is quite uncertain.

The Mariner V radio occultation experiment has contributed data concerning yet another portion of the atmosphere: well-defined strongly attenuating layers, possibly cloud layers, have been found near the 350 and $400\,°K$ levels, at pressures of several bars.

Two reliable methods of determining the water vapor abundance are available to us. The one which gives positive detection of water is infrared spectroscopy, whereby it is found that the water vapor abundance in and above the visible clouds generally falls in the range .001 to .01% and is time-variable. Microwave spectroscopy of the H_2O rotational transition at $\lambda 1.35$ cm can be performed, albeit crudely, by searching for a local minimum in the radio brightness temperature of Venus near this wavelength. No such dip can be detected, and the H_2O abundance throughout the lower atmosphere can be restricted to $<0.8\%$. If the $400\,°K$ cloud layer is attributed to water condensation, the water abundance would have to be $\sim 50\%$ of the total atmospheric pressure. On the other extreme, a water abundance of .003% would prohibit condensation except at temperatures below $\sim 205\,°K$, presumably near the 50 mb level.

Current evidence suggests a haze of H_2O-HCl solution clouds at high altitude, with solution concentrations approaching 9 Molar. These clouds would be in equilibrium with the observed HCl and H_2O abundances if the local temperatures were as low as $205\,°K$. The freezing point of a eutectic solution of HCl in water is only $187\,°K$.

The visible cloud layer is another problem entirely. I prefer to think of it as merely the top surface of an extremely complex sequence of cloud layers dominated by volatilized materials from the surface. On the basis of extensive chemical equilibrium calculations and geochemical modeling of Venus, I would propose that the most likely candidate for the $240\,°$ cloud layer is Hg_2Cl_2, a white, crystalline compound. Clouds containing the less volatile materials Hg, Hg_2Br_2, Hg_2I_2, HgS, As_2S_3, Sb_2S_3, etc. are plausible candidates for lower-lying cloud layers. It is interesting to note that even the rare element Hg is abundant enough in the crust of Earth so that degassing would provide roughly 1 gram of mercury per cm^2 of surface area of the planet.

Upper atmosphere. At high altitudes the chemical behavior of the atmosphere is dominated by the photochemistry of CO_2. The primary photolysis reaction is

$$CO_2 + h\nu \rightarrow CO + O. \qquad (14)$$

The ultimate reaction for recombination of CO and O is

$$CO + O + M \rightarrow CO_2 + M, \qquad (15)$$

where M represents a third body capable of absorbing a portion of the recombination energy of CO + O as internal excitation. Important side reactions include the recombination of two O atoms to make O_2, the formation of O_3 (ozone) by reaction of O_2 with O, and the photolysis of O_2 and O_3.

One difficulty with this scheme is that the reconstitution of CO_2 from CO and odd oxygen (O or O_3) is quite slow, and net destruction of CO_2 should occur until an essentially $CO-O_2$ atmosphere is developed. Such is not the case, and some mechanism must be found either to recombine CO and O in situ in the upper atmosphere or to mix the atmosphere on a time scale of hours all the way down to the cloud tops, where surface-catalysis may provide adequate recombination rates.

Of observed species in the Venus atmosphere, HCl is by far the most photolabile. It may contribute most of the hydrogen observed in the exosphere by the Mariner V Ly-α photometer. Photolysis of water vapor greatly complicates the picture, since numerous species containing hydrogen must be considered at all altitudes. Study of the upper atmosphere of Venus has reached the point where a new round of spacecraft observations of the turbopause region are needed. Dual-frequency ratio occultation experiments are capable of returning much useful data on this difficult region where the electron number density is decreasing exponentially with depth and where the refraction of radio waves by the neutral atmosphere is still exceedingly small. In situ mass spectrometric probing of the upper atmosphere, although still several years in the future, must be regarded as a necessity.

Conclusions

Since the principal reason for studying the Venus atmosphere is to gain a general knowledge of those effects which are obscured and complicated by various mechanisms peculiar to Earth, it is important to point out that study of Venus also introduces some new complications. Fortunately the basic information we are after is not likely to be obscured by any known or anticipated complexities, and we may confidently expect parallel study of the atmospheres of Earth, Venus, Mars, and Jupiter to be fruitful. In particular, Venus may assist us in understanding the geochemistry of the volatile elements, the origin of planetary atmospheres, the dynamics of a possible planet-wide Hadley cell circulation regime, the physical significance of the much-used but little-understood concept of eddy diffusion, and mechanisms for recombination of photolysis products which may be of use in understanding Earth's upper atmosphere.

Perhaps the most intriguing aspect of Venus exploration is the prospect of discovering more about the origin and evolution of the terrestrial planets, particularly why Earth and Venus diverge so profoundly in nature. Theories of the origin and history of our solar system must be found which account not only for the overall compositions of these planets but also for the origin and stability of their atmospheres, their surface conditions, rotational angular momentum, possession or lack of satellites, and so on. Many features of a general theory of planetology can now be anticipated with some degree of certainty, but the formulation of such a theory is still many years in the future.

Within the next few years we may anticipate further scientific investigations of Mars (by the Mariner Mars 1971 spacecraft), Jupiter (by Pioneer F and G and Grand Tour missions), and Mercury (by the 1973 Mercury-Venus flyby). To date, exploration of the atmosphere of Venus has been largely the work of Soviet scientists, whose Venera IV, V, and VI spacecraft have entered the Venus atmosphere, and whose Venera VII probe has recently succeeded in landing on the planet's surface. Crucial areas for future Venus research by both the Soviet Union and the United States include: (1) detailed chemical analyses of the lower atmosphere by a deep-entry probe; (2) mapping of the planetary-scale circulation by balloon-borne "floaters" carrying radio transponders; (3) mass spectroscopic analyses of the upper atmosphere over a wide altitude range; and (4) geochemical and geophysical investigations of the surface of the planet.

Selected Bibliography

Review Papers

Briggs, M. H. *J. Brit. Interplanet. Soc.* 19:45–52 (1963).

Bronshtén, V. A. *Astron. Vestnik* 1:4–27 (1967).

Hunten, D. M., and R. M. Goody. *Science* 165:1317–23 (1969).

Johnson, F. S. *J. Atmos. Sci.* 25:658–62 (1968).

Kellogg, W. W., and C. Sagan. *The Atmospheres of Mars and Venus.* Washington, D.C.: National Academy of Sciences—National Research Council (1961).

Sagan, C. *Science* 133:849–59 (1961).

Vetukhnovskaya, Y. N., and A. D. Kuz'min. *Astron. Vestnik* 4:8–23 (1970).

Rotation, radius, surface features, etc.

Ash, M. E., R. P. Ingalls, G. H. Pettengill, I. I. Shapiro, W. B. Smith, M. A. Slade, D. B. Campbell, R. B. Dyce, R. Jergens, and T. W. Thompson. *J. Atmos. Sci.* 25:560–63 (1968).

Goldstein, R. M., and R. L. Carpenter. *Science* 139:910–11 (1963).

Goldstein, R. M., *Radio Sci.* 5:391–96 (1970).

Kliore, A., and D. L. Cain. *J. Atmos. Sci.* 25:549–54 (1968).

Kuz'min, A. D., and A. E. Salomonovich. *Sov. Astron.* 7:116–18 (1963).

Martynov, D. Y., *Astron. Vestnik* 3:82–84 (1969).

Rogers, A. E. E., and R. P. Ingalls. *Radio Sci.* 5:425–34 (1970).

Sinclair, A. C. E., J. P. Basart, D. Buhl, W. A. Gale, and M. Liwshitz. *Radio Sci.* 5:347–54 (1970).

Smith, W. B., R. P. Ingalls, I. I. Shapiro, and M. E. Ash. *Radio Sci.* 5: 411–24 (1970).

Interior of Venus

Bullen, K. E. *Monthly Notices Roy. Astron. Soc.* 109:457–61 (1949).

Mac Donald, G. J. F., *Space Sci. Rev.* 2:473–557 (1963).

Majeva, S. V., *Astrophys. Lett.* 1: 11–16 (1969).

McCrea, W. H., *Nature* 224:28–29 (1969).

Lower atmosphere

Avduevsky, V. S., M. Y. Marov, A. I. Noykina, V. A. Polezhaev, and F. S. Zavelevich. *Radio Sci.* 5: 36–72 (1970).

Bareth, F. T., A. H. Barrett, I. Copeland, D. C. Jones, and A. G. Lilley. *Science* 139: 908 (1963).

Danilov, A. D., and S. P. Yatsenko. *Geomagnetism and Aeronomy* 3: 475–83 (1963).

Goody, R. M., and A. R. Robinson. *Astrophys. J.* 146: 339–55 (1966).

Hall, R. W., and N. J. B. A. Branson. *Monthly Notices Roy. Astron. Soc.* 151:185–90 (1971).

Hansen, J. E., and S. Matsushima. *Astrophys. J.* 150: 1139 57 (1967).

Kuz'min, A. D., and Y. N. Vetuknovskaya. *J. Atmos. Sci.* 25:546–48 (1968).

Kuz'min, A. D., *Radio Sci.* 5:339–46 (1970).

Mintz, Y., *Planet. Space Sci.* 5:141–52 (1961).

Muhleman, D. O., *Radio Sci.* 5:355–62 (1970).

Sagan, C., *Icarus* 1:151–69 (1962).

Sagan, C., and J. B. Pollack. *Icarus* 10:274–89 (1969).

Spinrad, H., *Publ. Astron. Soc. Pacific* 74: 187–201 (1962).

Stone, P. H., *J. Atmos. Sci.* 25:644–57 (1968).

Thaddeus, P., *J. Atmos. Sci.* 25:665 (1968).

Vetuknovskaya, Y. N., and A. D. Kuz'min. *Astron. Vestnik* 1:85–89 (1967).

Wildt, R., *Astrophys. J.* 91:266–67 (1940).

Young, A. T., and L. D. Gray. *Icarus* 9:74–78 (1968).

Atmospheric composition: data

Adams, W. S., and T. Dunham, Jr. *Publ. Astron. Soc. Pac.* 44:243–47 (1932).

Anderson, R. C., J. G. Pipes, A. L. Broadfoot, and L. Wallace. *J. Atmos. Sci.* 26:874–88 (1969).

Avduevsky, V. S., M. Y. Marov, and N. K. Rozhdestvensky. *J. Atmos. Sci.* 25:537–45 (1968).

Avduevsky, V. S., M. Y. Marov, N. K. Rozhdestvensky, N. F. Borodin, and W. Kerzhanovich. *J. Atmos. Sci.*, in press (1971).

Barth, C. A., J. B. Pearce, K. K. Kelly, L. Wallace, and W. G. Fastie. *Science* 158: 1675–78 (1967).

Barth, C. A., *J. Atmos. Sci.* 25:564–71 (1968).

Barth, C. A., L. Wallace, and J. B. Pearce. *J. Geophys. Res.* 73:2541–45 (1968).

Belton, M. J. S., and D. M. Hunten. *Astrophys. J.* 146: 307 (1966).

Belton, M. J. S., *J. Atmos. Sci.* 25:596–609 (1968).

Belton, M. J. S., A. L. Broadfoot, and D. M. Hunten. *J. Atmos. Sci.* 25:582 (1968).

Belton, M. J. S., and D. M. Hunten. *Astrophys. J.* 156:797 (1969).

Bottema, M., W. Plummer, and J. Strong, *Ann. d'Astrophys.* 28:225–28 (1965).

Chamberlain, J. W., *Astrophys. J.* 141:1184–1205 (1965)

Connes, J., and P. Connes. *J. Opt. Soc. Am.* 56:896–910 (1966).

Connes, P., J. Connes, W. S. Benedict, and L. D. Kaplan. *Astrophys. J.* 147:1230–34 (1967).

Connes, P., J. Connes, L. D. Kaplan, and W. S. Benedict. *Astrophys. J.* 152:963–66 (1968).

Dollfus, A., *Compt. Rend.* 256:3250–53 (1963).

Gillett, F. C., F. J. Low, and W. A. Stein. *J. Atmos. Sci.* 25:594–95 (1968).

Gray, L. D. *Icarus* 10:90–97 (1969).

Gray, L. D., R. A. Schorn, and E. Barker. *Applied Optics* 8:2087–93 (1969).

Ho, W., I. A. Kaufman, and P.Thaddeus. *J. Geophys. Res.* 71:5091–5108 (1966).

Hunten, D. M., M. J. S. Belton, and H. Spinrad. *Astrophys. J.* 150:L125 (1967).

Kaplan, L. D. *Planet. Space Sci.* 8:23–29 (1961).

Moroz, V. I. *Sov. Astron.* 7:109–15 (1963).

Owen, T., *Astrophys. J.* 150:L121–23 (1967).

Owen, T., *J. Atmos. Sci.* 25:583–85 (1968).

Pollack, J. B., and D. Morrison. *Icarus* 12:376–90 (1970).

Prokofyev, V. K., and N. N. Petrova. *Mem. Soc. Roy. Sci. Liège* 7:311–21 (1963).

Schorn, R. A., E. S. Barker, L. D. Gray, and R. C. Moore. *Icarus* 10:98–104 (1969).

Schorn, R. A., L. D. Gray, and E. S. Barker. *Icarus* 10:241–57 (1969).

Schorn, R. A., L. Gray Young, and E. S. Barker. *Icarus* 12:391–401 (1970).

Sinton, W. M., *J. Quant. Spectr. and Radiative Transfer* 3:551–58 (1963).

Spinrad, H., *Publ. Astron. Soc. Pac.* 74: 156–158 (1962).

Spinrad, H. *Icarus* 1:266–70 (1962).

Spinrad, H., and E. H. Richardson. *Astrophys. J.* 141: 282–86 (1964).

Spinrad, H., and S. J. Shawl. *Astrophys. J.* 146:328 (1966).

Strelkov, G. M. *Astron. Vestnik* 2:217–28 (1968).

Vinogradov, A. P., U. S. Surkov, and C. P. Florensky. *J. Atmos. Sci.* 25:535–36 (1968).

Wildt, R., *Nachr. Ges. Wiss. Gottingen* 1: 1–9 (1934).

Young, L. G., *Icarus* 11:66–75 (1969).

Young, L. G., R. A. Schorn, E. S. Barker, and M. Mac-Farlane. *Icarus* 11:390–407 (1969).

Young, L. G. *Icarus* 13:270–75 (1970).

Young, L. G., R. A. Schorn, and E. S. Barker *Icarus* 13:58–73 (1970).

Young, L. G., R. A. Schorn, and E. S. Barker *Icarus* 13:74–81 (1970).

Atmospheric composition: interpretation

Adamcik, J. A., and A. L. Draper. *Planet. Space Sci.* 11: 1303–07 (1963).

Dayhoff, M. O., R. V. Eck, E. R. Lippincott, and C. Sagan. *Science* 155:556–59 (1967).

Fricker, P. E., and R. T. Reynolds. *Icarus* 9:221–30 (1968).

Holland, H. D. In *The Origin and Evolution of Atmospheres and Oceans*, P. J. Brancazio and A. G. W. Cameron, eds. N.Y.: Interscience (1963).

Ingersoll, A. P. *J. Atmos. Sci.* 26:1191–98 (1969).

Knudsen, W. C., and A. D. Anderson. *J. Geophys. Res.* 74:5629–32 (1969).

Lewis, J. S. *Icarus* 8:434–56 (1968).

Lewis, J. S. *Earth Planet. Sci. Lett.* 10:73–80 (1970).

Lingenfelter, R. E., W. N. Hess, and E. H. Canfield. *J. Atmos. Sci.* 19:274–76 (1962).

Lippincott, E. R., R. V. Eck, M. O. Dayhoff, and C. Sagan. *Astrophys. J.* 147:755 (1967).

Mueller, R. F. *Science* 141:1056–57 (1963).

Mueller, R. F. *Icarus* 3:285–98 (1964).

Mueller, R. F. *Nature* 203:625–26 (1964).

Mueller, R. F. *Icarus* 4:506–12 (1966).

Rasool, S. I. *J. Atmos. Sci.* 25: 663–64 (1968).

Rasool, S. I., and C. de Bergh. *Nature* 226:1037–39 (1970).

Smith, L. L., and S. H Gross. *Bull. Amer. Astron. Soc.* 1:363 (1969).

Urey, H. C. *The Planets*, New Haven: Yale University Press (1952).

Walter, L. S. *Science* 143:1161 (1964).

Clouds

Arking, A., and J. Potter. *J. Atmos. Sci.* 25:617–28 (1968).

Arking, A., and C. R. Nagaraja Rao. *Nature* 229:116–17 (1971).

Bottema, M., W. Plummer, J. Strong, and R. Zander. *Astrophys. J.* 140:1640–41 (1964).

Bottema, M., W. Plummer, J. Strong, and R. Zander. *J. Geophys. Res.* 70:4401–02 (1965).

Coffeen, D. L. *J. Atmos. Sci.* 25:643 (1968).

Dollfus, A., and D. L. Coffeen. *Astron. Astrophys.* 8:251–66 (1970).

Fukuta, N., T.-L. Wang, and W. F. Libby. *J. Atmos. Sci.* 26:1142–45 (1969).

Gehrels, T., and R. E. Samuelson. *Astrophys. J.* 134: 1022–23 (1961).

Goody, R. *Icarus* 3:98–102 (1964).

Goody, R. *Planet. Space Sci.* 15:1817–20 (1967).

Hansen, J. E., and H. Cheyney. *J. Atmos. Sci.* 25:629–33 (1968).

Hansen, J. E., and A. Arking. *Science* 171:669–71 (1971).

Irvine, W. M., *J. Atmos. Sci.* 25:610–16 (1968).

Kuiper, G. P., *Comm. Lunar Planet. Lab.* 6:229–50 (1969).

Lewis, J. S. *Astrophys. J.* 152:L79–83 (1968).

Lewis, J. S. *Icarus* 11:367–85 (1969).

Lewis, J. S. *J. Atmos. Sci.* 27: 333–34 (1970).

Martynov, D. Ya., and M. M. Pospergelis. *Sov. Astron.* 5: 419–20 (1961).

Menzel, D. H., and F. L. Whipple. *Publ. Astron. Soc. Pac.* 67:161–68 (1955).

Murray, B. C., R. L. Wildey, and J. A. Westphal. *J. Geophys. Res.* 68:4813–18 (1963).

Murray, B. C., R. L. Wildey, and J. A. Westphal. *Science* 140:391–92 (1963).

Pollack, J. B., and C. Sagan. *J. Geophys. Res.* 70:4430–26 (1965).

Pollack, J. B., and C. Sagan. *Icarus* 4:62–103 (1965).

Rasool, S. I. *Radio Sci.* 5:367–68 (1970).

Rea, D. G., and B. T. O'Leary. *J. Geophys. Res.* 73:665–75 (1968).

Sagan, C., and J. B. Pollack. *J. Geophys. Res.* 72:469–77 (1967).

Samuelson, R. E. *J. Atmos. Sci.* 25:634–43 (1968).

Tolbert, C. W., and A. W. Straiton. *J. Geophys. Res.* 67: 1741–44 (1962).

Westphal, J. A., R. L. Wildey, and B. C. Murray. *Astrophys. J.* 142:799–802 (1965).

Westphal, J. A. *J. Geophys. Res.* 71:2693–96 (1966).

Wildt, R. *Astrophys. J.* 96:312–16 (1942).

Upper atmosphere

de Vaucouleurs, G., and D. H. Menzel. *Nature* 188:28–33 (1960).

Donahue, T. M., *J. Atmos. Sci.* 25:568–73 (1968).

Eshleman, V. R. *Radio Sci.* 5:325–32 (1970).

Goody, R., and T. B. McCord. *Planet. Space Sci.* 16:343–52 (1968).

Goody, R. *Comments Astrophys. Space Phys.* 2:7–11 (1970).

Harteck, P., R. R. Reeves, Jr., and B. A. Thompson. *NASA TN D-1984* (1963).

Hogan, J. S., and R. W. Stewart. *J. Atmos. Sci.* 26:332–33 (1969).

Hunten, D. M. *Can. J. Chem.* 47:1875 (1969).

Kliore, A., G. S. Levy, D. L. Cain, G. Fjeldbo, and S. I. Rasool. *Science* 158:1683–88 (1967).

Kurt, V. G., S. B. Dostovalow, and E. K. Sheffer. *J. Atmos. Sci.* 25:668–71 (1968).

Levine, J. S. *Planet. Space Sci.* 17:1081–88 (1969).

Marmo, F. F., and A. Engelman. *Icarus* 12:128–30 (1970).

McElroy, M. B. *J. Geophys. Res.* 73:1513–21 (1968).

McElroy, M. B. *J. Atmos. Sci.* 25:574–77 (1968).

McElroy, M. B., and D. M. Hunten. *J. Geophys. Res.* 74:1720–39 (1969).

McElroy, M. B., and D. M. Hunten. *J. Geophys. Res.* 75:1188–1201 (1970).

Newkirk, G. *Planet. Space Sci.* 1:32–36 (1959).

Prinn, R. G. *J. Atmos. Sci.* (in press).

Shimizu, M. *Icarus* 9:593–97 (1968).

Shimizu, M. *Icarus* 10:26–36 (1969).

Stewart, R. W. *J. Atmos. Sci.* 25:578–82 (1968).

Warner, B. *Monthly Notices Roy. Astron. Soc.* 121:279–83 (1960).

The Geology of Mars

Michael H. Carr

Volcanic, tectonic, and fluvial features on the surface of Mars record a long and varied geologic history

Mars, the fourth planet from the sun after Mercury, Venus, and Earth, has been studied more thoroughly than any other planet except Earth. Spacecraft exploration began in 1964, when Mariner 4 flew by Mars and sent back to Earth several indistinct pictures of a cratered surface. This was followed by two additional U.S. flybys in 1969 and by the Mariner 9 orbiter in 1971. The Soviets sent two spacecraft to Mars in 1971 and four additional vehicles in 1974. Then, in the summer of 1976, the Viking project placed two elaborately instrumented vehicles on Mars's surface and left two others in orbit to make global observations. At the time of writing, in mid-1980, one lander is still operating, and the orbiter missions had just ended after 4 years of almost continuous data-taking. Of particular importance for understanding the geology of Mars are the orbiter pictures, which cover the entire planet at a resolution of around 200 m, and which also cover extensive areas at higher resolutions, ranging down to about 10 m.

Michael H. Carr received his B.S. from the University of London and his M.S. and Ph.D. from Yale University. In 1962 he joined the U.S. Geological Survey, where he was first engaged in mapping the moon and developing electron-microscope and electron-microprobe techniques for analyzing particulate debris in the atmosphere. Since 1970, as leader of the Viking Orbiter Imaging Team, he has been responsible for the development of the cameras on the Viking orbiters and the data-handling systems on the ground. He has supervised the acquisition of more than 60,000 pictures of Mars and has been involved in their analysis—particularly as it pertains to volcanic and fluvial processes. Address: Branch of Astrogeologic Studies, USGS, 345 Middlefield Road, Menlo Park, CA 94025.

Mars has followed a distinctively different evolutionary path from those of the moon or Earth. On the moon, volcanic activity ceased around 3 billion years ago, and since that time, the surface has remained essentially unchanged except for the occasional impact crater. Mars appears, however, to have been volcanically active throughout its history, and its large volcanoes are probably still active. Moreover, the surface has been modified to varying degrees by the wind, and possibly by water and ice also.

Despite these superficial resemblances to Earth, the differences between Earth and Mars are enormous. Earth's geology is dominated by the effects of plate tectonics. The motions of the plates control the positions of continents, mountain chains, and ocean deeps and affect the style of crustal deformation and the type, periodicity, and location of volcanic activity. Furthermore, the surface of the earth is extensively modified by running water, which redistributes materials and eliminates extremes of relief. In contrast, the crust of Mars is relatively stable: the mountain chains, linear troughs, and transcurrent fault zones characteristic of plate tectonics are absent, and while it appears that there have been episodes of fluvial erosion, its effect on the redistribution of surface materials has been trivial. The combination of sustained internal activity, no plate tectonics, and little fluvial erosion and deposition has resulted in a planet of enormous surface relief— with volcanoes towering 27 km above the surface and vast canyons several kilometers deep—on which relatively ancient features are preserved in almost pristine condition.

Before discussing the geology in more detail, I shall review some properties of Mars that might affect geologic processes. The equatorial radius of 3,390 km is a little over half that of the earth and close to twice that of the moon. The axis of rotation is inclined 25° to the ecliptic, which means that the planet has seasons; however, because of the relatively high orbital eccentricity (0.097), there is a seasonal asymmetry, with summers in the south being shorter and hotter than those in the north.

The atmosphere is composed of 95.3% CO_2, 2.7% N_2, and 1.6% Ar, with lesser amounts of O_2, CO, H_2O, and noble gases other than argon (Owen et al. 1977). It is thin, and a significant fraction of the CO_2 condenses on the polar caps in winter, which results in variations in atmospheric pressure from 7 millibars at zero elevation in southern winter to 9 mb at zero elevation in southern summer. Because the atmosphere is thin, temperatures at the surface have wide diurnal and seasonal ranges. They vary from as low as 140°K in winter on the southern polar cap to as high as 290°K at midday in summer in mid-southern latitudes. At no time or place is liquid water stable at the surface; it either evaporates or freezes.

At midsummer in the south, local dust storms are common, and a feedback mechanism between the amount of dust in the atmosphere and the magnitude of the tidal winds can result in a runaway effect, causing most of the planet to become engulfed in gigantic dust storms. The dust raised by the storms generally settles out in about 3 months, but the atmosphere always retains a significant dust component (optical depth ~0.8).

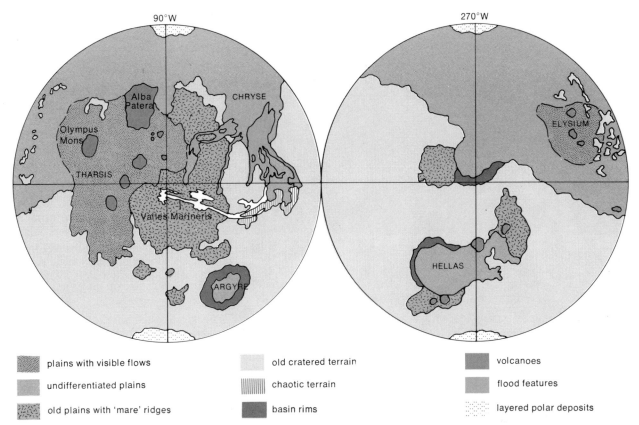

plains with visible flows

undifferentiated plains

old plains with 'mare' ridges

old cratered terrain

chaotic terrain

basin rims

volcanoes

flood features

layered polar deposits

Figure 1. As this geologic map of Mars shows, most of the southern hemisphere is densely cratered, whereas much of the northern hemisphere is sparsely cratered. Volcanoes are concentrated in Tharsis, Elysium, and Hellas. Most large floodlike features drain northward toward Chryse Planitia from the region north and east of the vast canyon system, Valles Marineris. (After Mutch et al. 1976.)

Cratered terrain and plains

Mars is markedly asymmetric in the distribution of its surface features and in its geologic evolution. It can be divided into two hemispheres by a plane dipping 50° to the equator and oriented so that it intersects the 50°N latitude parallel at 330°W. This division results in nearly all the most ancient and most densely cratered surfaces falling on the more southerly hemisphere and in most of the lightly cratered plains and large volcanoes falling on the northerly hemisphere (Fig. 1). In many parts of the northern hemisphere, however, remnants of old craters protrude through the plains to the surface, which indicates that the old cratered terrain, albeit considerably disrupted, underlies the plains.

The elevation of the lightly cratered plains in the north is generally 2–3 km lower than that of the cratered terrain in the south, the main exceptions being the anomalously high plains in the volcanic regions of Tharsis and Elysium. As a result, most of the southern hemisphere is higher than most of the northern hemisphere. The cause of this hemispheric difference is not known, but a plausible explanation is that low-density crustal materials are unequally distributed. Some support for this view is the lack of a gravity anomaly along the boundary between the plains and the uplands, which suggests a low-density crust beneath the high-standing uplands that is thicker than that beneath the adjacent plains.

The large number of craters in the uplands implies that the surface has survived from an early era of intense bombardment. On the moon, the cratering rate was very high around 4 billion years ago but shortly thereafter declined to its present low rate. There are good reasons to believe that all the terrestrial planets have a similar cratering history, the era of intense bombardment coinciding with the early period in the evolution of the solar system, when objects in nearly circular heliocentric orbits were being swept up by the planets.

The cratered uplands of the moon and Mars are significantly different in detail, however. The lunar highlands are saturated with craters of all diameters down to several tens of meters, which implies that the surface is in equilibrium and that old craters are destroyed as fast as new ones are formed. On Mars, the uplands (Fig. 2) are saturated only with craters larger than about 30 km in diameter, and these craters tend to be highly degraded. The distribution curves for smaller sizes fall far short of saturation, so much so that the density of craters 10 km in diameter in the martian uplands is over a factor of ten less than that in the lunar highlands. Moreover, the small craters that are present tend to look younger than the larger craters. These observations led Hartmann (1973) to suggest that the early era of intense bombardment was accompanied by enhanced crater destruction, which resulted in modification of the large craters and elimination of the smaller ones. According

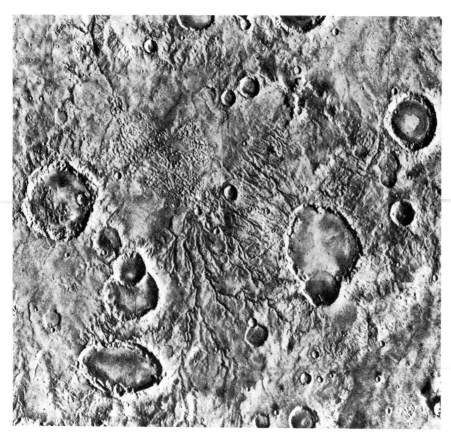

Figure 2. In this typical old cratered terrain at 230°S, 8°W, the largest crater is 80 km across. There are relatively few craters less than 10 km in diameter, and the large craters have flat floors and low rims. The dissection of the surface by branching valleys is characteristic of much of this old terrain.

to this theory, the small craters that exist today were formed after the early period of intense obliteration and the decline in the cratering rate. The cause of the enhanced obliteration is unclear. It could not be cratering alone, for this would result in a saturated distribution. High rates of volcanism could have been a contributing factor, since large areas of the uplands appear to be partly covered with a thin veneer of volcanics. A denser atmosphere may also have had an effect. Many of the old upland craters are channeled and gullied as though they had been modified by the slow erosion of running water. The scarcity of channels in the younger surfaces may indicate that the atmosphere later became thinner and temperatures colder.

Thus, the origin of the densely cratered terrain is almost certainly complex, with volcanism and eolian and fluvial processes occurring simultaneously with the intense bombardment. As a result, materials near the surface are probably interbedded units of differing origins that have frequently been disrupted by impacts. When the impact rate declined around 4 billion years ago, the landscape stabilized, and it has been only slightly modified since then.

The plains, in contrast, show a wide range in age, from those that are relatively heavily cratered and probably formed shortly after the early decline in the impact rate to some in the volcanic region of Tharsis that are geologically recent and are almost devoid of impact craters. The sim-

Figure 3. The ridges in these plains at 23°S, 79°W closely resemble those of the lunar maria. They appear to have formed as a result of deformation caused by the presence of the Tharsis bulge to the northwest. The distribution of crater ejecta in the form of lobate flows with distal ridges is typical of plains within 30° of the equator and suggests a highly fluid consistency. The frame is 240 km across.

plest and most easily understood plains are those within 30–40° of the equator. They are mostly of two types. The first, generally found around volcanoes, consists of successions of clearly identifiable flows, one on top of the other. Individual flows can commonly be traced for many tens—even hundreds—of kilometers. Many have lobate flow fronts a few tens of meters high, and some have leveed channels like those in terrestrial lava flows, although the Mars channels are usually far larger. The second type of equatorial plain is almost featureless except for superposed craters and low ridges that resemble those of the lunar maria (Fig. 3). In some areas, lines of cratered cones, low circular shields, and spatter ramparts are identifiable. The origin of all these plains is relatively unambiguous: they give every indication of having been formed by repeated eruptions of fluid lava, and, apart from the occasional impact crater and subsequent volcanic eruptions, they have undergone little modification.

The high-latitude plains, on the other hand, show a wide variety of puzzling features and appear to have had a complex history of erosion and deposition, which largely destroyed any primary volcanic characteristics. The present surface commonly has a bewildering array of topographic and albedo features, such as polygonal fractures, striped ground, craters on pedestals (Fig. 4), closely packed dimple-shaped depressions, and irregular-shaped hills. Some of the features may result from repeated deposition and partial removal by wind of debris blankets rich in volatiles such as H_2O and CO_2 (Soderblom et al. 1973). Others may be related in some way to the accumulation and removal of ground ice (Carr and Schaber 1977). The 30–40° latitude band is a region of transition between those areas nearer the equator, where surface materials are permanently dehydrated because water-ice is unstable with respect to the present atmosphere, and those areas nearer the poles, where ice is stable at the surface for part of the year and at relatively shallow depths for the entire year (Farmer and Doms 1979). The stability boundaries shift back and forth in latitude in response to long-term climate changes caused by variations in eccentricity and obliquity. Dissipation and accumu-

lation of ice caused by these changes could have affected the erodibility of the surface materials and had some influence on eolian erosion. A further complication arises from the fact that the northern plains are probable sites of accumulation of fluvial debris derived from large flood features to the south. The complex characteristics of the northern plains are thus in all likelihood the result of the interaction of a variety of processes, none of which is very well understood.

One obvious difference between the volcanic plains on Mars and those on the moon is the ejecta patterns around impact craters (see Fig. 3). The craters on Mars are surrounded by sheets of debris that have numerous indications of having been emplaced by flow (Carr et al. 1977), rather than simply from ballistic trajectories, as is generally the case for the moon (Oberbeck 1975). So-called rampart craters have petal-like arrangements of flows in which each ejecta petal has a distal ridge; other craters are surrounded by an annulus of material with a buckled and crumpled surface; still others have ejecta with a strong radial fabric.

Figure 4. Plains at high latitudes are generally more complex than those at low latitudes. In this plain region at 48°N, 349°W, are pedestal craters, around which remnants of a former, thick cover appear to have been retained. The frame is 78 km across.

The peculiar morphologies have been attributed to entrainment of large amounts of water and water-ice in the ejecta. The flow patterns occur only for craters larger than about 1–5 km in diameter. Such large craters may penetrate the permafrost that exists everywhere on Mars, excavating water-rich materials from below, which then become emplaced almost like a mud-flow. The very fluid-appearing ejecta patterns become less common toward the poles, which is consistent with the fact that the permafrost is thicker in the polar regions. The fluid patterns are not restricted to the plains, although that is where they are most obvious; they occur on all features, even atop the large volcanoes.

Volcanism and tectonism

In addition to the extensive volcanic plains, Mars has many discrete volcanoes. The most spectacular are large shield volcanoes several hundreds of kilometers across, but numerous smaller volcanoes a few hundred meters across are also present. Most of the large volcanoes are in two broad provinces: Tharsis, centered on the equator at 112°W, and Elysium, centered at 25°N, 214°W. Tharsis is at the center of a broad rise in the martian crust, about 10 km high and 6,000 km across, depending on how its base is defined. Toward the summit of the rise are three large volcanoes with summits 27 km above the Mars datum level and 17 km above the surrounding plains. These volcanoes are enormous by terrestrial standards. Olympus Mons (Fig. 5), at the northwest edge of the rise, towers 25 km above the surrounding plains and is circled by a peripheral cliff 550 km across; lavas drape over the cliff and extend far beyond, and thus the true diameter of Olympus Mons is closer to 700 km. Alba Patera, a large volcano north of Tharsis, is over 1,500 km across. By comparison, the largest volcanoes on Earth—those in Hawaii—only grow to 120 km across and 9 km above the ocean floor.

The volcanoes at the summit of the Tharsis rise have all had a similar growth cycle. Construction of the main shields by repeated eruptions of fluid, probably basaltic lava, at the summit and on the flanks was accompanied by repeated subsidence at the summit to form central calderas several tens of kilometers across. In

Figure 5. Olympus Mons, the tallest volcano on Mars, is 27 km high and is circled by a peripheral cliff 550 km in diameter. In several places (e.g., the lower left), lavas drape over the cliff and extend over the surrounding plains. The volcano has probably been accumulating for a long period of time—possibly billions of years.

the late stages of shield-building, eruption was confined mainly to the northeast and southwest flanks, where parasitic vents developed. The vast quantities of lava that erupted through these vents spread out over the surrounding plains and buried the lower flanks of the shields (Crumpler and Aubele 1978). Olympus Mons and Alba Patera appear to have had somewhat similar histories, although a northeast-southwest asymmetry never developed.

The large size of some of the martian volcanoes is probably the result of a thick, stable lithosphere (Carr 1973). Hawaiian volcanoes grow only for a limited length of time—a few hundred thousand years—because movement of the Pacific plate on which they sit carries them away from the magma source. The result is a line of extinct volcanoes that extends northwest across the Pacific from the source of magma beneath the cur-

rently active volcanoes on the island of Hawaii. On Mars, because there is no plate motion, a volcano remains stationary over a magma source and so can continue to grow as long as magma is available. Although crater counts on the flanks and calderas of the Tharsis shields suggest that martian volcanoes are relatively young, the flows that originate in the peripheral vents have an age range indicative of activity that has been sustained for a considerable length of time—possibly billions of years.

Even if magma is available, however, a volcano can grow in height only if the magma can be forced to the volcano summit. The rise is achieved largely by the hydrostatic pressure head created by the difference in density between the magma and the rocks through which it passes on its way to the surface. The three Tharsis shields have the same summit elevation of 27 km, which is unlikely to be

a coincidence. It is more likely that this represents the limit to which magma can be pumped on Mars. For terrestrial values of magma and mantle densities, the 27-km elevation implies a source depth of 200 km. By comparison, the source depth for the Hawaiian magma is thought to be around 60 km. It appears likely, therefore, that the martian lithosphere is thicker than the earth's, a conclusion that is consistent with most models of the evolution of the interior and with models of deformation of the lithosphere under the loads created by the volcanoes. Volcanoes older than those in Tharsis have lower elevations, but whether this is because the earlier lithosphere was thinner or because of different eruptive styles is not known.

The history of volcanism on Mars is one in which activity has become progressively more restricted with time. Crater densities on the plains suggest that most plains were formed during the first half of the planet's history. Activity in this era appears to have been almost global in extent, occurring within the old cratered terrain as well as in those areas now completely covered with volcanic plains (Greeley and Spudis 1978). Activity during the second half of the planet's history has, however, been confined largely to the province of Tharsis and possibly Elysium. In the last billion years or so, activity has occurred almost exclusively at Olympus Mons and the large shield volcanoes at the summit of the Tharsis rise. This progressive decline in activity is probably connected with a general global cooling and a thickening of the lithosphere.

Spectroscopic evidence and direct analyses of the surface materials suggest that most of the volcanics are iron-rich basalts. Rocks were not analyzed at the Viking landing sites, but analyses of the fine-grained debris that is moved around the surface by the wind showed low Al and K and high Fe and Mg contents, which suggests that the source rocks are mainly basaltic, and that rocks of granitic composition are rare at the surface, if present at all (Toulmin et al. 1977). This conclusion is consistent with the spectral reflectivity of the surface, the morphology of the volcanic features, the inferred density of the mantle, and the origin of Earth's granitic rocks, which were generally formed by alteration of subducted sedimentary sequences, which are not present on Mars.

The mineralogy of the fine-grained debris on the surface is not known, but it is widely accepted that it is composed mostly of iron-rich clays. Such clays could form underground by hydrothermal processes or by slow weathering of iron-rich rocks by groundwater, or at the surface by weathering induced by ultraviolet light. Those that form underground may be brought to the surface by a variety of processes, including impact and fluvial and eolian erosion.

As previously indicated, most of the large volcanoes are associated with the Tharsis bulge. Formation of the bulge was clearly a major event in the history of the planet. Surrounding it are numerous radial fractures, which can be traced outward for about 3,000 km in almost all directions from the bulge center at 0° latitude, 110°W, and affect almost a third of the planet's surface. Centered over the bulge is a large, positive, free-air gravity anomaly. Calculations of surface stresses based on the topography and gravity that now exist show that the present directions of maximum stress (Phillips and Ivins 1979) are almost uniformly oriented perpendicular to the fractures. The fractures appear to result from the presence of the Tharsis bulge, rather than from stresses associated with its formation. Old fractures and old lava flows with flow directions consistent with present slopes indicate that the bulge formed early in the planet's history, possibly before the decline in the impact rate around 4 billion years ago. A possible cause is mantle convection triggered by separation of the core. Whether the rise is still being actively supported by convection or is passively supported by a thick, rigid lithosphere is unclear, but the volcanic activity in the region suggests that the lithosphere may be too thin to support the Tharsis loads for billions of years and that active support may be required.

On the eastern flanks of the Tharsis bulge, and oriented radially to its center, is a vast system of canyons, Valles Marineris (Blasius et al. 1977). Throughout most of its length, the system is multiple, consisting of parallel canyons up to 200 km across (Fig. 6), lines of pits, crater chains,

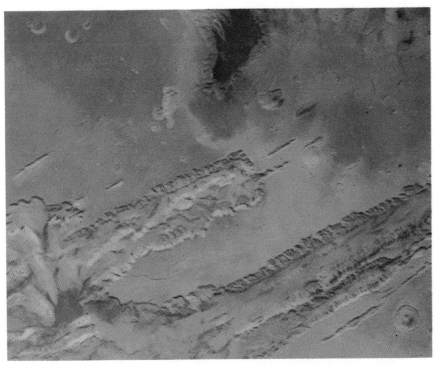

Figure 6. Extending eastward from the volcanic region of Tharsis is a vast system of canyons, Valles Marineris. Each of the larger canyons shown here is about 200 km across and, in this section, 4–7 km deep. The canyon floors are slightly bluer and darker than the surrounding plateau but contain some layered sediments that have a lighter tone. The color was reconstructed at USGS, Flagstaff, Arizona, from pictures taken by the Viking orbiters with different color filters.

and grabens. The system is widest and deepest in the central section between 65° and 77°W, where three huge, parallel troughs merge to form an almost continuous depression 600 km across and over 7 km deep.

The canyons appear to be largely tectonic in origin. Many of the walls are clearly scarps formed by faults radial to Tharsis, and parts of the floor appear downfaulted. But faulting has also triggered gigantic landslides, some of which are over 100 km across (Fig. 7). In addition, most of the walls are gullied or channeled to some degree, although whether this is the result of mass wasting or fluvial action is unclear. One of the least understood features of the canyon system is a series of layered sediments found within individual canyons, particularly in the central section. This has led some observers to speculate that lakes may have existed temporarily within the system at some time in the past.

The canyons extend from the summit of the bulge—at an elevation of 10 km—down the flanks until they merge with large areas of chaotic terrain 3,000 km to the east-southeast, at an elevation of 1 km. In these areas, the surface seems to have collapsed to form a jumble of jostled blocks 1–2 km below the surrounding areas. From these chaotic regions, large channels commonly emerge and extend northward, down the regional slope toward Chryse Planitia. The same sequence of canyon, chaos, and channels is seen to the north of the main canyons, where several box canyons have chaotic terrain on their floors and give rise to large channels. Clearly, the canyons, although largely tectonic in origin, are related in some way to the large channels, which appear to be predominantly fluvial.

Fluvial and eolian processes

The numerous channels on the surface of Mars present some of the most perplexing problems of martian geology, for, as we have seen, liquid water is unstable at the surface under present conditions. The channels can be divided into three main types: runoff, fretted, and outflow (Sharp

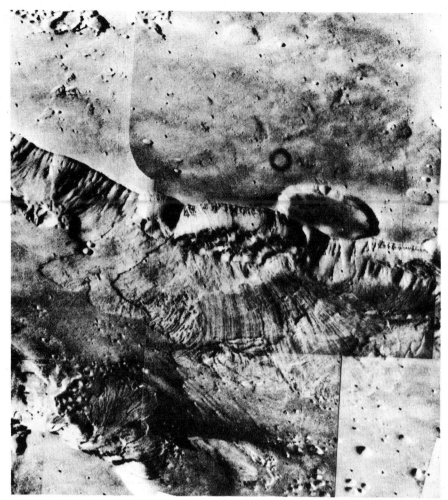

Figure 7. This oblique view across an eastern section of Valles Marineris shows several landslides on the far wall. The canyon at this point is ~150 km across and 2 km deep.

volcanic melting of ice (Masursky et al. 1977), eruption of artesian water trapped under high pressure (Carr 1979), and instantaneous liquefaction of fine-grained, water-charged sediments. Another plausible hypothesis is that these floodlike features were not formed by floods at all, but are rather the result of slow erosion by artesian-fed glaciers (Lucchitta 1980).

The outflow channels cut the lightly cratered plains, which are seldom dissected by runoff channels, and are thus mostly younger than the runoff channels. A possible explanation is that much of the water that cut the runoff channels earlier supplied a vast artesian system. Subsequent global cooling resulted in the development of a thick permafrost, and the trapped water migrated to low areas where substantial pore pressure developed. The water then either escaped catastrophically to form large flood features or leaked slowly to supply glaciers that have long since disappeared.

Although the surface of Mars has been exposed to the wind for billions of years, during which there have probably been hundreds of millions of violent global dust storms like those that were closely observed in 1971, 1977, and 1979, the effect of wind on surface topography in most regions is quite small. Many of the volcanic plains, for example, have crater-size frequencies similar to the lunar maria for crater sizes down to as small as 100 m. Apparently, the accumulated erosion on the lava plains over the last 2–3 billion years has not been sufficient to destroy craters of this size. Estimates are that the average erosion rate is no more than 10^{-3} μm/yr (Arvidson et al. 1979).

Features associated with wind erosion are, however, present in places. Extensive areas of scoured and pitted ground are found, for example, southwest of Olympus Mons (Ward 1979), and, as we have seen, many phenomena at high latitudes are suggestive of wind erosion. Localization of wind erosion in these areas appears to be due largely to the presence of friable, easily erodible materials at the surface.

The main action of the wind appears to be the continuous redistribution of the particulate debris that is already

and Malin 1975). Runoff channels resemble terrestrial river valleys and appear to have been formed by the slow erosion of running water. They start small, increase in size downstream, and commonly have tributaries (see Fig. 2). Most are less than 1 km across and are up to a few tens of kilometers long. They generally stop abruptly, as though the flow had disappeared underground.

Runoff channels occur almost exclusively in the old, heavily cratered uplands, which implies an early era of temperate climatic conditions, during which a slow flow of water across the martian surface was possible. Although runoff channels are ubiquitous in the cratered terrain, the drainage system is immature, the amount of material eroded is relatively small, and erosion has not been sustained sufficiently for individual streams to dominate large areas.

Fretted channels, which resemble runoff channels except for their wide flat floors, are probably formed as a result of secondary enlargement of runoff channels by mass wasting, since they always occur in areas that abound in features associated with mass wasting.

Outflow channels start full size (Fig. 8), mostly from areas of chaotic terrain to the east and north of the canyons. They have many features (Fig. 9) that bear a striking resemblance to the large Pleistocene flood features of eastern Washington, which suggests formation by catastrophic floods (Baker and Milton 1974). Estimates of discharge rates, based on channel dimensions, range as high as 10^9 m^3/sec—ten thousand times the mean annual discharge of the Amazon. The cause of the supposed floods has been the subject of considerable controversy. Suggestions include

at the surface. Addition of new debris to the generally circulating mass is extremely slow. Dunes are common (Fig. 10); they are present within many large craters in high southern latitudes, and there is an almost uninterrupted array around the north pole. Crater streaks, which form as a result of selective erosion and deposition in the lee of craters, provide additional evidence of the mobilization of surface debris. The streaks are stable for most of the year but are destroyed or modified during the great dust storms and re-form during the storms' waning phases, thereby preserving a record of wind directions at these times (Thomas and Veverka 1979).

Polar features

At the poles are some of the youngest features on the planet. A sequence of layered deposits at least 1–2 km thick extends from each pole out to the 80° latitude circle. The deposits are almost devoid of impact craters (Fig. 11), suggesting either a young age or some efficient self-annealing process. The succession almost certainly records variations in deposition, and hence climates, in the recent geologic past, and in fact, the layering and partial dissection of the deposits suggest that periods of deposition have alternated with periods of erosion.

Under present climatic conditions, global dust storms occur in the southern hemisphere when the northern seasonal cap is accumulating. As a result, dust is probably incorporated into the northern cap as it grows. In contrast, when the southern cap forms, the atmosphere is relatively clear, and therefore little dust is incorporated. Pollack and co-workers (1979) estimate that the accumulation rate is 4×10^{-2} cm/yr; at this rate, 30 m-thick layers, such as those observed at the poles, would take almost 10^5 years to accumulate.

Precession of the planet's rotation axis and the normal to the orbital plane causes climatic conditions to alternate between the two hemispheres on a 50,000-year cycle (Murray et al. 1973). The caps therefore alternate in accumulating dust, and this may be one cause of the observed layering. The actual situation must be considerably more complicated,

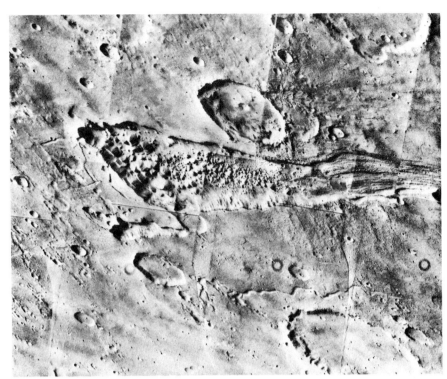

Figure 8. In this chaotic terrain at 1°S, 42°W, a 20-km-wide channel arises abruptly from a 40-km depression containing jumbled debris. The channel continues eastward, where it connects with other channels. A possible explanation of the channel's origin is that water erupted under great pressure from a confined aquifer.

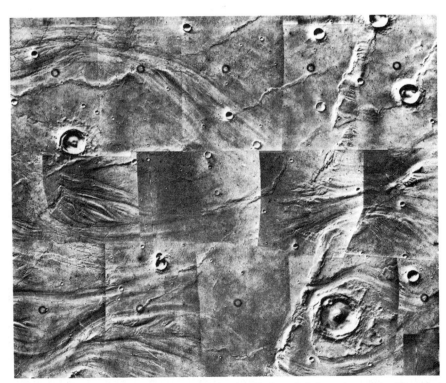

Figure 9. In this area, in the western part of Chryse Planitia, a vast flood apparently diverged widely over some lava plains and was locally funneled through gaps in ridges that obstructed the flow. The largest crater is 16 km across.

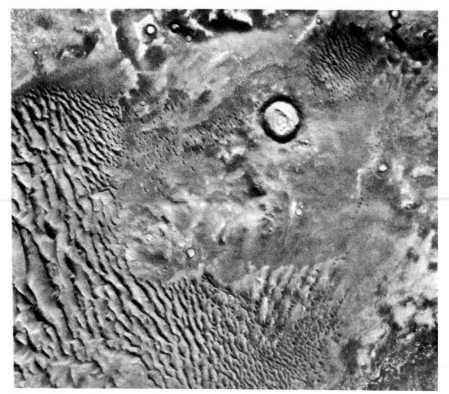

Figure 10. Dunes are particularly common in high latitudes. Here, at 47°S, 340°W, we see transverse dunes and also some isolated crescent-shaped dunes toward the top of the frame. The picture is 60 km across.

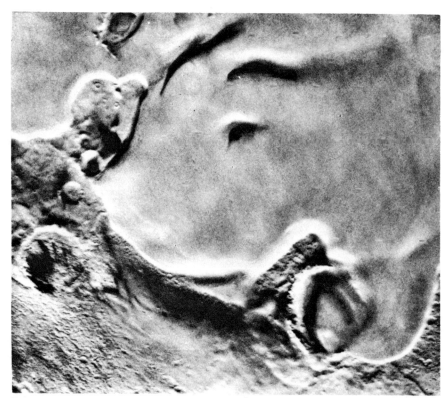

Figure 11. Close to the south pole, relatively young layered deposits rest on old densely cratered terrain. The deposits are, in all likelihood, accumulations of volatiles and eolian debris. As climatic conditions change, erosion and deposition probably alternate to produce the dissected, layered sequence observed. Picture width is 215 km.

however, since orbital eccentricity and obliquity also change with time. Variations in eccentricity modulate the amplitude of the hemispheric differences: during periods of low eccentricity, the two hemispheres have similar climates, whereas during periods of high eccentricity, the hemispheric differences are large.

Variations in obliquity control the latitudinal distribution of solar radiation. This affects not only the stability of volatiles and the general circulation of the atmosphere, but can also cause adsorption and desorption of volatiles in the regolith. Fanale and Cannon (1979) estimate that movement of volatiles in and out of the regolith could cause variations in atmospheric pressure of as much as a factor of ten. Thus, the polar deposits are probably the result of deposition of dust and condensation of volatiles, and their layering and dissection are brought about by a complicated interplay of various factors that affect the location and frequency of global dust storms, the stability of volatiles near the surface, and global wind patterns. They are among the youngest geological features recognized on the planet.

From this short overview, we see that Mars, like Earth, is a volcanically and tectonically active planet, whose surface has been affected by the action of wind, water, and ice. The earth's surface is dynamic, however; the lithosphere is continually being recycled through subduction and spreading of the ocean floor, and materials within the lithosphere are being altered, transported, and reformed by processes of weathering, erosion, and metamorphism. In contrast, while there is volcanism on Mars, the products simply accumulate, almost unaltered, at the surface. Fluvial erosion has occurred but has had only a trivial effect on the redistribution of materials across the surface. Weathering takes place but probably only extremely slowly because of inefficient removal of the weathered products. Mountains form, as do canyons, basins, and impact craters, but because of the thick, rigid lithosphere and the almost total absence of running water at the surface, they survive almost unchanged, even though they are billions of years old. The result is a spectacular planet on which geologic features of enormous scale and a wide variety of origins and ages are almost perfectly preserved.

References

Arvidson, R. E., E. A. Guiness, and S. W. Lee. 1979. Differential aeolian redistribution rates on Mars. *Nature* 278:533–35.

Baker, V. R., and D. J. Milton. 1974. Erosion by catastrophic floods on Mars. *Icarus* 23:27–41.

Blasius, K. R., J. A. Cutts, J. E. Guest, and H. Masursky. 1977. Geology of Valles Marineris: First analysis of imaging from the Viking Orbiter primary mission. *J. Geophys. Res.* 82:4067–91.

Carr, M. H. 1973. Volcanism on Mars. *J. Geophys. Res.* 78:4049–62.

———. 1979. Formation of martian flood features by release of water from confined aquifers. *J. Geophys. Res.* 84:2995–3007.

Carr, M. H., K. R. Blasius, R. Greeley, J. E. Guest, and J. E. Murray. 1977. Observations on some martian volcanic features as viewed from the Viking Orbiters. *J. Geophys. Res.* 82:3985–4015.

Carr, M. H., and G. G. Schaber. 1977. Martian permafrost features. *J. Geophys. Res.* 82:4039–65.

Crumpler, L. S., and J. C. Aubele. 1978. Structural evolution of Arsia Mons, Pavonis Mons, and Ascraeus Mons, Tharsis region of Mars. *Icarus* 34:496–511.

Fanale, F. P., and W. A. Cannon. 1979. Mars: CO_2 adsorption and capillary condensation on clays—significance for volatile storage and atmosphere history. *J. Geophys. Res.* 84:8404–14.

Farmer, C. B., and P. E. Doms. 1979. Global seasonal variation of water vapor on Mars and the implications for permafrost. *J. Geophys. Res.* 84:2881–88.

Greeley, R., and P. Spudis. 1978. Volcanism in the cratered terrain hemisphere of Mars. *Geophys. Res. Lett.* 5:453–55.

Hartmann, W. K. 1973. Martian cratering IV: Mariner 9 initial analysis of cratering chronology. *J. Geophys. Res.* 78:4096-116.

Lucchitta, B. K. 1980. Martian outflow channels sculpted by glaciers. In *Lunar and Planetary Sciences*, vol. 11, pp. 634–36. Houston: Lunar and Planetary Science Institute.

Masursky, H., J. M. Boyce, A. L. Dial, G. G. Schaber, and M. E. Strobell. 1977. Formation of martian channels. *J. Geophys. Res.* 82:4016–38.

Murray, B. C., W. R. Ward, and S. C. Young. 1973. Periodic insolation variations on Mars. *Science* 180:638–40.

Mutch, T. A., R. E. Arvidson, J. W. Head, K. L. Jones, and R. S. Saunders. 1976. *The geology of Mars*. Princeton Univ. Press.

Oberbeck, V. R. 1975. The role of ballistic erosion and sedimentation in lunar stratigraphy. *Rev. Geophys. Space Sci.* 13:337–62.

Owen, T., K. Biemann, D. R. Rushneck, J. E. Biller, D. W. Howarth, and A. L. Lafleur. 1977. The composition of the atmosphere at the surface of Mars. *J. Geophys. Res.* 82:4635–39.

Phillips, R. J., and E. R. Ivins. 1979. Geophysical observations pertaining to solid-state convection in the terrestrial planets. *Phys. Earth Planet. Inter.* 19:107–48.

Pollack, J. B., D. S. Colburn, M. Flasar, R. Kahn, C. E. Carlston, and D. Pidek. 1979. Properties and effects of dust particles suspended in the martian atmosphere. *J. Geophys. Res.* 84:2929–45.

Sharp, R. P., and M. C. Malin. 1975. Channels on Mars. *Geol. Soc. Am. Bull.* 86:593–609.

Soderblom, L. S., T. J. Kriedler, and H. Masursky. 1973. Latitudinal distribution of a debris mantle on the martian surface. *J. Geophys. Res.* 78:4117–22.

Thomas, P., and J. Veverka. 1979. Seasonal and secular variations of wind streaks on Mars: An analysis of Mariner 9 and Viking data. *J. Geophys. Res.* 84:8131–46.

Toulmin, P., A. K. Baird, B. C. Clark, K. Keil, H. J. Rose, R. P. Christian, P. H. Evans, and W. C. Kelliher. 1977. Geochemical and mineralogical interpretation of the Viking inorganic chemical results. *J. Geophys. Res.* 82:4625–34.

Ward, A. W. 1979. Yardangs on Mars: Evidence of recent wind erosion. *J. Geophys. Res.* 84:8147–66.

R. Smoluchowski

Jupiter 1975

The spectacular successes of the Pioneer 10 and 11 missions have brought Jupiter to the forefront of planetary studies today

There are numerous reasons why Jupiter now claims the primary attention of planetologists. It contains 75 percent of the mass and 64 percent of the angular momentum of all planets and has a higher magnetic field—with a huge Van Allen belt—and a more powerful source of internal energy than any other planet. It also has a unique magnetodisc that extends beyond its four major, or Galilean, satellites: Io, Europa, Ganymede, and Callisto. It is a source of powerful, continuous decimetric synchrotron emission and deca- and hectometric pulsed gyroemission, some of which is coupled to the motion of its satellites. Its interior consists mostly of metallic hydrogen, a form that does not exist on Earth. Its clouds show colorful bright zones, dark belts, and mysterious red and white spots, the most famous of which, the Great Red Spot, was discovered in 1665 (Figs. 1–3).

In addition, Jupiter is the best pre-

Dr. Smoluchowski, Professor of Solid State Sciences at Princeton University, was born in Austria and obtained his Ph.D. degree in physics and mathematics at the University of Groningen. Over the years his research has dealt with solid state physics and, in particular, with defects produced by radiation in nuclear reactors. This work led to his study of the lunar dust layer which showed that its sintering by protons in the solar wind assured support of the weight of the future manned-spacecraft landings on the moon. Besides his long-standing interest in Jupiter and especially in its highly condensed interior, he has also written on such topics as lunar magnetism, retention of ice on Mars, and starquakes on pulsars. He is a member of the Space Science Board of the National Academy of Sciences–National Research Council. Address: Materials Laboratory, Engineering Quadrangle, Princeton University, Princeton, NJ 08540.

served sample of the early solar nebula from which all the solar system originated, and we can speculate that some of the clouds may contain organic molecules with a presumptive relation to some sort of "life." If Jupiter were about 70 times more massive, it could have been a star in its own right, like our sun (which is about 1,000 times more massive than Jupiter). Given this long list of facts, it becomes obvious why Jupiter presents fascinating territory for scientists from all areas: cosmology, physics, chemistry, condensed matter, convection, magnetism, heat flow, atmospheric dynamics, optics, ionosphere, magnetosphere, plasma, and biology. Among the sciences usually associated with planetary studies, only geology so far seems to have been excluded from studies of Jupiter, owing to the almost certain absence on the planet of a solid mantle or crust. This discipline may, however, enter the picture when our exploration and knowledge of the Jovian satellites, some of which resemble Mars and our moon in size and density, has progressed further.

Composition and structure of Jupiter

The solar system was formed as a result of cooling and condensation in a rotating primeval gaseous solar nebula, in which density and temperature decreased outward from high values at the center. The compounds with the highest melting and sublimation temperatures formed grains the size of a few microns, some of which tended to gravitate toward the sun, so that the outer parts of the nebula, including the location of the present

Jovian orbit, were enriched in the more volatile gases, such as hydrogen and helium. The large distance of Jupiter and of other giant planets from the sun permitted them to compete more effectively with the sun for the primeval nebula matter; such competition was much more difficult for the inner planets.

According to some theories (1), through collisions, the grains formed larger clumps, of the order of centimeters, which in turn led to the formation of a large, rocky core composed of silicone dioxide, magnesium oxide, and ice. When the core became large enough, its gravity attracted gases such as hydrogen, helium, water vapor, ammonia, and methane in solar proportion, forming a protoplanet. This protoplanet had its own co-rotating gaseous and particulate nebula which led to the formation of the major Jovian satellites in a process analogous to the formation of the planet itself. According to this theory, the abundance of the heavier constituents in the planet, particularly of water, is higher than in the solar composition. This is in contrast to theories (2, 3) which assume that a suitably large region within the solar nebula condensed and gravitationally collapsed to form the planet without first forming a rocky core, thus retaining the solar abundance of all elements. Nevertheless, in all theories the important hydrogen-to-helium ratio has the solar value—3.4 ± 0.1 by mass.

In most theories various stages of condensation and gravitational contraction of the protoplanet are followed by a relatively brief period of very high central temperature—$\sim 50,000$K—and luminosity—about

Figure 1. The Great Red Spot of Jupiter and other details of the cloud structure. The light-colored zones are warmer and are rising; the darker belts are colder and are de- scending. (Figs. 1–3 taken in December 1974 by Pioneer 11 at a distance of approximately 1 million km; courtesy of NASA.)

10^{-3} of the present solar luminosity. The decrease in density of the Galilean satellites, 3.5, 3.1, 2.0 and 1.5 ± 0.1 g cm^{-3}, with increasing distance from the planet is considered as evidence of this flare-up, which led to a more rapid evaporation of the volatile constituents from the nearer

satellites. Figure 4 illustrates how the very early stage of the condensation process leads into the later stage and how the final result compares with various models of the present planet. While there are still considerable differences in the details of the various theories, the

general agreement is rather satisfactory.

The low density of Jupiter—1.33 g cm^{-3}—combined with its huge mass—318 times the mass of Earth—clearly indicates that helium and hydrogen must be its main constitu-

Figure 2. The northern latitudes of Jupiter, with the north pole at the top. Note the white plume on the left, the Great Red Spot near the right lower edge, and the absence of bands and zones at the higher latitudes.

Figure 3. The southern latitudes of Jupiter. Note the white "plume" above the equator.

ents. This conclusion agrees, of course, with the evolutionary history described above. It follows that any viable model of the Jovian interior must be based on equations of state of an approximately solar mixture of hydrogen and helium that is valid at high pressures and high temperatures (4, 5). On theoretical grounds it is expected that at pressures around 2–4 million bars (Mb), solid or liquid molecular hydrogen (H_2) becomes a metal somewhat analogous to such metals as lithium, sodium, etc.

The uncertainty in the value of the transition pressure stems from an amusing fact: the equation of state of metallic hydrogen seems to be theoretically well established but as yet has no experimental verification. The equation of state of condensed H_2, on the other hand, is poorly known theoretically; however, data obtained at high pressures in implosion experiments permit the deduction of suitable effective intermolecular interactions valid up to the transition pressure (6). Actually, a transition observed (7) in implosion experiments at 2.8 Mb is accompanied by a density increase of 20–30 percent, which is in reasonable agreement with theory. The phase rule requires that, in a hydrogen-helium mixture, such a phase change must be accompanied by a discontinuity in the helium concentration. One would also expect the H_2 region to be richer in He than the metallic region (8).

Until quite recently most models of the Jovian interior assumed that helium is completely miscible in hydrogen and that an equation of state of the mixture can be approximated by suitable averaging of the equations of state of the two components (4). Qualitative arguments (9) pointing to a limited solubility of He in metallic H have been recently confirmed in detailed calculations (10), and an analogous situation for He in H_2 has been experimentally demonstrated (11). The so-called miscibility gap of He in H disappears at sufficiently high pressures and temperatures, and the same may apply to He in H_2, although in the latter case the theoretical extrapolation from laboratory conditions is rather risky.

If the radial pressure and temperature profiles in Jupiter are such that both these miscibility gaps exist, then we would expect four distinct layers in the H-He region of Jupiter, each with a different H/He ratio, as illustrated in Figure 5 (12). Thermally driven convection would keep the composition within each region fairly uniform, apart from a small gradient reflecting the pressure and temperature dependence of solubility. Thus, for instance, the H/He ratio in the atmosphere could be different from that in the deep interior, provided that the overall ratio for the whole planet were close to the solar ratio. This turns out to be an important conclusion because, as we shall see, the atmosphere appears to be enriched in helium.

There is little doubt that the metallic H–He solution is fluid on Jupiter, a necessary condition to explain the presence of a strong magnetic field. (This solution may actually be fluid even at absolute zero because at that temperature the amplitude of vibrations of the atoms appears to be larger than that which normally results in melting.) A satisfactory model of the interior must give the correct mass and radius of the planet, and the model's radial density gradient must also account for the fact that neither the rapidly rotating planet nor its gravitational field is spherically symmetrical. The deviation of this field from spherical symmetry (4) is expressed by so-called gravitational harmonic coefficients, which were first obtained from an analysis of the motion of the Galilean satellites; later (13), more precise values were deduced from a detailed knowledge of the paths of Pioneer 10 and 11. The oblateness of the planet, that is, the ratio of the difference between the equatorial and polar radii to the equatorial radius (0.06)—which can be measured directly from Earth or by other methods such as occultation (14) of prominent stars— agrees with the value deduced from gravitational coefficients, indicating that Jupiter, unlike Earth, is in hydrostatic equilibrium. Interestingly enough, in a first approximation, Jupiter can be described as a body in which density ρ depends on pressure p according to the relationship $p = \alpha \rho^n$, where α is a constant and n is close to 2 (15). This situation simplifies many calculations.

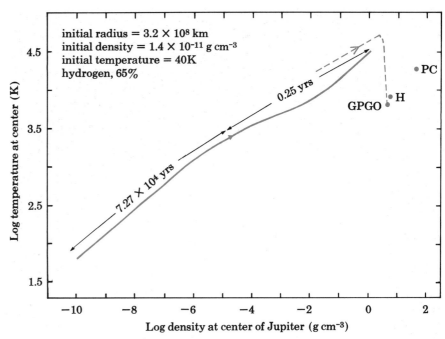

Figure 4. The evolution of Jupiter from the solar nebula (3) showing the sudden flare-up. The solid line is the very early long stage of the condensation process; the dashed line (GPGO) is the later stage of the evolution leading to the present planet (calculated using somewhat different parameters, 2). H and PC indicate other models of the present planet (4, 16, 17). (Courtesy of P. Bodenheimer.)

An important feature of Jupiter discussed below is that it emits more than twice as much energy as it receives from the sun. Thus, knowing the thermal outward flux, we can calculate the required temperature gradient and the central temperature, T_c, the latter being comparable to the temperature expected on the basis of evolutionary theories, as illustrated in Figure 4. The heat flux is so high that, in order to account for it by conduction in a solid interior or through a crust, the temperature gradient and the temperatures themselves would have to be so high that everything would be liquid. We conclude, therefore, that there is no solid surface on Jupiter. This does not preclude the possibility of the existence of isolated islands of solid H_2 at larger radii. As mentioned below, there is no liquid surface either.

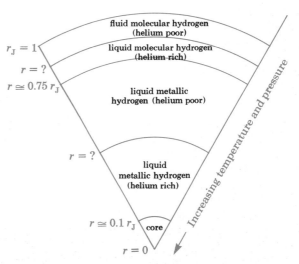

Figure 5. The various immiscible layers in the Jovian interior are schematically represented. As noted, the position of two boundaries is still uncertain; the proximity of the upper of these two is the possible location of He-poor, solid H_2 floating in He-rich liquid H_2 (12).

The many models of the Jovian interior can be roughly divided into three groups: (1) those in which, besides a small rocky core, the H/He ratio is uniform and lower than solar (4); (2) those in which the core is larger, the H/He ratio is uniform but solar (16), and there is a considerable excess of H_2O; and (3) those in which the core is relatively small, the limited solubility of He in metallic H is taken into account (10), and no excess H_2O may be needed. In all these models, the molecular-metallic transition of the H-He mixture occurs at 0.7 to 0.8 r_J, and the core has a radius between 0.1–0.2 r_J, where r_J = 71,000 km is the radius of the planet. Differences between these models stem not only from a variety of assumptions concerning the solubility of He in H and their ratio but also from differences in the equations of state of the various components. Thus, while the models do have many important common features, there are still considerable discrepancies between them. It may be added that a judicious use of the best values of the gravitational coefficients and of temperature at 1 bar led to a detailed analysis (15, 17) of the outer 3 percent of the radius of the planet, i.e. about 3,000 km.

Heat flux and convection

Ever since the discovery of the huge excess heat flux from Jupiter (of the order of 10^4 erg cm^{-2} sec^{-1}, which is 10^3 times higher than the terrestrial heat flux), many proposals have been made (4) concerning its source. Only three of them merit further consideration: (1) primeval heat, i.e. heat evolved during the early formation and contraction of the planet; (2) continued gravitational contraction—at a rate of about 0.1 cm per year—caused, for instance, by an outward motion of the metallic-molecular phase boundary (9); and (3) precipitation of excess He from the limited He-H solution upon slow cooling and subsequent gravitational drift of these precipitates to deeper levels of the interior (12, 18). The applicability of the first mechanism depends critically on the assumptions and details of the evolutionary calculations discussed earlier, and the answer is not certain. In the second mechanism, the rate-controlling process could be helium diffusion.

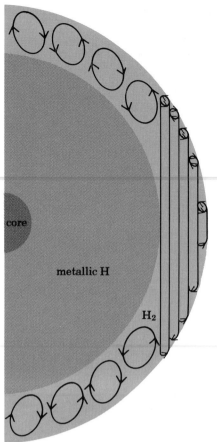

Figure 6. Convection patterns in the H_2 layer. In the equatorial region, the movement in the cells is perpendicular to the axis of rotation of the planet (even mode); the cells themselves are parallel to the axis. In the polar regions, the movement in the cells is more parallel to the axis of rotation (Bénard, or odd, mode). It is assumed here that no miscibility gaps exist. (From F. H. Busse, 20.)

The last mechanism is very attractive, but it requires temperatures which are not too high, so as to permit a limited solubility of the He in the metallic hydrogen, or in H_2, or in both.

Convective heat flow in a rapidly rotating planet (period around 9 hrs 50 mins) is a challenging theoretical problem. In the deep metallic interior, the pattern of convection is undoubtedly strongly affected by the presence and generation of the magnetic field, and little is known about it. There is no convection across the metallic-molecular phase boundary (19), and, similarly, convection may be obstructed by concentration gradients and by the miscibility boundaries shown in Figure 5. It should be stressed that, since the whole atmosphere of Jupiter is supercritical—that is, since there is no liquid-gas phase boundary—there is no "tangible" liquid or solid surface. If the miscibility gap within the H_2 region does not exist, then convection can be described (20) as shown in Figure 6. In the equatorial region the convection cells are parallel to the axis of rotation of the planet (even mode) while in the polar regions they are of the usual Bénard kind (odd mode), which is often seen in liquids heated from below.

It is interesting to note that the

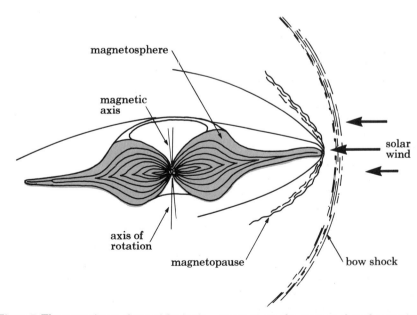

Figure 7. The approximate shape of Jupiter's magnetosphere (magnetodisc) and its interaction with the solar wind. The magnetopause and bow shock refer to different stages of perturbation of the flow of the plasma. Pioneer 11 results suggest that the magnetodisc is thicker at large radii than indicated in this diagram and that it extends farther in the direction away from the sun. (From 5.)

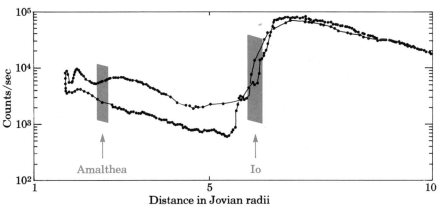

Figure 8. Perturbation of the density of protons (expressed in counts per second, as measured by Pioneer 11) in the Jovian electron flux by Amalthea and by Io, Jupiter's two closest satellites. 0.61 Mev < E < 3.41 Mev. (Courtesy of J. A. Van Allen.)

even mode of convection is bounded on the surface of the planet near the 40° latitudes, which are also the boundaries of the prominent belts-and-zones pattern of the visible Jovian clouds. It is not yet known whether this is a causal relationship or an accident, because Pioneer 11 data indicate a most remarkable uniformity (21) of temperature (125 ± 1.5K) of the visible surface all the way from the equator to the poles. Since the equator of the planet lies almost in the plane of the orbit, it gets much more solar energy than the poles, and a temperature difference of over 20K would be expected (22). We conclude that there may be temperature-equalizing meridional currents and also that the strong insolation at the equator may decrease the internal thermal gradient at low latitudes, with a resulting decrease of the internal heat flux in the equatorial direction as compared to the polar flux (21).

Magnetic field and atmosphere

Probably the most remarkable results of the Pioneer 10 and 11 exploration concern the magnetic field of Jupiter and its interaction with the surrounding plasma and the nearer satellites. Just as on Earth, the field is undoubtedly generated in the fluid metallic interior of the planet by the hydromagnetic dynamo mechanism (4), which is driven by the convection and rotation of the planet. The magnetic field of a planet is usually described as if there were a powerful magnetic moment located near its center. There are two sets of magnetic measurements, made with different instruments on Pioneers 10 and 11, which in the following will be separated by a hyphen (23, 24).

The effective dipole magnetic moment makes an angle of 9°-11° with the rotational axis and is displaced from the center of the planet by about 0.1 r_J (for Earth these parameters are 12° and 0.07 r_E). The field strength at the north and south poles is 13-14 and 10-11 gauss, respectively. What is particularly interesting is the presence of strong quadrupole and even octopole magnetic moments whose strengths are respectively 23-20 and 20-15% of the dipole moment. The presence of

such strong higher-field components may be related to the fact that, below a radius of about 0.92 r_J, the H_2 layer exhibits metallic conduction produced in the normally insulating condensed H_2 by the sufficiently high pressures and temperatures (8). This effect permits much of the magnetic field to be generated close to the surface, as required by recent hydromagnetic dynamic considerations (25), with the result that the quadrupole and octopole magnetic fields are strong outside of the planet in spite of their rapid decrease with distance. Analogous conclusions can be drawn by comparing the higher magnetic moments of Earth and Jupiter (26) and taking into account the difference in scale.

Outside of the planet, up to a radius of about 12 r_J, the magnetic field is

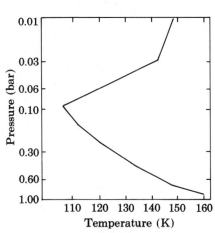

Figure 9. Typical variation of temperature with altitude (expressed as pressure) in the equatorial region of Jupiter. Inversion occurs near 0.1 bar. (From 29.)

essentially dipolar (27), but at larger distances it is elongated and concentrated along the equatorial plane, as shown in Figure 7. This appears to be caused by centrifugal effects of co-rotation of the ionosphere, which near the equatorial plane reaches temperatures corresponding to several KeV. Since the magnetic and rotational equators are not co-planar, the magnetodisc is bent slightly toward the rotational plane at higher radii and terminates at about 50 to 100 r_J on the sunward side because of dynamical interaction with the solar wind. The outer parts of the magnetosphere are far from being rotationally symmetrical, and their shape and extent fluctuate considerably because of the variability of the pressure of the solar wind. Nevertheless, all five nearest satellites are always immersed in the magnetosphere, and some of them interact strongly with the plasma along their orbits.

In analogy with the well-known Van Allen belt of electrons and ions which is kept around Earth by its magnetic field, Jupiter has a similar but much more powerful belt. This belt was the cause of much early worry about the detrimental effect it might have on the Pioneer 10 and 11 spacecraft instrumentation. As it turned out, the damage, though significant, was not as bad as feared and was essentially limited to the imagery mechanism. The intensity of the electronic and protonic flux in the magnetosphere shows a 10-hour period associated with the rotation of the planet. Below about 20 r_J, the intensity of stable trapped electrons and protons is

Figure 10. Cloud structure and cyclonic flow patterns in the northern hemisphere of Jupiter. *Left,* the two halves show the qualitative agreement between the theoretical and observed structure of the clouds (from *31,* permission of the American Meteorological Society). *Right,* theoretical model of the details of cyclonic flow patterns; compare with Figure 11 (from G. P. Williams).

very high, reaching a maximum in the magnetic equatorial plane around 3 r_J. The effect of the two nearest satellites on particle intensity is shown in Figure 8, as measured by Pioneer 11. It is interesting to note also that in the interplanetary space well outside of the magnetosphere, 3- to 30-MeV electrons coming from Jupiter have been observed. They show the characteristic 10-hour period in intensity and have been observed traveling along interplanetary magnetic field lines (*28*) even in terrestrial surroundings.

While the cloud structure of Jupiter—its belts, zones, and the famous Great Red Spot (Figs. 1–3)—is readily observable, the higher and the deeper layers of the atmosphere are not easily studied. Information about the deep layers comes from Pioneer 11 and ground-based 20–45 μ infrared data (*29*), while observations of Pioneer radio occultation give insight into the structure of the ionosphere (*30*). As mentioned earlier, the infrared radiometers have shown that the thermal flux is such that the temperatures at the poles and at the equator are essentially the same.

The thermal profile of the atmosphere near the equator shown in Figure 9 indicates a sharp temperature inversion near 0.1 bar. The most consistent fit to various data is obtained for a H/He ratio close to 3.0, which is lower than the solar value of about 3.4. At other latitudes the thermal profiles are probably similar, the small differences in intensity of the observed radiation being accounted for by an optically thick cloud layer near 0.7 bar. The striking bright zones below about 45° latitude, which are about 2K warmer than the dark belts, appear to be the result of large-scale organized convection (*31*) (elongated Bénard rolls) driven in a 500-km deep layer by the internal heat source of the planet. Because there is differential rotation within each belt and zone, the poleward edges of zones and equatorward edges of belts have a positive (prograde) motion, i.e. in the direction of rotation of the planet and vice versa. Also there is a strong equatorial positive jet stream of about 100 m sec^{-1}. Thus, within each belt and zone, the motion is actually three-dimensional (see Fig. 10). Nearer the poles, where the gravitational acceleration is nearly parallel to the axis of rotation, the large-scale convective patterns are suppressed. Actually the breakdown of the large-

Figure 11. Cloud formation near 51° north latitude showing the scalloped edges of the belts and zones. Note the zonal and cyclonic flow patterns. (Photograph taken 2 December 1974 by Pioneer 11 at a distance of 6 million km; courtesy of NASA.)

scale pattern is gradual, and with increasing latitude the edges of bands become more and more scalloped, as shown in Figure 11.

The Great Red Spot (GRS) (Figs. 12 and 13) is a fascinating phenomenon, and many attempts have been made to explain it. Its most intriguing characteristic, apart from its uniqueness and existence for over 300 years, is its large longitudinal (east-west) motion with almost complete absence of latitudinal (north-south) displacement. The GRS, which is about 40,000 km long and 13,000 km wide, shows small periodic changes in size which appear to be very closely related to small variations in its longitudinal velocity.

Two models for the GRS are much discussed at present. According to one, it is nothing but an enormous hurricane (32), similar to those that occur on Earth, the main difference being the absence of an underlying solid or liquid surface, which is considered critical for terrestrial hurricanes. In other versions of this theory, the GRS is a huge localized belt or zone with a strong up- or downflow of matter (31, 33). As yet no quantitative formulation of this theory has been given. The other model is based on the theoretically (34, 35) and experimentally established fact that an irregularity deep in a rotating fluid produces a vertical column of fluid which does not participate in the rotation but remains attached to the obstacle. This phenomenon, called a Taylor column, can reach the surface of the fluid. Experiment indicates also that solid H_2 containing some He can be in equilibrium with liquid H_2 containing more He and will float in it at a suitable pressure level (11).

Such a "floating island" could well act as the base of a Taylor column in a negative vorticity flow at the boundary between powerful prograde and retrograde currents, as required by theory and confirmed by experiment (36). This rather unique requirement is indeed found to be well satisfied at 22°S latitude, where the GRS exists. There are only two other latitudes (30 and 40°N) at which these conditions appear to be marginally favorable. Although many basic features of the GRS and of its motion can be nicely

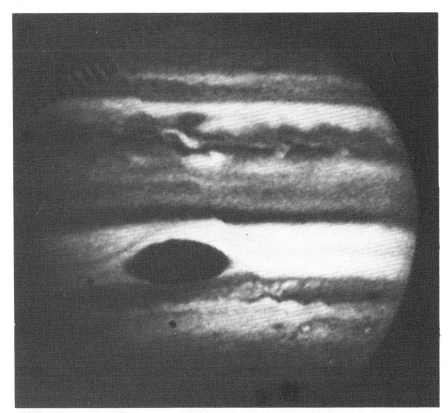

Figure 12. The Great Red Spot and surrounding area, photographed on 2 December 1973 by Pioneer 10 at a distance of approximately 2 million km. Compare with Figure 13, noting the change in the cloud structure. (Courtesy of NASA.)

Figure 13. The Great Red Spot photographed a year later, on 3 December 1974, by Pioneer 11 at a distance of approximately 540,000 km. (Courtesy of NASA.)

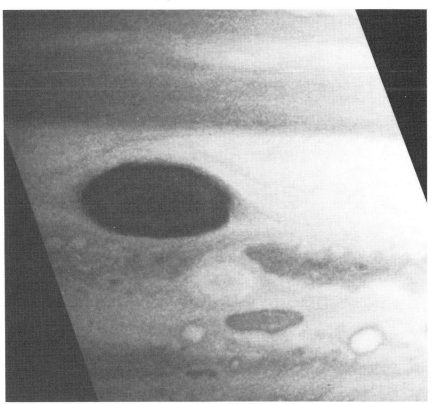

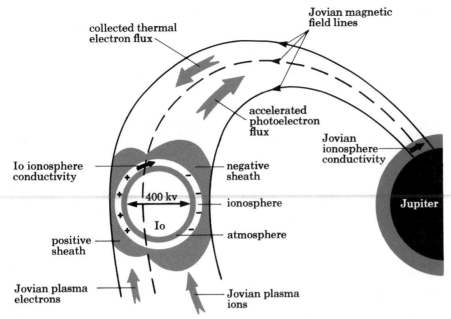

Figure 14. Variations in the radiation (8–10 m wavelength) from Jupiter appear to be related to the motion of its satellite Io. This diagram, which is not drawn to scale, shows the expected plasma sheath formed around Io, the magnetic flux tube reaching toward the planet, and the paths of bombarding and emitted ions and electrons. The motion of Io through the Jovian magnetic field produces an electric potential of about 400 kv across the satellite. The power carried down the tube is estimated at 10^{13} watts. (From 47.)

nomical (rather than terrestrial) measurement of the D/H ratio, which is of fundamental importance in the cosmological theories of the universe. The result seems to support the theory that the universe is of low density and is open rather than closed (45) in the cosmological sense.

Radiation and satellites

Twenty years ago it was found that Jupiter is the source of powerful electromagnetic radiation in the decimetric region (10–100 cm wavelength), and shortly thereafter the origin of this radiation was very satisfactorily explained in terms of synchrotron radiation coming from Jupiter's Van Allen belt—that is, from electrons which spiral back and forth around magnetic field lines of the planet. The observed location of the source of maximum intensity of the radiation is in good agreement with the Pioneer observations of the density of electrons in the Van Allen belt and with the inclination and position of the dipole moment, as described above.

In addition to the continuous decimetric radiation, there is also gyroradiation (46) which extends over the decametric and hectometric ranges of the spectrum (10–1,000 m wavelength), with a peak at about 8 MHz, or 40 m. The high-frequency part of this spectrum is pulsed as if there were discrete sources on the planet which produce directed beams. Curiously enough, near 30 to 36 MHz these pulses depend not only on the longitude of Jupiter but also on the position of the satellite Io. At somewhat lower frequencies, the effect of Io weakens, and near 10 MHz the longitudinal dependence fades out. Some of these surface sources are independent of Io; others are strongly activated by it, with a simultaneous, nearly twofold increase in the frequency of the radiation in a rather narrow bandwidth.

There are no completely satisfactory explanations of these phenomena, although there seems little doubt that a tube of magnetic flux lines attached to Io moves through Jupiter's atmosphere and ionosphere and during its passage activates some of the sources (Figure 14). Indeed, certain Pioneer 11 re-

explained (37) on this basis, the model is rather unusual and has not been generally accepted.

Pioneer missions and terrestrial observations have revealed the existence of many other much smaller red spots, and four of them have been carefully observed (38). Their lifetime is presumably shorter than that of the GRS, but since all of them were discovered in 1974, not enough pertinent data have been collected. There are many other fascinating details visible on Jupiter (39), such as plumes and white ovals, but none is as yet properly understood. The red or orange color of the GRS and of the smaller spots has been variously interpreted as due to certain mercury compounds, red phosphorus (40), polymeric sulfur, or aerosols. Some of these compounds can be produced by intense ultraviolet irradiation (41) of gases present in the Jovian atmosphere or by electrical discharge.

Also, amino acids and high-carbon straight-chain alkanes (42) can be obtained in this manner, and it has been suggested that, if there is dust or other particulate matter suspended in the atmosphere at a level where the temperature is near 300K, then some very primitive form of life—including balloon-type organisms—may have developed (43). This hypothesis is questioned by those who believe that the rate of vertical turnover (about once a month) is so fast that the particles, droplets, or organic molecules would reach deeper, much hotter regions and decompose before having a chance to attain higher levels of complexity.

From a detailed study of absorption bands, emission lines, and polarization, we can obtain a large amount of information about the composition and structure of the Jovian clouds. Between 40 and 140 km, there is aqueous ammonia, ice, ammonium hydrogen sulfide, and solid ammonia (44). At higher altitudes in the neutral atmosphere (30), there are large amounts of hydrogen, helium, methyl radicals, methane, acetylene, and ethane. These molecules are being continuously ionized by the solar extreme ultraviolet radiation, while various recombination reactions between positive ions and electrons tend to establish an equilibrium. It should be pointed out that some of the Jovian methane contains deuterium (CH_3D), permitting the only astro-

sults concerning trapped radiation (47) suggest that, as the spacecraft passed within 6,000 km of the Io flux tube, an anomalous intensity spike of 500 KeV particles was detected. This is just about the value of the potential calculated to form across Io during its motion through the Jovian magnetic field.

Among the 12 or 13 Jovian satellites, only the four originally discovered by Galileo are reasonably well known, even though most of our information about them is quite recent. Io and Europa are about the size of our moon, while Ganymede and Callisto are about the size of Mercury, which is smaller than Mars. It is interesting to note that, if it were not for the brightness and proximity of Jupiter, they are big enough to be visible to the unaided eye. The fifth satellite, Amalthea, is nearer to the planet than Io, which makes it very difficult to observe, and little is known about it except that it is small. As mentioned earlier, the density of these satellites drops with increasing distance from the planet, and because it is rather low, they appear to be composed of ice, water, and silicates (48). The proportion of water depends critically on the ambient temperature and is, therefore, not well known. Other ices such as methane and ammonia are not stable on these bodies. The satellites presumably contain enough radioactive elements, such as uranium, thorium and potassium, to have caused melting and partial differentiation of the interiors of some of them. Europa and Ganymede have a silicate core covered with a mantle of water and ice.

Strong tides produced by Jupiter have caused the five nearest satellites to be locked facing the planet in a manner analogous to our moon. Their leading hemispheres often differ in color and composition from their trailing hemispheres, very likely because their orbits lie well inside the Jovian plasma and magnetosphere, and thus the leading hemispheres are more exposed to impinging ions and to interplanetary meteoroids. Callisto has the darkest surface and, in fact, is presumably covered with a thin crust of silicates. Very recent radar studies have shown that Ganymede has an extremely rough surface made of rocky or metallic material embedded in ice; according to Pioneer 11 photographs it is covered with craters and large flat plains like our moon. The presence of ice on Ganymede's surface has been confirmed by infrared observations, and it may have a very thin atmosphere of methane and ammonia. Most of Europa is covered with water, ice, or frost, which accounts for its high brightness.

Io is perhaps the most intriguing of the Jovian satellites: it has an atmosphere, it interacts very strongly with the planet's plasma and radiation environment, and it is responsible for the regular variations in the pulsation of the gyroemission, as described above and in Figure 8. Its main constituents are partially hydrated silicates (there is no evidence for ice on its surface), and it shows an exceedingly high reflectivity in the 1–3 μ infrared region. Its pole caps and one side have a reddish color that has been associated with the formation of color centers in the surface materials by radiation. If there ever was ice on Io, it has evaporated, leaving dissolved salts on the surface that are rich in sodium (49). This would account for the strong sodium-deuterium line observed in emission on many occasions. The presence in the "evaporite" of elemental sulfur as well as ammonia ice with dissolved sodium and calcium are other possibilities.

The strong bombardment of Io by the Jovian plasma causes the ions of the evaporite to be sputtered off, producing an atmosphere and a torus rich in sodium which extends partially along Io's orbit (50). There is also a hydrogen torus along this orbit. The huge number of ions that must be emitted from Io's surface in order to explain these observations (10^7 Na and 10^{11} H atoms per cm^2 per sec) are an indication that the sputtering is due to impinging plasma ions 23–28 times more massive than protons, which are much less efficient. As yet, we are unable to account for the observed ratio of these ions and for the escape of other constituents from the satellite. The other, more distant eight satellites are almost certainly captured bodies formed outside of the Jovian nebula. Four of them move around Jupiter in the direction of rotation of the planet, and four move in the opposite direction, thus suggesting separate capturing and perhaps fragmentation events.

This brief summary of our mid-1975 ideas about Jupiter clearly indicates that much work remains to be done, even without new observations from the ground or from a spacecraft. In the 1978–79 missions to Saturn and Uranus, the spacecraft will rely on Jupiter for an additional velocity boost and will gather some new information about that planet. However, the spacecraft's greater distance from Jupiter than that of Pioneers 10 and 11 will make observations correspondingly more difficult, although a few of the satellites, the heat balance of the planet, and the magnetosphere will perhaps be better studied than has so far been possible.

The real turning point in the Jovian exploration will be the launching of an orbiter (a Pioneer or, preferably, a Mariner spacecraft) in 1981—a particularly favorable year—and its arrival near Jupiter about 20 months later. Nearly every question touched upon in the present survey will require a new study and may receive a new, probably even more fascinating answer. The crucial advantage of orbiters over fly-bys is the possibility of obtaining systematic data about the temporal variation of various phenomena and a coverage of the whole planet at all phase angles. The lack of this information is the major obstacle to a satisfying understanding of the greatest of all planets.

References

Many of the papers listed below were presented at a conference on Jupiter held in Tucson, Arizona, in May 1975 and will be published in *Jupiter: The Giant Planet*, edited by T. Gehrels (University of Arizona Press).

1. A. G. W. Cameron. 1973. *Space Sci. Rev.* 14: 383; and A. G. W. Cameron and J. B. Pollack. In *Jupiter: The Giant Planet*, forthcoming.

2. H. C. Graboske, J. B. Pollack, A. S. Grossman, and R. J. Olness. 1975. *Astroph. J.* 199:265.

3. P. Bodenheimer. 1974. *Icarus* 23:319; and in *Jupiter: The Giant Planet*, forthcoming.

4. W. B. Hubbard and R. Smoluchowski. 1973. *Space Sci. Rev.* 14:599 (a pre-Pioneer review paper).

5. R. Smoluchowski. 1975. *Icarus* 25:1 (a post-Pioneer summary).

6. M. Van Thiel, M. Ross, B. L. Hord, A. G. Mitchell, W. H. Gust, M. J. D'Addario, R. N. Keller, and K. Boutwell. 1973. *Phys. Rev. Lett.* 31:982.

7. F. V. Grigoriev, S. B. Kormer, O. L. Mikhailova, A. P. Tolochko, and V. D. Urlin. 1972. *JETP Letters* 16:201.

8. R. Smoluchowski. 1975. *Astroph. J. Lett.* 200: L119.

9. R. Smoluchowski. 1967. *Nature* 215:691.

10. J. D. Stevenson and E. E. Salpeter. In *Jupiter: The Giant Planet*, forthcoming.

11. W. B. Streett. 1969. *J. Atmosph. Sci.* 26:924.

12. R. Smoluchowski. 1973. *Astroph. J.* 185:L95.

13. J. D. Anderson, G. W. Null, and S. K. Wong. 1974. *J. Geoph. Res.* 79:3661; and in *Jupiter: The Giant Planet*, forthcoming.

14. W. B. Hubbard and T. C. Van Flandern. 1972. *Astron. J.* 77:65.

15. W. B. Hubbard. 1974. *Icarus* 21:157.

16. M. Podolak and A. G. W. Cameron. 1974. *Icarus* 22:123; and *Icarus*, in press.

17. W. B. Hubbard, W. L. Slattery, and C. De Vito. 1975. *Astroph. J.* 199:504; and in *Jupiter: The Giant Planet*, forthcoming.

18. E. E. Salpeter. 1973. *Astroph. J.* 181:L83.

19. E. E. Salpeter and D. J. Stevenson, 1975. *CRSR* (Cornell University Center for Radiophysics and Space Research), Pub. No. 595.

20. F. H. Busse. In *Jupiter: The Giant Planet*, forthcoming.

21. A. P. Ingersoll. In *Jupiter: The Giant Planet*, forthcoming.

22. P. H. Stone. 1972. *J. Atmosph. Sci.* 29:405.

23. E. J. Smith, L. Davis, Jr., D. E. Jones, P. J. Coleman, Jr., D. S. Colburn, P. Dyal, C. P. Sonett, and A. M. A. Frandsen. 1974. *J. Geoph. Res.* 79:3501; and in *Jupiter: The Giant Planet*, forthcoming.

24. M. H. Acuña and N. F. Ness. 1975. *Nature* 253:327.

25. F. H. Busse. In *Jupiter: The Giant Planet*, forthcoming.

26. M. H. Acuna and N. F. Ness. In *Jupiter: The Giant Planet*, forthcoming.

27. J. A. Van Allen, D. N. Baker, B. A. Randall, and D. D. Sentman. 1974. *J. Geoph. Res.* 79: 3559; and in *Jupiter: The Giant Planet*, forthcoming.

28. D. L. Chenette, T. F. Conlon, and J. A. Simpson. 1974. *J. Geoph. Res.* 79:3551.

29. G. S. Orton. *Icarus*, in press; and in *Jupiter: The Giant Planet*, forthcoming.

30. S. K. Atreya and T. M. Donahue. In *Jupiter: The Giant Planet*, forthcoming.

31. G. P. Williams and J. B. Robinson. 1973. *J. Atmosph. Sci.* 30:684.

32. G. P. Kuiper. 1972. *Sky and Telescope* 43:4.

33. T. Maxworthy and L. G. Redekopp. In *Jupiter: The Giant Planet*, forthcoming.

34. R. Hide. 1969. *J. Atmosph. Sci.* 26:841.

35. J. W. Cottrell. 1971. *Proc. of the Woods Hole Oceanographic Institute Summer Program.*

36. C. W. Titman, P. A. Davies, and P. M. Hilton. 1975. *Nature* 255:538.

37. W. B. Streett, H. I. Ringemacher, and G. Veronis. 1971. *Icarus* 14:319.

38. P. C. Crump, R. M. Kellerman, and D. P. Cruikshank. In *Jupiter: The Giant Planet*, forthcoming.

39. J. W. Fountain, D. L. Coffeen, L. R. Doose, T. Gehrels, and M. G. Tomasko. 1974. *Science* 184:1279.

40. R. G. Prinn and J. S. Lewis. 1975. *Bull. Am. Astron. Soc.* 7:381.

41. B. N. Khane and C. Sagan. 1975. *Science* 189: 722.

42. B. N. Khane and C. Sagan. 1973. *Icarus* 20:311.

43. S. L. Miller and L. Orgel. 1973. *The Origins of Life.* Englewood Cliffs, N.J.: Prentice-Hall.

44. J. S. Lewis. 1969. *Icarus* 10:365.

45. R. Beer and F. W. Taylor. 1973. *Astroph. J.* 179:309.

46. M. D. Desch and T. D. Carr. 1974. *Astroph. J. Lett.* 194:L57; and in *Jupiter: The Giant Planet*, forthcoming.

47. S. D. Shawhan, C. K. Goertz, R. F. Hubbard, D. A. Gurnett, and J. Glenn. 1974. In *Proc. of the Frascati Conference on the Magnetospheres of the Earth and Jupiter.* Dordrecht: D. Reidel; and in *Jupiter: The Giant Planet*, forthcoming.

48. D. Morrison and J. A. Burns. In *Jupiter: The Giant Planet*, forthcoming.

49. F. P. Fanale, T. V. Johnson, and D. L. Matson. 1974. *Science* 186:922.

50. R. A. Brown and Y. L. Yung. In *Jupiter: The Giant Planet*, forthcoming.

James B. Pollack

The Rings of Saturn

Recent observations have provided important clues about the structure of Saturn's rings, the nature of the particles, and the origin of this intriguing feature of the solar system

Soon after he had discovered the four largest moons of Jupiter, now named in his honor, Galileo pointed his telescope at Saturn and became the first person to view the rings of Saturn (1). In the fashion of his times, he laid claim to this discovery, while hedging his bets a bit, by sending an anagram to his friends in Italy and Germany. The solution to the anagram read, "I have observed the most distant planet to be a triple one." Thus, Galileo interpreted his initial observations as indicating not the presence of a ring, but rather that of two satellites in orbit about Saturn. This failure to realize that the rings were in fact rings can be attributed in part to a combination of the relatively poor image quality of his telescope,

James B. Pollack is a staff scientist at NASA's Ames Research Center. For the last 15 years he has engaged in a variety of theoretical and experimental research projects concerning the solar system. His research activities dealing with the rings of Saturn have included theoretical analyses of infrared, radar, and radio observations; studies of the early history of the Saturn system when the rings may have first formed; and consulting with the Pioneer spacecraft team about the targeting of the Pioneer 11 spacecraft with respect to the rings. Dr. Pollack received his Ph.D. in astronomy from Harvard University in 1965 and was a research scientist with the Smithsonian Astrophysical Observatory from 1965 to 1968. He was a senior research scientist at Cornell University from 1968 to 1970, before moving to Ames Research Center. A member of the TV imaging team of the Mariner 9 spacecraft mission to Mars in 1971, he was in charge of obtaining the first close-up photographs of the Martian satellites. Currently, he is responsible for studies of particles in the Martian atmosphere, which are being conducted with the Viking Lander imaging cameras. Last year he was awarded NASA's Medal for Exceptional Scientific Achievement for his research work in the planetary area. Address: Ames Research Center, NASA, Moffett Field, CA 94035.

the unlikely circumstance that the rings were viewed when they were only narrowly open, and finally the obvious analogy to the Galilean satellites of Jupiter.

Subsequent observations by Galileo made him question his initial interpretation and cast widespread doubt about the nature of these strange appendages of Saturn. In the first place, they did not show a daily shift in their apparent position relative to Saturn, as exhibited by the Galilean satellites and as expected for any close satellite moving about its primary. Second, observations obtained in late 1612, when the rings were much narrower as viewed from the Earth and close to edge-on with respect to the Sun, failed to reveal anything other than Saturn.

For the next 50 years, astronomers recorded the alternating appearance of the rings of Saturn and puzzled over its interpretation. Some bizarre hypotheses were advanced that cried out for elimination by Occam's razor. For example, Honore Fabri, S. J., a French professor of philosophy and mathematics, proposed that the phases of Saturn could be explained by the movement of two large, dark satellites located close to the planet and two large bright ones somewhat further away. However, these satellites were required to go only behind Saturn and not in front of it so that they would not cast shadows on the disk of Saturn, something never observed to occur.

The correct nature of the rings was discerned in 1655 by 26-year-old Christiaan Huygens. In the same year, this skillful observer, brilliant theoretician, and master craftsman also discovered Titan, the largest

satellite of Saturn. Huygens showed that the complete set of appearances of Saturn and its appendages (see Fig. 1) could be explained by a thin flat ring that was completely detached from Saturn and lay in Saturn's equatorial plane.

During the next 250 years, studies of the ring system led to both deeper insights into the nature of the rings and the development of general astrophysical theories. Initially, the ring system was thought to be a single solid disk of material. But in 1675 Cassini discovered a dark gap, subsequently named in his honor, which separated the rings into two parts (see Fig. 2), and in 1850 the Bonds and Tuttle at Harvard College Observatory and Dawes in England inferred the existence of a faint ring, the crepe or C ring, located inside the main rings.

Theoretical studies showed that there were serious stability problems presented by rings made of continuous solid material and led to the concept that the rings consisted of numerous small bodies in orbit about Saturn. In 1785 Laplace showed that wide rings rotating at a constant angular velocity would not be mechanically stable since at only one radial position could gravitational forces due to Saturn be balanced by centrifugal forces due to their rotation. To eliminate this problem, Laplace proposed that the rings were subdivided into numerous, very narrow ringlets. Furthermore, to make the ringlets stable against small horizontal displacements, he hypothesized the presence of sizable density variations in the azimuthal direction along the ringlets. But Laplace never fully abandoned the concept that the ringlets were continuous in structure.

The concept that the rings consisted of many independent tiny satellites was first advanced by James Clark Maxwell in an essay that won the Adams prize at Cambridge University in the year 1856. This conclusion was based upon a thorough stability analysis for a variety of ring models. Maxwell's conclusion, derived on strictly theoretical grounds, was empirically verified in 1895 by James Keeler of Allegheny Observatory. In one of the first astrophysical applications of Doppler's principle, Keeler measured the spectral shift of reflected solar absorption lines due to the motion of the rings and showed that the velocity distribution across the rings was in accord with that expected from satellites in independent orbits about Saturn, but in disagreement with that predicted for solid body rotation.

During the late eighteenth century, Kant and Laplace developed their nebular model for the origin of the solar system. The analogy of the rings entered prominently into the development of this theory, according to which the Sun and planets formed out of a primeval rotating nebula. Upon its contraction, the nebula broke into a series of rings that subsequently gravitationally condensed into the Sun and planets. The rings of Saturn were viewed as a fossil record of this process, carried out only part way. An alternative model for the origin of the rings was advanced in 1847 by Roche, who suggested that a liquid satellite wandered too close to Saturn and was tidally disrupted. In modified form, the above two hypotheses are still the leading contenders to explain the origin of the rings.

Here I shall examine more modern observations of the rings and discuss their implications for the structure of the rings and the physical characteristics of the ring particles. Toward the end of this article, I shall discuss theories of the origin and subsequent evolution of the rings, and look to the future to see what new studies are being planned.

Ring structure

As illustrated in Figure 2, the main portion of the rings consists of an inner, brighter ring B and an outer, somewhat fainter ring A. Separating the two rings is a narrow dark zone,

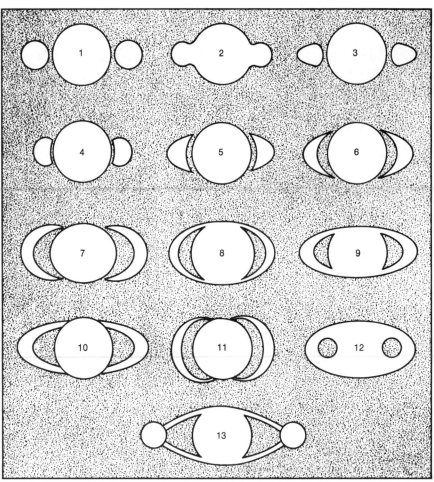

Figure 1. These early drawings show various interpretations of Saturn and the rings. 1, Galileo (1610); 2, Scheiner (1614); 3, Riccioli (1641 or 1643); 4–7, Hevel (theoretical forms); 8, 9, Riccioli (1648–50); 10, Divini (1646–48); 11, Fontana (1636); 12, Biancani (1616), Gassendi (1638–39); 13, Fontana and others at Rome (1644–45). The variations in the form are due in part to variations in the elevation angle of the ring plane with respect to the line of sight from the Earth and in part to limitations on the optical quality of the early telescopes. (Adapted from ref. 1.)

Cassini's division. Limitations in the resolving power of ground-based telescopes, due in part to atmospheric turbulence, have made it difficult to determine whether there is some residual, nonzero brightness within Cassini's division and therefore whether a small number of particles occupy this gap. Interior to ring B is a much fainter ring, the C or crepe ring, which extends part way toward the surface of Saturn. There have been occasional reports of still fainter rings interior to the C ring and exterior to the A ring, but their existence remains controversial. Table 1 summarizes the horizontal dimensions of the three principal rings.

An amazing property of the ring system is its extreme thinness. Estimates of the ring thickness can be made from measurements of the brightness of the rings as the Earth passes through the ring plane, thereby al-

lowing an edge-on view of the rings. Such an estimate is tricky because it involves an extrapolation of brightness measurements made on either side of ring passage to the exact moment of ring passage. Values of several kilometers for the ring thickness have been obtained in this manner (2, 3); but a spread of almost a factor of two between independent estimates implies that such values may be upper limits rather than actual values of the ring thickness.

The thinness of the rings can be attributed to the cumulative effect of collisions between the ring particles, which rapidly dampen random components of their motion. The high surface brightness of the rings implies that incident sunlight has a high probability of encountering and being scattered by a ring particle before it passes through the rings. Similarly, a ring particle, particularly one with a

Figure 2. Photograph of Saturn and rings A and B, the two brighter rings. The inner ring B is separated from the outer ring A by a dark zone, Cassini's division. This picture was taken with the 120 inch telescope at Lick Observatory, Mt. Hamilton, California. (Photograph courtesy Lunar and Planetary Laboratory, University of Arizona.)

and will collide with the second particle if the separation in their orbital distances is less than the average size of these two particles. The resulting partially elastic collision will lead to the conversion of a small amount of their orbital velocities into vertical motions. Thus, at present the ring thickness may be determined by a balance between collisions that dampen random motions and ones that produce such motions. It has been suggested that if the size distribution of ring particles is very broad the thickness of the rings is determined principally by the gravitational scattering of the smaller particles by the larger ones (4). This type of scattering occurs when the two types of particles come close but do not actually collide. In this case, most of the small particles will be located within a height of several times the size of the larger ones, with the number density of small particles rapidly diminishing at greater altitudes above the central plane.

We have already seen that a major division separates the A and B rings. As illustrated in Figure 3, the brightness of the individual rings exhibits considerable fine structure and variability in the radial direction (5). The presence of gaps, the location of the ring boundaries, and the brightness variations within individual rings are due, mainly, to gravitational perturbations by the large satellites of Saturn's system, especially Mimas. Because Saturn's mass is so much larger than that of any of its satellites, generally the orbital motion of a ring particle is totally dominated by Saturn's gravitational field. However, when the orbital period of a ring particle lies close to some simple fraction of the orbital period of a large satellite, random occurrences of closest approach between the two bodies happen on a number of consecutive orbits and the small perturbing influence of the satellite on the ring particle's motion can be amplified into a significant disturbance. For example, a particle traveling in a circular orbit within Cassini's division would have an orbital period quite close to $1/2$ that of Mimas' orbital period. Calculations of this "three-body problem"—Saturn, Mimas, and a ring particle—show that there are no stable orbits possible within a zone around the $1/2$ resonance position, with the width of the zone depending on the ratio of

significant random component to its motion, has a high probability of encountering another particle within a single orbital period. Suppose that the ring particles initially had orbits with a large spread in orbital eccentricity and inclination to the ring plane. Such motions would have large random components of velocity within and perpendicular to the ring plane, respectively. Collisions would quickly dissipate these random motions and lead to nearly circular orbital motions that lay close to the central ring plane, i.e. a thick ring would quickly become a very thin ring.

A point of present controversy is whether the end state of collisional processes would be a ring that was only one particle thick or a ring that was at least several particles thick. The latter possibility is suggested by the continued occurrence of collisions between particles due to the high density of particles within the rings: a particle traveling in a circular orbit will overtake a second particle traveling in an orbit slightly outside of it

Table 1. Dimensions of the rings of Saturn (from ref 17)

| | | Radial distance from the center of Saturn | | |
	Radial extent (km)	Angular distance* (arc sec)	km	Units of Saturn's equatorial radius
Outer edge of ring A		19.82	137,050 ± 1000	2.293
Inner edge of ring A		17.57	121,490 ± 1000	2.032
Cassini's division	4850 ± 2500			
Outer edge of ring B		16.87	116,650 ± 1000	1.951
Inner edge of ring B		13.21	91,350 ± 1000	1.528
Inner edge of ring C†		11.1	76,760 ± 3000	1.284

* With Saturn at a distance of 9.5388 A.U. from the Earth
† This is a very approximate value. It is currently uncertain as to whether there is a small amount of ring material closer to Saturn.

Mimas' mass to that of Saturn (6). However, because this ratio is very small (~10^{-7}), the width of the zone of avoidance is only about 20 km, which is much less than the size of Cassini's division (see Table 1).

To understand why Cassini's division is much larger than the size of the zone of avoidance, we need to consider the characteristics of the stable orbits that lie close to this zone. These orbits have significant eccentricities due to Mimas' perturbations. Thus particles traveling in these orbits collide with other ring particles at greater relative velocities than those that collide further from the $\frac{1}{2}$ resonance position. The result of such collisions may be a reduction in the number density of ring particles close to the resonance position. The outer edge of ring A and the inner edge of ring B, respectively, lie close to the $\frac{2}{3}$ and $\frac{1}{3}$ resonance positions of Mimas' orbit. Thus, these boundaries could also be due to resonant perturbations. Calculations for a collisionless, one-particle-thick ring yield brightness patterns similar to the observed one (6). These results represent a start toward understanding the relationship between satellite perturbations and radial brightness variations. However, many uncertainties still remain, and analogous calculations that incorporate collisions between ring particles are needed.

When account is taken of the inclination of the satellite orbits relative to the ring plane, an out-of-plane component is found for the gravitational perturbation by the satellites. This component leads to a warping of the ring plane that tracks the motion of the perturbing satellite. Calculations of this effect indicate that the dominant perturber is Titan, the most massive of Saturn's satellites, which causes a warping with a vertical amplitude of about 10 meters near the outer portion of ring A and 3 m near the inner edge of ring B (7).

Recent observations (8, 9) have confirmed a rather unexpected type of brightness variation. The brightness of the A ring varies in the azimuthal direction, i.e. along the direction of a particle's orbit about Saturn. Peak brightness occurs in the quadrants before a particle's orbit reaches the line between the Earth and Saturn on the near and far sides of Saturn ("trailing quadrants"), with mini-

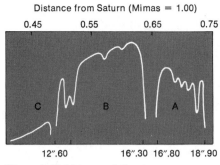

Figure 3. Brightness of the ring system as a function of radial distance from Saturn. The letters along the horizontal axis denote the names and positions of the three major ring segments. (From ref. 5.)

mum brightness occurring approximately symmetrically in the two "following" quadrants. However, no such variation is seen for ring B.

Symmetry considerations would have led us to predict a priori that there would not be any azimuthal variations: in the first place, we would have expected collisions to produce a random distribution of spin directions and hence to eliminate any brightness variations due to the nonspherical shape of the particles. Second, collisions would quickly smooth out any temporary density fluctuation in the azimuthal direction. Reconciliation of these a priori expectations and the observational data can be achieved by finding mechanisms that either cause preferential alignment or persistently cause density fluctuations. Here we consider such a model.

The most promising explanation of the azimuthal brightness variation invokes an increase in the density of small ring particles in the vicinity of large ring particles. According to this model (10), the gravitational field of large particles creates intense trailing density wakes that are analogous to ones found in theoretical studies of the spiral structure of galaxies. The differential rotation of the rings shears this wake and causes its long axis to be rotated. When the large particles are within the trailing quadrants the long axes of their wakes will be oriented approximately perpendicular to the line of sight from the Earth, and the enhanced density within the wakes leads to a statistical brightening of the ring in these quadrants. Such a mechanism is most effective when the rings are only partially filled with particles, as viewed along the line of sight from the Earth. The absence of an azimuthal

brightness variation for the brighter B ring may be due to its having a higher density of particles than the A ring, rather than to its lack of density wakes. Alternative models invoke tidally aligned particles that produce the azimuthal brightness variation either because of a nonuniform pattern of brightness across their surfaces (11) or because of their nonspherical shape (10).

Still another brightness variation exhibited by the rings is one with phase angle—the angle between the direction from the Sun to the rings and the direction from the rings to the Earth. At zero degrees phase angle, the Sun and Earth are located along the same line to the rings. Observations (12) show that the brightness of the rings increases with decreasing phase angle more rapidly close to zero degrees phase angle than at larger angles. This nonlinear increase in brightness at small phase angles, called the "opposition effect," is due either to shadowing within the surfaces of individual particles or to shadowing of particles by particles in front of them: at zero phase angle, when the direction to the Sun and Earth coincide, we view only places that are being directly illuminated by the Sun. But at larger phase angles, we see an increasing proportion of places that are in shadow.

If the opposition effect is due chiefly to interparticle shadowing, we can use its magnitude to estimate the packing fraction or volume density of ring particles. This ability to estimate the volume density can be understood as follows: on the one hand, when the mean distance between ring particles is large, the probability of a given particle lying within the shadow of another particle is small (because of the finite angular size of the Sun at the rings, shadows cast by particles cover conical volumes of finite length). Thus, the magnitude of the opposition effect becomes smaller as the mean separation becomes larger. On the other hand, when the mean separation is small, the opposition effect gets spread over a large range of phase angles. Calculations (13) indicate that the observed magnitude and shape of the opposition effect can be matched if the ring particles occupy 1% of the volume of the rings, i.e. if the mean separation between ring particles is about five times their size.

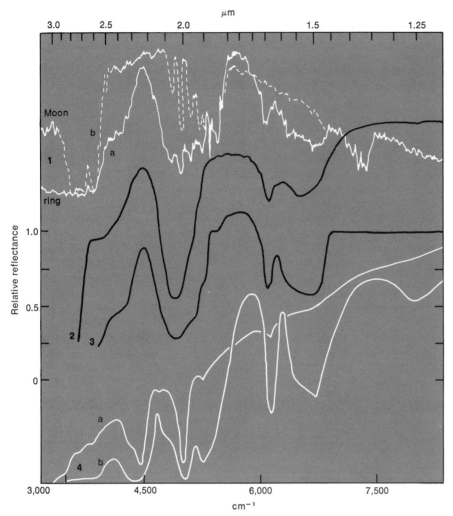

Figure 4. Comparison of the reflectance spectra for H$_2$O and NH$_3$ frosts and Saturn's rings. 1a. The Saturn ring spectrum of Kuiper et al. (14). 1b. Lunar comparison spectrum. 2. Fine-grained H$_2$O frost spectrum. 3. Spectrum of the rings divided by that of the moon. 4. NH$_3$ frost, fine-grained a; coarse-grained b. (From ref. 15.)

The alternative explanation of the opposition effect involves shadowing within the surfaces of individual particles. While very dark objects, such as the Moon, exhibit an opposition effect comparable in magnitude to that of the rings, bright solar-system bodies, such as several of the Galilean satellites of Jupiter, show a significantly smaller opposition effect. As the brightness of the ring particles is more comparable to the latter set of objects, intraparticle shadowing may be insufficient to explain the opposition effect and therefore mutual shadowing may also need to be invoked. If we accept this line of reasoning, not only does the opposition effect provide an estimate of the volume density of the ring particles but it also shows that the rings are at least several particles thick. These inferences should be considered as tentative, owing to the uncertainty in how faithfully the behavior of the ring particles can be

modeled with much larger solar-system bodies.

Particle properties

The application of modern infrared, radio, and radar techniques to the study of the rings has yielded information on the composition and size of the ring particles. Many materials have strong absorption bands in the near-infrared portion of the spectrum. The wavelength location of such bands in the reflectivity spectrum of an object offers a means of obtaining data concerning the composition of that object. Using a Fourier-transform spectrometer, Kuiper and his associates (14) measured the spectrum of the rings and the Moon, as shown by curves 1a and b of Figure 4, respectively. By ratioing the spectra of the rings with that of the Moon, the wavelength sensitivity of the instrument and absorption in the Earth's atmosphere are eliminated

and the desired reflectivity curve is obtained. This ratio spectrum is shown by curve 3 of Figure 4.

We see that the reflectivity spectrum of the rings exhibits absorption bands centered near 1.5, 1.6, and 2.0 microns. Comparing these features with laboratory spectra of water and ammonia ices, as shown by curves 2 and 4 of Figure 4, respectively, Pilcher and his associates (15) deduced that the features in the ring spectrum were due to water ice. Because infrared light at these wavelengths penetrates to depths only on the order of 100 microns, these results provide information on the composition of the surface of the ring particles, but not necessarily on their interior composition.

Further information on the composition of the surfaces of the ring particles is provided by their spectral behavior at visible wavelengths. The reflectivity of the rings shows a marked decline from a high value in the red portion of the spectrum to steadily decreasing values at shorter wavelengths (16). Because water ice is quite transparent throughout the visible region, this spectral behavior implies the presence of some additional material in the surfaces of the ring particles. Candidates for the blue-absorbing material include small rock particles intimately mixed with the water ice, and compounds generated by the exposure of the ring particles to solar ultraviolet light (16). The former might be the result of meteoroid bombardment (17). An example of the latter is hydrogen polysulfide, which is generated when water ice samples, containing simple sulfur compounds, are exposed to ultraviolet irradiation (18).

Bodies emit radiation over the entire electromagnetic spectrum simply as a result of their nonzero temperature. At radio wavelengths, the amount of thermal radiation emitted by solar-system bodies is proportional to their temperature, and thus radio observations can be used to determine a body's temperature. The results of such measurements are expressed in terms of brightness temperature, which is the temperature of a blackbody of the same size that produces the same amount of radio emission. The actual physical temperature of a body is generally slightly higher than its radio brightness temperature since its emissivity is usually slightly less than unity. Temperature can also be determined from measurements of the radiation emitted at infrared wavelengths.

Observations of the rings at infrared wavelengths show that the ring particles have a physical temperature of about 95°K (17, 19). However, early observations at radio wavelengths failed to detect the rings and implied that their brightness temperature was less than about 20°K (17). In turn, this upper bound indicates that the emissivity of the ring particles is less than about 20% at radio wavelengths. Such a low emissivity can be contrasted with values closer to 90% that typify large bodies of the inner solar system, such as our Moon. This comparison led people to believe that the low emissivity of the rings at radio wavelengths was a size effect. As illustrated in Figure 5, the efficiency of a body for interacting with radiation diminishes rapidly as the size of the body becomes small compared to the wavelength of the radiation. In this figure, X is the ratio of the circumference of the body to the wavelength, and the efficiency factor Q is the ratio of the interaction cross section to the body's geometric cross section. Thus, the failure to detect radio emission from the rings at wavelengths of several centimeters could be interpreted as owing to the ring particles being much less than a centimeter in size.

This hypothesis was immediately ruled out when Goldstein and Morris (20) detected strong echoes from a 12 cm-wavelength radar beam directed at the rings. The strength of the echoes implied a reflectivity comparable to the rings' high value in the red portion of the visible spectrum. We may now use the size-efficiency argument given above to deduce from these observations and Figure 5 that the ring particles are at least a few centimeters in size (17, 21).

How then can we reconcile the radio observations with the radar measurements? Particles that are comparable to or larger than a wavelength can be poor emitters of radiation under two very different circumstances. A low emissivity characterizes dielectric particles that are transparent to radiation passing through them at the wavelength of interest, as well as metallic particles that reflect almost all the radiation

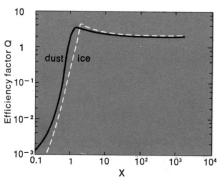

Figure 5. Efficiency factor Q, as calculated from Mie scattering theory, as a function of X, the ratio of a particle's circumference to the wavelength of incident radiation. Q is the ratio of the particle's interaction cross section to its geometric cross section. For large values of X, Q is about 2 rather than 1, because the particle's diffraction cross section is included in the calculation of Q. (From ref. 21.)

incident on their external boundaries. Both types of particles satisfy the radio and radar observations because they efficiently scatter microwave radiation but are poor absorbers and therefore poor emitters at these wavelengths.

Some preliminary estimates of particle size and composition can be made from a comparison of these two types of particles with the radar and radio observations. Consider first the dielectric particles. In order for this type of particle to be sufficiently transparent at microwave wavelengths, the product of the size of the particle and its absorption coefficient per unit length must not exceed some upper bound. But the particle size alone must be larger than a few centimeters for the particles to interact efficiently with radar waves, and therefore the microwave absorption coefficient has to be less than another upper bound.

These considerations imply that some dielectrics can be ruled out because they absorb too strongly at microwave wavelengths. Theoretical calculations show that rings containing water ice particles can produce radar reflectivities comparable to the observed value, but rings containing rocky particles cannot (22). A similar conclusion is found from theoretical studies of the radio data (17). Furthermore, the mean size of the ice particles is about 10 cm, and the size distribution is quite broad (17, 22). Alternatively, the mean particle size is much larger, but the particles have

an inhomogeneous structure, with scattering centers having a characteristic length of about 10 cm.

No bound can be placed on the size of metallic particles other than a lower limit of a few centimeters, as discussed above. A distinction between metallic and dielectric particles can be made on the basis of the radio observations at very short wavelengths. Typical metallic materials retain their metallic characteristics until wavelengths well within the visible or near-infrared portion of the spectrum. Thus, the radio brightness temperature should remain very low at millimeter wavelengths. But dielectrics such as water ice become progressively more absorbing as the wavelength is decreased and therefore, at the shorter wavelengths in the millimeter region, the brightness temperature should significantly increase. Radio observations at wavelengths close to 1 mm apparently show a positive detection of the rings with a brightness temperature significantly larger than the upper bounds found at longer wavelengths (23). These results imply that the bulk composition of the ring particles is water ice (17).

Origin and evolution of the rings

At the beginning of this article we briefly cited two alternative hypotheses about the origin of the rings that date back to Roche in one case and Kant and Laplace in the other case, and now we shall consider modern versions of these hypotheses. As originally stated by Roche, the rings formed from the tidal disruption of a liquid satellite of Saturn, whose orbit decayed to within the ring area. He showed that, for a liquid satellite having the same density as Saturn, disruption occurs when its orbital radius is less than about 2.4 times the radius of Saturn. In its more modern guise, this hypothesis has been expanded to include a stray body that passed close to Saturn, as well as a prior satellite.

The tidal disruption model has several serious problems. One of these involves the logistics of placing the material in the region of the rings. If the precursor of the ring material was originally a satellite of Saturn which, of necessity, formed beyond the outer boundary for tidal disruption, it is difficult to find a way to bring the satellite within this boundary. The most plausible mechanism for such an orbital evolution involves tidal forces raised by the satellite on Saturn. But little tidal orbital evolution is expected for gaseous planets like Saturn (24). Furthermore, the initial orbital radius would need to lie within the region of the rings for such an evolution to cause a decrease rather than an increase in the satellite's distance (17). If on the other hand, the precursor object was a stray body, such as a comet or an asteroid, this body could have a close approach with Saturn. But it seems unlikely that some of the resulting fragments would have the needed large change in velocity from that of their parent body for them to go into orbit about Saturn at the ring's distance.

A second difficulty with the tidal theory is related to Roche's assumption that the parent body was in a liquid state. For plausible compositions, a small body at Saturn's distance from the Sun would be in a solid state. If this body is smaller than about 100 km, its tensile strength would play a more important role than its self-gravity in holding it together, and consequently it would not be tidally disrupted (17, 25). Similar considerations imply that the tidal disruption of a large object might lead to fragments whose size greatly exceeds the current thickness of the rings.

Let us now consider the alternative hypothesis that the ring particles formed as part of the same process that formed the large satellites of Saturn. It is thought that the planets formed within the outer portions of a disk of gas and small dust particles, with the Sun forming in the inner portion of this primordial solar nebula. The giant planets such as Jupiter and Saturn formed as a result of local increases in the density of this nebula, which enabled the gas and dust in these localized regions to contract into smaller-sized objects. Gaseous protoplanets formed at the center of these localized, disk-shaped regions, and satellites were constructed farther out from dust particles and ices that condensed from less volatile gases, such as water vapor. The larger satellites may have been formed, in part, as a result of gravitational instabilities that allowed the dust and ice particles to aggregate into much

larger objects (26). Since the planets no longer have gas disks about them, at some point these disks were dissipated and satellite formation ceased, perhaps as a result of a strong wind emanating from the early Sun.

Within the context of the above model, we can imagine a scenario by which the ring particles formed. The same tidal disrupting forces considered earlier would prevent the gravitational aggregation of dust and ice particles in the vicinity of the rings. Thus, these primordial particles would never form a large satellite. However, such small-sized particles would be in danger of being eliminated from the region of the rings because of drag by gas also present there. Gas drag would cause their orbital radii to decay until they were incorporated into the protoplanet. Thus, the dust particles that were initially present would be lost by this process. But ice particles that condensed shortly before the planetary gas disk was eliminated might remain in the region of the rings.

Theoretical calculations of the gravitational contraction history of Saturn indicate that it emitted a large amount of thermal radiation in its early history as a result of rapid contraction at this time (27). In fact, an ember of this early heat is shown by present-day Saturn (27). Currently, Saturn radiates to space more than twice the amount of sunlight it absorbs. During the first million or so years of contraction, Saturn's thermal radiation was initially large enough to keep the temperature of its nebula above the freezing point of any ice species in the vicinity of the rings, but subsequently this radiation decreased to the point where water ice particles could form (28). Such a model of Saturn's early history is therefore consistent with the requirements of the condensation model of the rings' origin, which requires ice particles to be formed in the region of the rings shortly before the planetary disk was eliminated.

The above scenario for the origin of the rings of Saturn also suggests a reason for rings existing around Saturn, but not around Jupiter (28). As Jupiter has three times more mass than Saturn, it was about ten times as luminous as Saturn at the times of satellite formation. Hence, the region within Jupiter's tidal disruption

boundary may not have become cool enough for water vapor to condense prior to the dissipation of its nebular disk. In addition, it would have lost any dust particles initially present there, just as Saturn did. According to this model, no rings would therefore be expected to be present around Jupiter. These cosmogonic considerations also imply that the rings of Saturn are made of water ice. This deduction is consistent with the composition of the ring particles obtained from analyses of infrared, radio, and radar observations, as described above.

Once the rings formed they evolved in several ways. First, collisions led to a rapid thinning of the rings, for reasons cited earlier. Also, grazing collisions between particles having slightly different orbital radii promoted an exchange of angular momentum between the particles that caused the outer particle to move farther out and the inner one to move farther in (29). In this manner, the rings spread out in the radial direction. The rings also widened rapidly because of a gravitational instability that occurred when the ring particles were packed too closely together (17).

Meteoroid bombardment also caused some important evolutionary changes. When small meteoroids hit the ring particles, tiny craters were produced on the surfaces of the ring particles and material was ejected from these holes with a mass several thousand times greater than the mass of the impacting meteoroids (17). The tiny fragments contained in the ejected material did not have a large enough velocity to escape from Saturn's gravitational field, but rather went into orbit about Saturn in the region of the rings. Subsequent collisions with other ring particles quickly circularized their orbits and caused them to lie very close to the ring plane. The amount of meteoroid material at Saturn's distance from the Sun is probably large enough so that over the lifetime of the solar system many more small ejecta particles were formed than the number of original ring particles. In fact, the cumulative surface area of these secondary particles is probably much larger than that of their parent bodies, and thus, while most of the rings' mass resides in the primaries, most of the light scattered by the rings is due to these secondaries. At the present time there may be a bimodal distribution of particle sizes. The mean particle size of about 10 cm suggested by the microwave observations, if the particles are water ice, could refer to these secondary particles, while the larger-sized primaries could be identified with the objects causing the density wakes and azimuthal brightness asymmetry of the rings.

Future observations

Several important ground-based, as well as spacecraft, observations of the rings will be made over the next five years. Observations of the variation of their radar reflectivity as a function of the tilt of the ring plane will reveal whether the rings are one particle thick or at least several particles thick (22). On October 19, 1977, and January 8, 1978, Saturn's satellite Iapetus will be eclipsed by the rings. Observations of the diminished brightness of Iapetus during this period will provide a good determination of the transparency of the rings as a function of radial distance. In 1979 the Pioneer 11 spacecraft, which flew by Jupiter in 1975, will enter Saturn's system and obtain the first close-up view of the rings. A year later, a Mariner spacecraft, carrying more sophisticated equipment, will gather additional data on the rings. The rings, which have intrigued astronomers since their discovery over 350 years ago, continue to provide a challenge for the future.

References

1. A. F. O'D. Alexander. 1962. *The Planet Saturn.* Macmillan. Much of the material in the introduction of this article was derived from Alexander's book.

2. J. H. Focas and A. Dollfus. 1969. Optical characteristics and thickness of Saturn's rings observed on the ring plane in 1966. *Astron. Astrophys.* 2:251.

3. R. I. Kiladze. 1969. Observations of Saturn's rings at the moments of the Earth's transit through their plane. *Byull. Abastuman. Astrofiz. Obs.* 37:151.

4. J. Cuzzi, J. Burns, R. Durisen, and P. Hamill. 1977. On the physical thickness of Saturn's rings. Presented at the annual meeting of the Division of Planetary Sciences/AAS.

5. A. Dollfus. 1970. Visual and photographic studies of planets at the Pic Du Midi. In *Surfaces and Interiors of Planets and Satellites,* ed. A. Dollfus, p. 46. Academic Press.

6. F. A. Franklin and G. Columbo. 1970. A dynamical model for the radial structure of Saturn's rings. *Icarus* 12:338.

7. J. Burns, P. Hamill, R. Durisen, and J. Cuzzi. In press. Warping of the ring plane by Saturn's satellites.

8. H. J. Reitsema, R. F. Beebe, and B. A. Smith. 1976. Azimuthal brightness variations in Saturn's rings. *Astr. J.* 81:209.

9. K. Lumme and W. M. Irvine. 1976. Azimuthal brightness variations of Saturn's rings. *Astrophys. J. Lett.* 204:L55.

10. G. Columbo, P. Goldreich, and A. W. Harris. 1976. Spiral structure as an explanation for the asymmetric brightness of Saturn's A ring. *Nature* 264:344.

11. R. Beebe, H. Reitsema, and B. Smith. 1977. A model of the azimuthal variation of the brightness of Saturn's A-ring. Presented at the annual meeting of the Division of Planetary Sciences/AAS.

12. F. A. Franklin and A. Cook 1965. Optical observations of Saturn's rings II. Two color phase curves of the two bright rings. *Astron. J.* 70:704.

13. Y. Katawa and W. M. Irvine. 1974. Models of Saturn's rings which satisfy the optical observations. In *Exploration of the Solar System,* ed. A. Woszczyk and C. Iwaniszewska, I.A.U. Symposium 65:441.

14. G. P. Kuiper, D. P. Cruikshank, and U. Fink. 1970. The composition of Saturn's rings. *Sky Telesc.* 39:14, 39:80.

15. C. B. Pilcher, C. R. Chapman, L. A. Lebofsky, and H. H. Kieffer. 1970. Saturn's rings: Identification of water frost. *Science* 167:1372.

16. L. A. Lebofsky, T. V. Johnson, and T. B. McCord. 1970. Saturn's rings: Spectral reflectivity and compositional implications. *Icarus* 13:226.

17. J. B. Pollack. 1975. The rings of Saturn. *Space Sci. Rev.* 18:3.

18. L. A. Lebofsky and M. B. Fegley, Jr. 1976. Laboratory reflection spectra for the determination of chemical composition of icy bodies. *Icarus* 28:379.

19. D. Morrison. 1974. Infrared radiometry of the rings of Saturn. *Icarus* 22:57.

20. R. M. Goldstein and G. A. Morris. 1973. Radar observations of the rings of Saturn. *Icarus* 20:260.

21. J. B. Pollack, A. Summers, and B. Baldwin. 1973. Estimates of the size of the particles in the rings of Saturn and their cosmogonic implications. *Icarus* 20:263.

22. J. Cuzzi and J. B. Pollack. In press. Saturn's rings: Particle composition and size distribution as constrained by microwave observations. I. Radar observations.

23. J. Rather, P. Ade, and P. Clegg. 1975. Planetary brightness temperature measurements at 1 mm wavelength. *Icarus* 23:448.

24. P. Goldreich and S. Soter. 1966. Q in the solar system. *Icarus* 5:375.

25. H. Jeffreys. 1947. The relationship of cohesion to Roche's limit. *Monthly Not. Roy. Astron. Soc.* 107:260.

26. P. Goldreich and W. R. Ward. 1973. The formation of planetesimals. *Astrophys. J.* 183:1051.

27. J. B. Pollack, A. S. Grossman, R. Moore, and H. C. Graboske, Jr. 1977. A calculation of Saturn's gravitational contraction history. *Icarus* 30:111.

28. J. B. Pollack, A. S. Grossman, R. Moore, and H. C. Graboske, Jr. 1976. The formation of Saturn's satellites and rings, as influenced by Saturn's contraction history. *Icarus* 29:35.

29. H. Jeffreys. 1947. The effects of collisions on Saturn's rings. *Monthly Not. Roy. Astron. Soc.* 107:263.

"Sure it's dark out here, but still they're going to very surprised to find that it's Mercury, Venus, Earth, Mars, Jupiter, Saturn, Uranus, Pluto, and Neptune."

PART 4 *Other Objects in the Solar System*

Meteorites

Brian Mason

Research on meteorites, i.e., extraterrestrial rocks, has its frustrating aspects. Whereas the Earth-bound geologist has few limitations on the number and variety of rocks he may study, and can always go into the field and collect more, researchers on meteorites are dependent on what has been randomly picked up, usually by nongeologists, over the last century and a half. Meteorite falls are few and unpredictable. Perhaps this is just as well, otherwise more of us might suffer the unique and distressing experience of Mrs. Hewlett Hodges of Sylacauga, Ala., who was relaxing on her sofa after lunch one November day in 1954 when an 8-pound meteorite broke through the ceiling, bounced off a radio, and hit her on what was delicately reported as her upper thigh! Mrs. Hodges thereby achieved immortality as the first person known to have been injured by an extraterrestrial flying object.

Those who date the Space Age from Sputnik I should recall that meteorites are the only tangible objects we yet have from the universe beyond our Earth, and that they have been objects of scientific investigation since the beginning of the last century. Meteorite research, however, has a history of ups and downs. The attitude of educated persons at the beginning of the nineteenth century was well expressed by the then President, Thomas Jefferson. When informed that two Yale professors had reported a meteorite fall at Weston, Conn., on December 14, 1807, he is said to have remarked, "It is easier to believe that two Yankee professors would lie than that stones would fall from heaven." Nevertheless, in the first decade of the nineteenth century the reality of stones falling from the sky was established beyond doubt, thanks to a number of well-authenticated meteorite showers. Collecting and preserving meteorites then became an approved activity of natural history museums. In 1834 Berzelius, the greatest chemist of his day, wrote a lengthy paper "Om meteorstenar" (concerning meteor stones) in which he provided good-quality quantitative chemical analyses of

Brian Mason, the Curator, Division of Meteorites, Smithsonian Institution, U.S. National Museum, Washington, D.C. 20560, was born in New Zealand in 1917, and took his Master's degree in the University of New Zealand in 1938. A Ph.D. of the University of Stockholm in 1943, he came to the U.S.A. in 1947. He has held academic appointments in geology and mineralogy in Canterbury College, N.Z., University of Indiana, The Museum of Natural History in New York City, and, since 1965, in Washington. A naturalized American since 1953, his books include Principles of Geology, Meteorites, The Literature of Geology *and, with L.G. Berry, in 1959,* Mineralogy.

several different types, and established their general mineralogical and chemical composition. In 1863 H.C. Sorby made a major step forward when he prepared thin sections of some stony meteorites and examined them with the newly devised polarizing microscope, thereby establishing a powerful technique for meteorite research. The significance of such research for geology was pointed out by Daubrée in 1866, who remarked that meteorites were probably fragments of a disrupted planet like the Earth, and by analogy suggested that the Earth has a nickel-iron core like the iron meteorites, and a mantle consisting largely of magnesium-iron silicates, like the stony meteorites. Before that time, the Earth was generally considered to be a great ball of granite, chemically homogeneous throughout.

For the next sixty years or so, meteorite research was largely descriptive, based on more and better chemical and mineralogical analyses. The modern upturn in meteorite research can be said to date from the thirties, when V.M. Goldschmidt and his co-workers made a large number of determinations of minor and trace elements in meteorites. On this basis Goldschmidt, in 1937, published the first table of cosmic or absolute abundances of the elements, based largely on the concept that meteorites provide us with the best sample we have of nonvolatile cosmic matter. Goldschmidt's ideas were brilliantly confirmed in later years, when theories of nucleosynthesis provided a remarkable fit between theoretically deduced elemental abundances and those provided by analyses of meteorites.

Since the end of World War II the pace and scope of meteorite research has shown an exponential increase. Much credit for this goes to Harold Urey, who became fascinated by these enigmatic objects and who has stimulated the development of many fruitful new lines of research. During the past 10–15 years extensive and reliable data have been obtained on many of the minor and trace constituents (minerals and elements alike), and these are frequently the ones that provide the most information on the past history of meteorites, and hence of the Solar System. For instance, the time scale for consolidation, cooling, and breakup of the parent body or bodies has been the subject of extensive investigations, many of which have been possible only because of the development of new, highly sensitive, and accurate analytical techniques. The enormously increased interest in space exploration has encouraged a more thorough and sophisticated examination of the only material we yet have from the universe beyond our planet—this material being received cost free to the recipient. As Edward Anders of the University of Chicago so neatly expressed it, a meteorite is the poor man's space probe! It carries within it a history, albeit imperfectly preserved and difficult to decipher, of events in the universe over the past five billion years.

Recovery of meteorites

Only a small fraction of meteorite falls provide specimens for study. It has been estimated that meteorite falls average about one per million square kilometers per year, or about 500 annually for the whole Earth. Of these, some 350 fall in the ocean and are probably lost forever. However, of the 150 which fall on land our recovery rate is very unsatisfactory, averaging recently about four per year; there has been no recovered fall in the U.S. since the Bells (Texas) meteorite in September 1961.

Statistics of observed meteorite falls by decades are quite intriguing. From 1860 to 1920 the number per decade was fairly constant at about 40–50; for 1920–1930 the figure was 61, for 1930–1940 it was 79. For 1940–1950 the number was halved to 41, but this was generally ascribed to the effects of war and revolution (although World War I had no obvious influence on meteorite recovery). However, this low rate of observed meteorite falls has continued, and it would seem that the Earth passed through a meteorite-rich region of the Solar System in 1920–1940 and since then its orbit has carried it through a region comparatively sparsely populated by these bodies.

Since meteorite falls are rare, and unpredictable as to time and place, planned observations are extremely difficult. The only precise information we have on a meteorite trajectory comes from the Pribram fall in Czechoslovakia in 1959, whose passage through the atmosphere was photographed simultaneously by two recording cameras some miles apart. From these photographs the direction and speed of the meteorite were determined; the orbit calculated therefrom was an ellipse near the plane of the ecliptic, between the orbits of Venus and Jupiter. Most asteroids have orbits between Mars and Jupiter, and the Pribram calculation tends to confirm the long-held view that meteorites are derived from the asteroidal belt. A system of recording cameras has been set up on the Great Plains by the Smithsonian Astrophysical Observatory, and over a longer period this should provide considerable information on meteorite orbits and improve our recovery rate.

On the basis of the above statements the search for recurring meteorites would seem to be an exercise in futility; nevertheless, at different times we do receive meteorites so similar in composition and structure that it is tempting to regard them as different specimens from the same parent body. However, an examination of the time and place of fall of such meteorites fails to show any systematic pattern. Consideration of the recovery rate indicates, however, that we probably recover less than one in every twenty meteorite falls; therefore any statistical pattern would be extremely difficult to distinguish, with nineteen out of every twenty falls unavailable for comparison. This is like trying to assemble a jigsaw puzzle with most of the parts missing.

Composition and classification

Like terrestrial rocks, meteorites are usually classified on the basis of composition and structure. The basic division into irons, stony-irons, and stones is simple and straightforward; however, most of the stones are characterized by the presence of small (~1 mm diameter) spheroidal aggregates known as chondrules (Fig. 1). Stones with chondrules are known as chondrites, those without as achondrites. The current fourfold grouping of meteorites into irons, stony-irons, chondrites, and achondrites, though sanctified by long acceptance and an apparent logic, may actually obscure significant genetic relationships. A genetic grouping would probably be into chondrites and nonchondrites, the latter being derived from the former by a variety of secondary processes, which will be discussed later.

About sixty minerals are known from meteorites at the present time, but many of these are rare accessories. The common and abundant minerals are listed in Table 1. Some contrasts to terrestrial mineralogy should be pointed out: nickel-iron is practically absent from terrestrial rocks; the common minerals in meteorites are largely magnesium-iron silicates, whereas in the Earth's crust the commonest minerals are quartz and aluminosilicates; the common meteorite minerals are anhydrous, whereas hydrated minerals are common and abundant on Earth. These features indicate that meteorites formed in a highly reducing environment, in which nickel and iron were largely in the metallic state. A small but remarkable class of meteorites, the carbonaceous chondrites, differ fundamentally; they consist mainly of serpentine $(Mg,Fe)_6 Si_4O_{10}(OH)_8$, their nickel is present largely in silicates and sulfides, and they contain considerable amounts of organic compounds of extraterrestrial origin. A notable feature of the over-all mineralogy of meteorites is the absence of phases, such as pyrope garnet and jadeitic pyroxenes, indicative of high pressures (i.e., large parent bodies); the origin of the diamond in the Canyon Diablo iron has been plausibly ascribed to the shock of impact with the Earth, which formed Arizona's Meteor Crater, and the presence of diamond in the small group of ureilites appears to be due to extraterrestrial shock effects.

The classification of meteorites based on mineralogy and structure is set out in Table 2; photomicrographs of some typical examples are given in Figures 1–8. Analysis of the observed falls shows that the populations of the different classes vary widely (the figures for observed falls are used as being the best approach to actual extraterrestrial abundances; irons dominate meteorite finds, since they are resistant to weathering and are readily recognized as meteorites or at least as very unusual objects). Over 80% of meteorite falls are chondrites, and over 90% of these belong to two classes, frequently referred to jointly as the common chondrites. Of the other classes of meteorites, some are represented by a single fall, evidence that there may well be additional classes as yet unknown. One of the scientific attractions of manned lunar landings will be the possibility of collecting a more extensive range of meteoritic material, accumulated over hundreds of millions of years.

Table 1. The Common Minerals of Meteorites.

Kamacite	α-(Fe,Ni)	(4–7% Ni)
Taenite	γ-(Fe,Ni)	(15–60% Ni)
Troilite	FeS	
Olivine	$(Mg,Fe)_2SiO_4$	
Orthopyroxene[1]	$(Mg,Fe)SiO_3$	
Pigeonite	$(Ca,Mg,Fe)SiO_3$	(About 10 mole per cent $CaSiO_3$)
Diopside	$Ca(Mg,Fe)Si_2O_6$	
Plagioclase	$(Na,Ca)(Al,Si)_4O_8$	

[1] Divided into enstatite, with 0–10 mole % $FeSiO_3$, bronzite, 10–20%, and hypersthene, >20%; these minerals are orthorhombic, and have monoclinic polymorphs known as clinoenstatite, clinobronzite, and clinohypersthene.

Not only are the chondrites the most abundant meteorites, but many features indicate a primary origin for them and a derivate origin for the other meteorite groups. Of the different classes of chondrites, the carbonaceous chondrites, in particular a subclass known as Type I, show a remarkable correspondence in elemental abundances with the solar photosphere (Fig. 9). If elemental abundances were the same in both, the points on Figure 9 would lie on the 45° line. The close approach to this line for most elements is the basis for considering the chondrites, and specifically the Type I carbonaceous chondrites, as approximating in composition to the primordial nonvolatile matter of the Solar System.

Chondrules and chondrites

Chondrules are unique to meteorites, being unknown in terrestrial rocks, either igneous, metamorphic, or sedimentary, which suggests that they originated by some exotic process. Many hypotheses have been offered as to their origin; the more plausible are as follows:

1. Chondrules originated as molten silicate droplets, a "fiery rain" (Sorby, 1877).
2. Chondrules are fragments of pre-existing meteorites, which have been rounded by oscillation and attrition (Tschermak, 1885).
3. Chondrules are products of magmatic segregation, formed by rapid, arrested crystallization in a molten mass (Brezina, 1885).
4. Chondrules condensed directly from the solar nebula as amorphous aggregates which later crystallized (Levin and Slonimsky, 1957).

We may pass over the more bizarre and outmoded ideas, such as Fermor's view (1938) that chondrules are metamorphosed garnets, and Mason's suggestion (1960) that chondrules were formed by the solid-state thermal conversion of serpentine into spheroidal aggregates of olivine and pyroxene.

Of the different hypotheses, the oldest one, that of Sorby, holds up best. The spherical form, together with the presence of glass and high-temperature minerals such as cristobalite and inverted protoenstatite in some chondrites, strongly favors an origin as molten silicate droplets. A significant feature, inadequately considered in most theories, is the evidence of rapid chilling—the survival of glass and the presence of minerals of the clinoenstatite-clinohypersthene series—followed in most instances by a reheating and partial or complete equilibration at temperatures usually of the order of 800–1000°C. However, the process which gave rise to the molten droplets has been the subject of considerable

Table 2. Classification of the Meteorites. (Figures in parentheses are the numbers of observed falls in each class.)

Group	Class	Principal Minerals
Chondrites	Enstatite (11)	Enstatite, nickel-iron
	Bronzite (227)	Olivine, bronzite, nickel-iron
	Hypersthene (303)	Olivine, hypersthene, nickel-iron
	Carbonaceous (31)	Serpentine, olivine
Achondrites	Aubrites (8)	Enstatite
	Diogenites (8)	Hypersthene
	Chassignite (1)	Olivine
	Ureilites (3)	Olivine, clinobronzite, nickel-iron
	Angrite (1)	Augite
	Nakhlites (1)	Diopside, olivine
	Eucrites and howardites (39)	Pyroxene, plagioclase
Stony-irons	Pallasites (2)	Olivine, nickel-iron
	Siderophyre (1)[1]	Orthopyroxene, nickel-iron
	Lodranite (1)	Orthopyroxene, olivine, nickel-iron
	Mesosiderites (6)	Pyroxene, plagioclase, nickel-iron
Irons	Hexahedrites (8)	Kamacite
	Octahedrites (30)	Kamacite, taenite
	Ataxites (1)	Taenite

[1] Find.

Figure 1. Chondrules in the Chainpur meteorite; the largest is 1.2 mm in diameter.

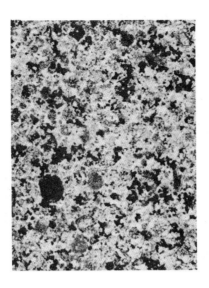

Figure 2. The Miller meteorite, a bronzite chondrite; chondritic structure is less marked than in Chainpur, probably because of partial recrystallization; the largest chondrule is 1.5 mm in diameter.

Figure 3. The Shaw meteorite, a hypersthene chondrite which is a granular aggregate of olivine, pyroxene, and a little plagioclase; chondritic structure is practically absent; average grain size is about 0.2 mm.

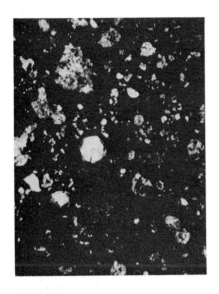

Figure 4. The Haripura meteorite, a typical carbonaceous chondrite with chondrules and small grains of olivine and pyroxene in an opaque groundmass of serpentine and carbonaceous matter; the chondrule in center is 0.5 mm in diameter.

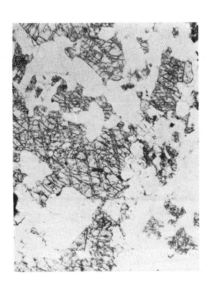

Figure 5. The Moore County meteorite, a pyroxene (gray) plagioclase (white) achondrite; the texture and composition resembles some terrestrial gabbros; average grain size is about 2 mm.

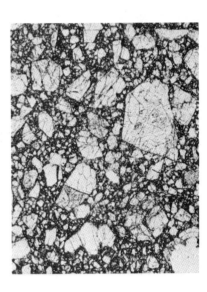

Figure 6. The Johnstown meteorite, a hypersthene achondrite, showing angular fragments of hypersthene, up to 5 mm across, in a groundmass of comminuted hypersthene.

diversity of opinion. Sorby suggested that they may have been derived directly from the surface of the Sun. Current ideas of their origin include volcanism on the meteorite parent bodies, splash drops formed in collisions between asteroids, condensation of liquid droplets from a hot gas of solar composition, and fusion of dust in the primordial solar nebula. The latter idea has been elaborated by Whipple (1966). He proposes that chondrules formed in the cooler part of the solar nebula through the fusion of dust by lightning. This would provide for the rapid high-temperature melting followed by instantaneous chilling, which appears to be required by the phase relations. This theory of origin also receives considerable support from recent work on I^{129}-Xe^{129} dating (Hohenberg et al., 1967). The results indicate that chondritic material crystallized essentially simultaneously (within 1–2 million years) some 4.7 billion years ago. Chondritic material is thus of primary, not secondary, crystallization, a point in favor of Whipple's proposal and against volcanic or impact theories.

Although most chondrules appear to have originated as molten droplets, I would hesitate to assert that everything referred to as chondrules originated in a single process. In the carbonaceous chondrites, for example, many of the rather irregular chondrules appear possibly to be ag-

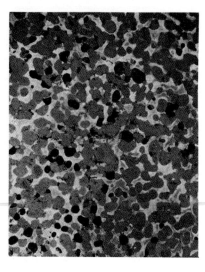

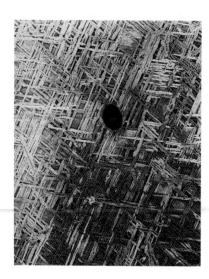

Figure 7. The Brenham meteorite, a pallasite, consisting of rounded crystals of olivine (gray) about 5 mm across, in a nickel-iron matrix (white); black spots are holes where olivine crystals have been plucked from the surface during polishing.

Figure 8. Polished and etched surface of the Carbo iron, showing three sets of parallel kamacite bands, each about 1 mm thick, defining the Widmanstatten structure; the round black inclusion is a nodule of troilite, FeS.

gregates of originally solid particles rather than molten droplets.

As long ago as 1878 Nordenskjöld commented on the remarkable uniformity of composition of chondritic meteorites, basing his comment on the close correspondence between analyses of the common (bronzite and hypersthene) chondrites. This uniformity is especially well illustrated when the composition is expressed as atomic percentages on an oxygen-free basis, since the most prominent difference between different classes of

chondrites is the degree of oxidation of their iron content. The uniformity of composition also extends to the carbonaceous chondrites when allowance is made for their content of combined water and organic compounds. In all the chondrites, Si + Mg + Fe + O >90%, both in weight and in number of atoms; in terms of numbers of atoms $Mg \approx Si \approx 0.5\text{--}0.8\ Fe \approx 3\text{--}4\ O$. The relationship between chemical and mineralogical composition between the carbonaceous, bronzite and hypersthene, and enstatite chondrites can be grossly represented as follows:

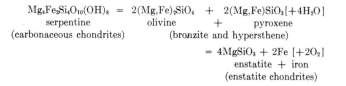

$$Mg_4Fe_2Si_4O_{10}(OH)_8 = 2(Mg,Fe)_2SiO_4 + 2(Mg,Fe)SiO_3[+4H_2O]$$

serpentine   olivine   +   pyroxene
(carbonaceous chondrites)   (bronzite and hypersthene)

$$= 4MgSiO_3 + 2Fe\ [+2O_2]$$
enstatite + iron
(enstatite chondrites)

On the basis of this relationship Prior (1920) proposed that the different classes of chondrites were formed by the progressive oxidation of a highly reduced magma, the earliest stage being represented by the enstatite chondrites. Mason (1960) and Ringwood (1961) have theorized that the parent material was highly oxidized, its original state being similar to the carbonaceous chondrites and the other classes of chondrites produced from this material by dehydration and progressive reduction.

However, theories deriving all chondrites from a common parent material of uniform composition are oversimplifications. In 1953, Urey and Craig made a critical review of all analyses of chondritic meteorites and found that they could be divided into two groups, one averaging 28.6% total iron (their high-iron or H group), and one averaging 22.3% total iron (their low-iron or L group). They brought out this division into two groups in the form of a diagram in which Fe combined in silicates (i.e., oxidized iron) was plotted against Fe in metal and FeS (i.e., reduced iron). On their original diagram the points

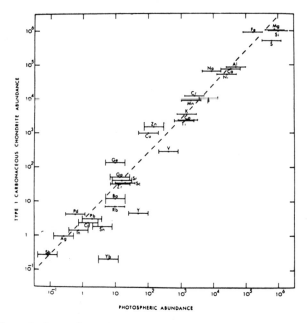

Figure 9. Comparison of elemental abundances in carbonaceous chondrites (Type I) with those in the solar photosphere (abundances normalized on the basis of Si = 10⁶ in the Sun and in the meteorites). Horizontal bars correspond to a twofold uncertainty in the solar abundances, vertical bars represent estimated errors in the chondritic abundances (Ringwood, 1966).

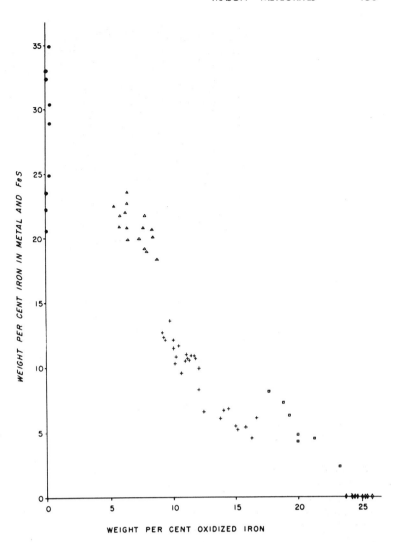

WEIGHT PER CENT IRON IN METAL AND FeS

WEIGHT PER CENT OXIDIZED IRON

Figure 10. Relationship between oxidized iron and iron as metal and sulfide in analyses of chondrites, illustrating the separation into distinct classes and the variation within the classes (● = enstatite chondrites; △ = bronzite chondrites; + = hypersthene chondrites; □ = carbonaceous chondrites, Type III; ◇ = carbonaceous chondrites, Types I and II).

representing the individual analyses showed a considerable scatter and some overlap between the two groups. A revised diagram based on more and better analyses is given as Figure 10. This diagram is exceedingly illuminating, in that it shows clearly that each of the four classes of chondrites occupies a specific composition field, and there is essentially no overlap. Another remarkable feature is the coincidence of Urey and Craig's L-group with the hypersthene chondrites—all hypersthene chondrites belong to the L-group, none to the H-group, whereas the reverse is true for the bronzite chondrites.

Significant and consistent variations in Mg and Si content also exist between the different classes of chondrites, and are illustrated in Figure 11. The Si/Mg ratio (atomic) is slightly less than 1.0 for the carbonaceous chondrites, between 1.0 and 1.1 for the bronzite and hypersthene chondrites (which plot in different areas of the diagram because of the difference in their total iron content), and averages about 1.3 for the enstatite chondrites.

Clearly, therefore, although all chondrites are remarkably similar in elemental composition, some fractionation of the elements had already taken place before their parent bodies accumulated, and the physico-chemical conditions of crystallization and accumulation differed widely, from cold, humid, and relatively oxidizing for the serpentinous matrix of the carbonaceous

chondrites to hot, anhydrous, and highly reducing for the enstatite chondrites. It is conceivable that this range of conditions may have existed in the primitive solar nebula in the extensive region between Mars and Jupiter now occupied in part by the asteroidal belt. Magnetic forces may have served to fractionate the iron in the ancestral nebula, either as metal particles or as magnetite. Mechanisms for the fractionation of magnesium and silicon are not so obvious—however, silicon is rather easily reduced to elemental state and then alloys with iron, whereas magnesium would remain in oxidic compounds under conditions in the outer part of the solar nebula. It is probably significant that the metal phase in the enstatite chondrites always contains silicon in solid solution.

Each of the classes of chondritic meteorites provides evidence for some degree of recrystallization following the accumulation of chondrules and matrix in the meteorite parent bodies. This recrystallization is shown not only by mineralogical changes such as the formation of plagioclase from glass and the inversion of clinoenstatite-clinohypersthene to enstatite-hypersthene, but also by partial or complete erasure of the chondritic structure (cf. Figs. 1–3). Van Schmus and Wood (1967) have divided the chondrite classes into subclasses based on the degree of recrystallization and equilibration, which they ascribe to thermal metamorphism in the meteorite parent bodies.

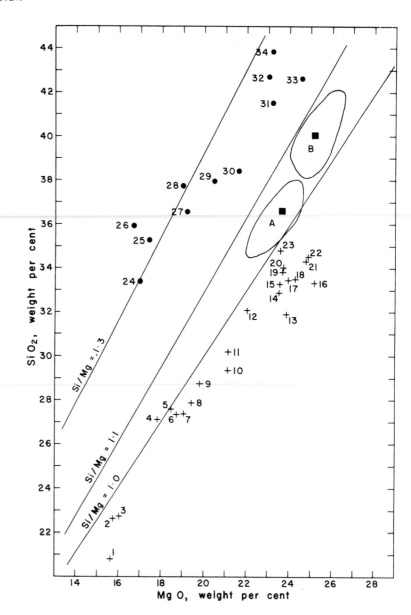

Figure 11. SiO₂ plotted against MgO (weight percentages) for chemical analyses of chondrites; the diagonal lines are for Si/Mg atomic ratios of 1.0, 1.1 , and 1.3. A is the field for 36 analyses of bronzite chondrites, B is the field for 68 analyses of hypersthene chondrites, the black squares being the means for each group. The analyses of carbonaceous chondrites (1–23) and enstatite chondrites (24–34) are plotted individually.

The sequence of recrystallization of chondritic material is most clearly demonstrated in the enstatite chondrites (Mason, 1966). These meteorites also present a rather simple system, the silicate material being essentially uniform in composition and falling within the ternary system Mg₂SiO₄–NaAlSi₃O₈–SiO₂ (approx. 55% Mg₂ SiO₄, 15% NaAlSi₃O₈, 30% SiO₂). The phase relations for this system have been elucidated by Schairer and Yoder (1960); the liquidus temperature for this composition is about 1500 °C, and the phase crystallizing is protoenstatite. Rapid chilling would produce the typical primitive enstatite chondrites, in which we find the association clinoenstatite (inverted proto)—cristobalite—glass. Reheating and recrystallization of this material would produce the association enstatite-albite-tridymite and/or quartz, characteristic of the "equilibrated" enstatite chondrites. In this connection the mineral association enstatite-albite-quartz in the St. Marks meteorite is particularly significant. The mineral association defines the equilibration temperature rather narrowly. The presence of enstatite (assuming that this mineral was produced by inversion from clinoenstatite) indicates a temperature of recrystallization above 620 °C, while the presence of quartz shows that the temperature

did not rise above 870 °C. Tridymite is the stable form of SiO₂ between 870° and 1470 °C. Studies of the common chondrites (Van Schmus and Koffman, 1967) also suggest equilibration temperatures of the order of 850 °C. The carbonaceous chondrites, however, have never been subjected to recrystallization under these conditions; serpentine decomposes around 500 °C, and their organic compounds would decompose at considerably lower temperatures. It therefore appears that the parent bodies of carbonaceous chondrites never heated up to any great extent, whereas parent bodies of the other chondrite classes did undergo a considerable degree of internal heating. Although most of the ordinary chrondrites have mineral compositions indicating equilibration at temperatures of the order of 850 °C, a considerable range of temperature is probably represented. The unequilibrated ordinary chondrites (Dodd, et al., 1967) contain clinoenstatite-clinohypersthene, indicating that they have not been reheated above about 600 °C, whereas in Shaw, a chondrite in which recrystallization has practically eliminated chondritic structure, the equilibration temperature was of the order of 1100–1200 °C (Fredriksson and Mason, 1967).

Organic compounds in carbonaceous chondrites

The comment that meteorite research has been characterized by ups and downs is well exemplified by the interest or lack thereof in the organic compounds, found especially in the carbonaceous chondrites. Among the meteorites analyzed by Berzelius was Alais, a carbonaceous chondrite which fell in France in 1806. Berzelius recounts how, at first, he refused to believe Alais was a meteorite; it looked so unlike the other meteorites he had examined and so like a piece of clay. He inquired of his French colleagues who had sent him the specimen, and they convinced him that it had indeed fallen from the sky. Berzelius clearly demonstrated the presence of volatile organic compounds, and of a "humus-like complex" in this meteorite, and he asked the prescient question—which we are still trying to answer today—does this material possibly indicate the presence of organisms on extraterrestrial bodies? (Berzelius thought not.)

An upsurge of interest in these organic compounds followed the fall of the Orgueil meteorite in southern France in 1864. This fall was well observed, and several stones, with a total weight of about 12 kg, were recovered, thereby providing adequate material for research. Within the next years several eminent scientists analyzed material of this meteorite and speculated on the origin of the organic compounds therein. The noted chemist, J. Lawrence Smith, investigated the organic compounds in this and other carbonaceous chondrites and published several papers on them around 1875.

Interest then lapsed for three-quarters of a century. Several carbonaceous chondrites fell in the interim, but their arrival failed to arouse much interest and they were entered in museum collections and largely forgotten. The modern interest may be said to date from the publication by George Mueller in 1953 of the results of a detailed examination of the organic material in the Cold Bokkeveld meteorite. In 1956 H.B. Wiik analyzed a number of the carbonaceous chondrites and showed that they could be grouped in three types—his Type I, Type II, and Type III—each characterized by the amount of organic material they contained and by other compositional features. Type I were richest in organic matter, with 3-5% combined carbon (Alais and Orgueil are Type I carbonaceous chondrites).

A major step forward was achieved by Nagy, Meinschein, and Hennessy, who, in March 1961, presented the results of their work on the organic compounds in the Orgueil meteorite. They applied the procedures of organic geochemistry, developed largely for the extraction and identification of the organic compounds in sediments and sedimentary rocks, to this meteorite. On the basis of their work they wrote, "The mass-spectrometric analyses reveal that the hydrocarbons in the Orgueil meteorite resemble in many important aspects the hydrocarbons in the products of living things and sediments on Earth. Based on these preliminary studies, the composition of the hydrocarbons in the Orgueil meteorite provides evidence for biogenic activity." Later in the same year, Claus (a microbiologist) and Nagy announced that they had discovered "microscopic-sized particles, resembling fossil algae, in relatively large quantities within the Orgueil and Ivuna carbonaceous chondrites."

These announcements aroused a hot controversy that is still far from resolved, the fundamental question being, of course, "Is this material of biological or nonbiological origin?" Since 1961, several man-years of work have been devoted to the organic material in carbonaceous chondrites, but real progress in understanding the significance of this material has been slow. We should not be too critical of this, however, since much of the organic matter is a black insoluble complex of high-molecular weight compounds reminiscent of asphalt or bitumen. Consider how little we know and understand of the constitution of coal, which has been investigated intensively for over a century.

The present situation has been reviewed by Hayes (1967) and can perhaps be summarized as follows. Hydrocarbons—alkanes, cycloalkanes, and aromatics—phenolic compounds, and fatty acids have been identified in one or more carbonaceous chondrites. Amino acids have been reported in minute amounts, but are probably terrestrial contaminants. Heterocyclic nitrogen compounds of no biological significance (melamine and ammeline) have been identified. Minute amounts of porphyrins have been found. Evidence for optical activity or the presence of extraterrestrial organisms is still inconclusive. The isoprenoid hydrocarbons pristane and phytane, possible degradation products of chlorophyll, have been identified by Meinschein and independently confirmed by Oró and his co-workers. The evidence for a nonbiological origin of meteoritic organic compounds is easily summarized: every compound found in meteorites, except pristane and phytane, has been made in some type of abiogenic synthesis.

A most intriguing thought is the possibility that the organic compounds in meteorites are not the products of life, but rather the precursors of life. It is conceivable that similar organic compounds were received by the Earth during the final stages of its aggregation from the material of the solar nebula, thereby providing the basic compounds on which life got started. The resemblance that Nagy and his co-workers noted between the hydrocarbons in the Orgueil meteorite and those in recent sediments would then reflect the self-replicating biological mechanisms using such compounds since the beginning of life on Earth.

Achondrites and stony-irons

Superficial inspection of a collection of achondrites and stony-irons reveals a great diversity and an apparent lack of any unifying features. Some of them, especially the mesosiderites, are obviously breccias made up of fragments of widely different chemical and mineralogical composition, cemented together by a nickel-iron matrix. However, a closer examination does indicate relationships that imply common processes of genesis for most of them. The composition and structure of the silicate material indicate original crystallization from a melt, similar to the magmas that gave rise to terrestrial basic and ultrabasic igneous rocks.

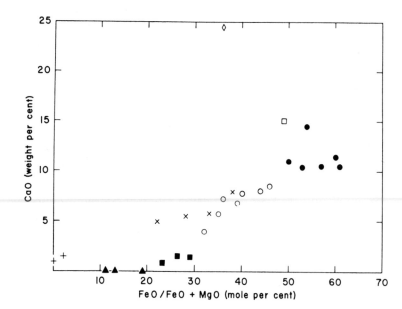

Figure 12. Plot of CaO (weight per cent) against FeO/FeO + MgO (mole per cent) for the achondrites and stony-irons (+ = enstatite achondrites; ▲ = pallasites; ■ = hypersthene achondrites;   × = mesosiderites; ○ = howardites; ● = eucrites; □ = nakhlite; ◇ = angrite).

These relationships can be most readily visualized in the form of a diagram (Fig. 12). This shows a regular trend, from calcium-poor, magnesium-rich compositions to compositions richer in calcium and ferrous iron. The enstatite achondrites can plausibly be accounted for by the partial melting of a parent body of enstatite chondrite composition, whereby metal and sulfide were melted and removed by gravitational forces, leaving a residue of coarsely crystallized enstatite and a little sodic plagioclase. Most of the other achondrites and stony-irons can be derived by the melting and fractional crystallization of one or more parent bodies with the over-all composition of the common chondrites. Such a melt would begin to crystallize at about 1500°C with the separation of magnesium-rich olivine, similar in composition to that found in pallasites. This olivine, together with the molten nickel-iron, would sink and eventually form a core of pallasitic composition. The crystallization of olivine would be followed by hypersthene, with a higher Fe/Mg ratio, as we find in the hypersthene achondrites. At a slightly later stage plagioclase would begin to crystallize, giving the association hypersthene-plagioclase characteristic of the howardites. This fractional crystallization would result in a steady increase in the concentration of calcium and ferrous iron in the melt, and eventually pigeonite, a pyroxene richer in iron and calcium than hypersthene, would be the stable ferro-magnesian silicate, giving the association pigeonite-plagioclase characteristic of the eucrites. Mesosiderites appear to be breccias of all these achondrite types, together with nickel-iron and olivine.

The whole process can be depicted as producing a differentiated asteroid (Fig. 13). The radius of the asteroid (300 km) was chosen to be consistent with the measured cooling rates of pallasitic meteorites (Buseck and Goldstein, 1967). The mineral compositions used for the calculations were those from the meteorites themselves. The result provides a gravitationally layered body consistent with laboratory data on silicate melts and petrological experience with igneous rocks. The over-all composition of this body shows a remarkable correspondence with the average composition of chondrites—except for sodium, which shows a tenfold depletion (the other alkali elements, except lithium, show a

similar depletion). It appears to me that the consistencies of this picture of the derivation of achondrites and stony-irons from chondritic material far outweigh the inconsistency in sodium content. There are several possible explanations for the depletion of sodium:

1. The sodium may have been concentrated into a relatively thin alkali-rich crust, specimens of which have not yet been received as meteorites.
2. The sodium may have been lost during the melting and crystallization sequence that produced the stony-irons and achondrites; selective volatilization is one process that has been suggested.
3. The original material, while of chondritic composition, may have been deficient in alkalis. (Some carbonaceous chondrites are much lower in these elements than the average chondrites.)
4. The achondrites and stony-irons were not derived from material of chondritic composition. (However, for all the other major elements the hypothesis of original material of chondritic composition is strongly supported by Figure 13.)

An unequivocal solution to this problem is not possible at this time; other factors not envisaged above may have played a part.

Irons

Iron meteorites may be grouped into a sequence, based on structure, and closely linked with the nickel content. This sequence is:

	Falls	Finds	Total
Hexahedrites (4–6% Ni)	8	39	57
Octahedrites (6–18% Ni)	30	412	442
Ataxites (>10% Ni)	1	41	42

Figure 14 illustrates some significant features of this sequence. No iron meteorite contains less than 4% Ni—an extremely useful diagnostic criterion when one is called on to determine whether a mass of iron is meteoritic or not! The next feature is the sharp peak at 5.5% Ni, which

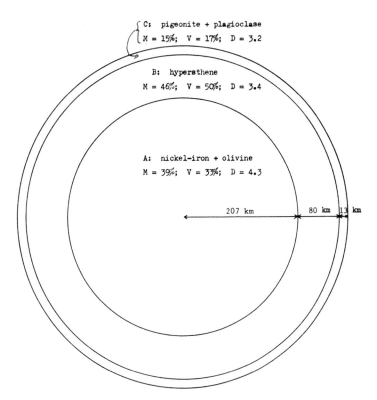

C: pigeonite + plagioclase
M = 15%; V = 17%; D = 3.2

B: hypersthene
M = 46%; V = 50%; D = 3.4

A: nickel-iron + olivine
M = 39%; V = 33%; D = 4.3

207 km   80 km   13 km

Figure 13. Cross-section of a hypothetical differentiated asteroid parent of pallasites and achondritic meteorites: nickel-iron, 14%; olivine, 25%; hypersthene, 46%; pigeonite, 9%; plagioclase, 6% (M = mass, V = volume, D = density).

	SiO_2	MgO	FeO	Al_2O_3	CaO	Na_2O	Fe	Ni
Composition	42.7	26.2	12.7	2.3	2.0	0.1	12.5	1.5
Av. chondrite[1]	41.2	26.3	13.1	2.5	2.0	0.9	12.5	1.5

[1]Omitting FeS.

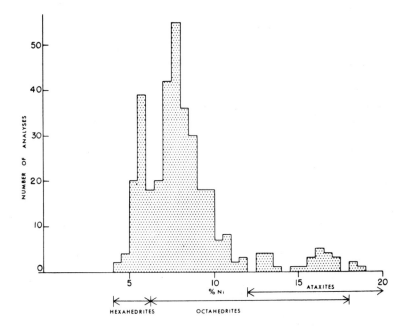

Figure 14. The frequency of distribution of nickel content in analyses of iron meteorites.

is the composition of many hexahedrites. Another peak at 8% Ni corresponds to a large number of octahedrites. Comparatively few irons have more than 10% Ni, and only seven contain more than 20%.

The hexahedrites are so-called because they normally consist of large crystals of kamacite, which has cubic structure (cube = hexahedron). The octahedrites, by far the commonest class of iron meteorites, show an oriented intergrowth of kamacite bands parallel to octahedral planes—the Widmanstatten structure (Fig. 8). The ataxites received this name originally because they appeared to be structureless, but many of them do show microscopic oriented plates of kamacite in a groundmass of fine-grained kamacite and taenite called plessite.

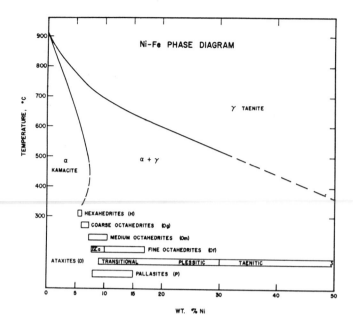

Figure 15. Subsolidus phase relations in the Fe–Ni system, correlated with the composition and structure of iron meteorites (after Goldstein and Ogilvie, 1965).

These relations are consistent with the subsolidus phase diagram for the Fe—Ni system (Fig. 15). Above 900°C, nickel-iron alloys exist as the gamma phase or taenite, with a face-centered cubic structure. On cooling, pure iron changes at 910°C into the alpha phase (kamacite), with a body-centered cubic structure. The addition of nickel decreases the temperature of the gamma-alpha transformation and introduces a two-phase region, within which the low-nickel alpha phase coexists with the gamma phase richer in nickel. This two-phase region broadens at lower temperatures; as the temperature falls, the nickel content of each phase increases, but that of the gamma phase increases much more rapidly than that of the alpha phase. The amount of each phase, for a specific composition, varies to accommodate the change in composition of the individual phases, and if during the cooling process the alpha phase reaches the same nickel content as the original gamma phase, the amount of the latter will diminish to zero.

The relation of this phase equilibrium diagram to the composition and structure of the iron meteorites is readily seen. The hexahedrites, consisting entirely of kamacite, fall in the field of the alpha phase at temperatures below 500°C; the majority of the iron meteorites (octahedrites and ataxites), consisting of intergrown kamacite and taenite, fall in the two-phase region at these temperatures; a few extremely nickel-rich ataxites consist entirely of taenite and fall in the field of the gamma phase. The Widmanstatten pattern of exsolved kamacite develops parallel to the octahedral planes in the host taenite. The finer scale of the Widmanstatten pattern that occurs with increasing nickel content reflects the lower temperature

for beginning exsolution and the lower diffusion rates at lower temperatures.

The recent application of the electron-beam microprobe to the study of Widmanstatten structure has greatly extended our understanding, and has provided estimates of cooling rates. Composition traces across the structure (Fig. 16) show that taenite is always strongly enriched in nickel at the contact with kamacite. This is a consequence of the diffusion rates in taenite falling off rapidly with decreasing temperature. Diffusion rates in kamacite are always much higher than in taenite at any given temperature. This is reflected in the essential uniformity of composition of kamacite. Cooling rates for iron meteorites, calculated from nickel gradients in taenite, range from 0.4° to 500°C per million years, consistent with parent bodies of about 25–300 km radius if the irons formed the cores (Goldstein and Short, 1967).

Studies of minor and trace elements in iron meteorites have provided significant information relative to their origin. Lovering et al. (1957) showed that gallium and germanium contents varied by orders of magnitude, and were roughly quantized into four distinct groups. This work has been refined and extended by later investigators (Fig. 17). The reasons for the distribution of the iron meteorites into specific Ga-Ge groups are not fully understood. It would appear that meteorites containing different Ga-Ge concentrations crystallized in different regions of a single parent body, or more probably in different parent bodies. Instead of coming from the core of a single parent body, as was once believed, iron meteorites may represent local metal segregations in a

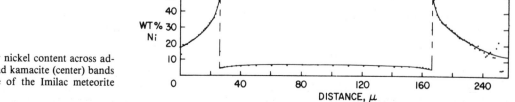

Figure 16. Microprobe trace for nickel content across adjoining taenite (left and right) and kamacite (center) bands in the Widmanstatten structure of the Imilac meteorite (Goldstein and Short, 1967).

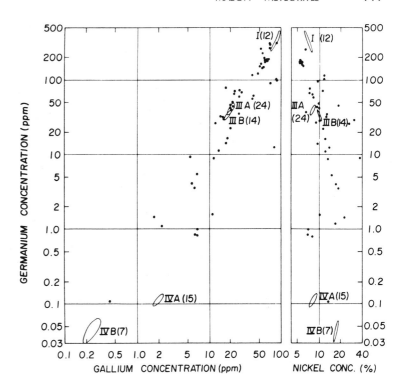

Figure 17. Gallium and germanium contents of iron meteorites; each dot represents the analysis of a single meteorite, the outlined fields the analyses of several meteorites (number in parentheses) of very similar composition (Wasson and Kimberlin, 1967).

variety of parent bodies. Such an origin would be consistent with the great range of cooling rates and of trace element concentrations. It is noteworthy that the slowest cooling rates yet recorded are for the metal in pallasites, suggesting that they represent well-insulated cores of relatively large asteroids, whereas the iron meteorites come from smaller bodies. Another significant fact is that silicate inclusions in iron meteorites resemble chondrites in chemical and mineralogical composition, indicating that these iron meteorites came from less differentiated bodies than those giving rise to achondrites and stony-irons.

The chronology of meteorites

Our knowledge of meteorites, and theories of origin of the Solar System, suggest the following sequence of events: (1) nucleosynthesis; (2) formation of the ancestral solar nebula, with a central protosun surrounded by a lens-shaped cloud of dust and gas, diminishing in temperature from the center to the margin; (3) formation of chondrules; (4) aggregation of chondrules and dust into asteroidal-sized bodies; (5) heating-up of these bodies to varying degrees, depending largely on their size, with partial or complete melting of some; (6) cooling and crystallization of molten material, with differentiation of nickel-iron and silicates; (7) breakup of meteorite parent bodies; (8) arrival of meteorites on the Earth.

The possibilities of dating these events are indicated in Table 3. Nucleosynthesis of the heavier elements and the formation of the ancestral solar nebula probably originated in a supernova outburst. Within a short time thereafter, geologically speaking, the whole sequence of events 3 through 6 must have taken place. Wasserburg *et al.* (1965), who established a 4700 m.y. Rb/Sr age for the Weekeroo iron, commented: "Our work further emphasizes the significance of the date of about 4.4–4.7 × 10⁹ years as the time of an event of major importance in the solar system. From the independent dating of stony meteorites and of iron meteorites and the coupled dating of the Earth and iron meteorites, it appears that the most significant period of chemical differentiation—coupled with the formation of solid objects within the solar system—occurred about 4.6×10^9 years ago within a narrow time band (less than 0.2×10^9 years). The significance of this period is enhanced by the observation of radiogenic Xe^{129} in meteorites from I^{129} decay ($t_{1/2}$, 1.6×10^7 years), and of $Xe^{131-136}$ and excess fission tracks attributed to the spontaneous fission of Pu^{244} ($t_{1/2}$, 7.6×10^7 years) . . . These observations thus indicate that the physical and chemical differentiation which took place about 4.6×10^9 years ago occurred soon after isolation of our solar system from nucleosynthetic processes in which heavy elements, including transuranics, were produced. Consequently, this time also dates the separation of the solar system from the interstellar medium, because the time scale for the evolution of the solar system, from the interstellar medium, is required to fall within a few half-lives of Pu^{244}."

The subsequent history of the meteorites can be deduced from a study of their exposure to cosmic rays since the time they were reduced to pieces approximately their present size. High-energy cosmic-ray particles colliding with meteorites in outer space produce an array of new and usually radioactive atoms by spallation. Cosmic-ray exposure ages have been determined for a large number of meteorites, with intriguing results. Stones give exposure ages ranging up to about 200 m.y., most ages, however, being of the order of 20 m.y.; irons give much longer exposure ages, about 100–2000 m.y.

Many of the hypersthene chondrites give U-He ages of 520 m.y., and about half of the octahedrites have similar exposure ages. This suggests the possibility of a collision between two asteroids, one of hypersthene chondrite composition and one of octahedrite composition, at this time. The impact effectively outgassed the chondritic material, thereby resetting the U-He "clock." Since stone

Table 3. Chronology of Meteorites. (After E. Anders, unpublished.)

Stage	Dating Method	Event Dated	Typical Results (m.y.)[2]
Nucleosynth.	I^{129}/Xe^{129}	Interval between nucleosynthesis and cooling of MPB[1] to low enough temperatures to retain $Xe^{129,136}$	t_0–60
	Pu^{244}/Xe^{136}		t_0–150
	Fission Track	Ditto, to retain Pu^{244} tracks	t_0–300
Melting	Pb^{207}/Pb^{206}	End of U/Pb fractionation	4550
	Rb^{87}/Sr^{87}	End of Rb/Sr fractionation	4700
	Re^{187}/Os^{187}	End of Re/Os fractionation	4400
Cooling	K^{40}/Ar^{40}	Low enough T to retain Ar^{40}, He^4	4400
	U^{238}/He^4		4200
Breakup I	U^{238}/He^4	Reheating MPB by major collision	520
Breakup II	He^3, Ne^{21}	Onset of cosmic-ray irradiation after reduction of meteorite to meter-sized fragments	20
	K^{40}/K^{41}		
Earth Impact	H^3, Ar^{39}	End of cosmic-ray irradiation	0–1
	C^{14}, Cl^{36}		

[1] MPB = meteorite parent bodies.
[2] m.y. = million years.

meteorites are more fragile than irons, they probably have been broken into smaller pieces by further collisions from time to time, hence explaining their lower exposure ages.

The origin of meteorites

As far as the origin of meteorites is concerned, the problem appeared solved when R.A. Daly of Harvard University published his authoritative paper "Meteorites and an Earth-Model" in 1943. From a detailed analysis of the analogies between meteorites and terrestrial rocks, coupled with deductions on the probable nature of the Earth's interior, Daly concluded that the meteorites originated by distruption of a planet once situated between Mars and Jupiter, in the region now occupied by the asteroids. He even estimated the dimensions of this planet: mean radius 3000 km, a volume 0.1 and a mass 0.07 that of the Earth, a mean density of 3.77, and an iron core of about 1000 km diameter. Daly's theory has been quoted approvingly by textbooks, including my own "Principles of Geochemistry" in the 1952 and 1958 editions; however, it is excised from the latest edition. It is now clear that, however the meteorites originated, they have not come from a single parent body; for one thing, there are too many chemical and mineralogical discontinuities between different groups of meteorites.

As a result of the great resurgence of interest since 1943 in meteorites as objects of scientific research, and a remarkable improvement and extension of the techniques for their investigation, we now have a vast accumulation of new data on these enigmatic bodies. Chondrites, especially carbonaceous chondrites, are now recognized to be relatively well-preserved samples of the nonvolatile matter of the Solar System. Compositional differences between different classes of chondrites indicate that the ancestral solar nebula probably underwent some chemical fractionation prior to the accumulation of the meteorite parent bodies. The extreme differences in mineralogy and oxidation state between the carbonaceous chondrites and the enstatite chondrites probably reflect differences in the environment of accumulation of their original material, and in the physicochemical

conditions within the meteorite parent bodies. The major events in the genesis of meteoritic material—the birth of the ancestral solar nebula, the formation of chondrules, the accumulation of meteorite parent bodies, and the subsequent differentiation of some of them to give the material of the nonchondritic meteorites—all took place within a relatively short time interval about 4.7 b.y. ago.

It appears to me that the wide variety of meteorites, and the contrast between chondrites and nonchondrites, may have their origin largely in different sizes of meteorite parent bodies. Some chondrites—the carbonaceous chondrites and the unequilibrated ordinary chondrites—have not been subjected to serious reheating since accumulation, and presumably are derived from relatively small parent bodies. Most chondrites show features indicating reheating in their parent bodies to temperatures of 800°C and higher. The higher the temperature the greater the loss of chondritic structure by recrystallization. Ultimately, if temperatures of 1400–1500°C were reached inside the parent bodies, the material melted. The degree of reheating was probably controlled in considerable part by the size of the parent bodies; the larger the parent body the greater the heating-up of the interior. Heating-up was probably a comparatively rapid process, brought about not only by the greater abundance of the long-lived radioactive elements 4.7 b.y. ago, but also by short-lived radioactivities such as Al^{26}, I^{129}, Pu^{244}, and others.

It is conceivable, of course, that each meteorite class might represent a single parent body, i.e., all the hypersthene chondrites, for example, might be samples from different levels in a single parent body, the highly chondritic unrecrystallized types being from cool near-surface layers, the progressively more recrystallized types coming from progressively greater depths. However, it appears more likely to me that a large number of potential parent bodies formed, perhaps from the size of footballs up to some hundreds of kilometers in radius (the largest asteroid, Ceres, has a radius of 385 km). The cooling rates for the pallasites indicate that they formed the core of a body with a maximum radius of about 300 km (Buseck and Goldstein, 1967). Since the pallasites have the lowest cooling rates for any class of meteorites, this seems to indicate an approximate upper limit for the size of meteorite parent bodies.

If the above hypothesis for the evolution of meteorite parent bodies has some validity, the populations of the different classes of meteorites may provide some relevant information. The abundance of the common chondrites indicates that most of the parent bodies must have had similar compositions. Parent bodies giving rise to the achondrites, irons, and stony-irons must be comparatively rare. This suggests that comparatively few of the parent bodies were sufficiently large to heat up to the melting point. The low abundance of the carbonaceous chondrites, their unique Si/Mg ratio, and their specific properties indicating a low-temperature origin for much of their material, suggest that the parent bodies for these meteorites accumulated farther from the center of the ancestral solar nebula, where temperatures were low. Their parent bodies never grew large enough for any thermal metamorphism to take place, perhaps because the proportion of solid matter to gas in that region of the nebula was considerably lower than where the other meteorite parent bodies accumulated. Within the space between Mars and Jupiter there is probably room to provide for the initial chemical and mineralogical differences and the divergent evolutionary paths of different parent bodies, whose results we observe today in meteorites.

Acknowledgments

I am indebted to J.H. Reynolds, J.I. Goldstein, W.G. Melson, and Miss Jane Nielson, who read the manuscript and made useful comments and suggestions.

Bibliography

Anders, E., 1964. Origin, age, and composition of meteorites. *Space Sci. Rev., 3,* pp. 583-714.

Krinov, E. L., 1960. Principles of meteoritics. Pergamon Press, New York, 535 pp.

Mason, B., 1962. Meteorites. John Wiley & Sons, New York, 274 pp.

Ringwood, A.E., 1966. Genesis of chondritic meteorites. *Rev. Geophysics, 4,* pp. 113-175.

Wood, J.A., 1963. Physics and chemistry of meteorites. Chapter 12 of "The moon, meteorites, and comets" (B. M. Middlehurst and G.P. Kuiper, eds.). University of Chicago Press.

References

Brezina, A., 1885. Die Meteoritensammlung des k. k. mineralogischen Hofka binettes in Wien. *Jahrb. k. k. Geol. Reichsanstalt, 35,* pp. 151-276.

Buseck, P.R., and Goldstein, J.I., 1967. Olivine compositions and cooling rate of pallastic meteorites. *Trans. Am. Geophys. Union, 48,* pp. 165-167.

Claus, G., and Nagy, B., 1961. A microbiological examination of some carbonaceous chondrites. *Nature, 192,* pp. 594-596.

Daly, R.A., 1943. Meteorites and an Earth-model. *Bull. Geol. Soc. Am., 54,* pp. 401-456.

Dodd, R.T., Van Schmus, W.R., and Koffman, D.M., 1967. A survey of the unequilibrated ordinary chondrites. *Geochim. Cosmochim. Acta, 31,* pp. 921-951.

Fermor, L.L., 1938. On khoharite, a new garnet, and on the nomenclature of garnets. *Rec. Geol. Surv. India, 73,* pp. 145-156.

Fredriksson, K., and Mason, B., 1967. The Shaw meteorite. *Geochim. Cosmochim. Acta, 31,* in press.

Goldstein, J.I., and Ogilvie, R.E., 1965. A re-evaluation of the iron-rich portion of the Fe-Ni system. *Trans. Metall. Soc. Am. Inst. Mining Eng., 233,* pp. 2083-2087.

Goldstein, J.I., and Short, J.M., 1967. The iron meteorites, their thermal heating, and parent bodies. *Geochim. Cosmochim. Acta, 31,* in press.

Hayes, J.M., 1967. Organic constituents in meteorites. *Geochim. Cosmochim. Acta, 31,* pp. 1395-1440.

Hohenberg, C.M., Podosek, F.A., and Reynolds, J.H., 1967. Xenon-iodine dating: sharp isochronism in chondrites. *Science, 156,* pp. 202-206.

Levin, B.Y., and Slonimsky, G.L., 1958. Question of the origin of meteoritic chondrules. *Meteoritika, 16,* pp. 30-36.

Lovering, J.F., Nichiporuk, W., Chodos, A., and Brown, H., 1957. The distribution of gallium, germanium, cobalt, chromium, and copper in iron and stony-iron meteorites in relation to nickel content and structure. *Geochim. Cosmochim. Acta, 11,* pp. 263-278.

Mason, B., 1960. Origin of chondrules and chondritic meteorites. *Nature, 186,* pp. 230-231.

Mason, B., 1966. The enstatite chondrites. *Geochim. Cosmochim. Acta, 30,* pp. 23-39.

Mueller, G., 1953. The properties and theory of genesis of the carbonaceous complex within the Cold Bokkeveld meteorite. *Geochim. Cosmochim. Acta, 4,* pp. 1-10.

Nagy, B., Meinschein, W.G., and Hennessy, D.J., 1961. Mass spectroscopic analysis of the Orgueil meteorite: evidence for biogenic hydrocarbons. *Trans. N.Y. Acad. Sci., 93,* pp. 25-35.

Prior, G.T., 1920. The classification of meteorites. *Mineral. Mag., 19,* pp. 51-63.

Ringwood, A.E., 1961. Chemical and genetic relationships among meteorites. *Geochim. Cosmochim. Acta, 24,* pp. 159-197.

Schairer, J.F., and Yoder, H.S., 1960. The system albite-forsterite-silica. *Ann. Rep. Geophys. Lab., 1959-60,* pp. 69-70.

Sorby, H.C., 1877. On the structure and origin of meteorites. *Nature, 15,* pp. 495-498.

Tschermak, G., 1885. Die mikroskopische Beschaffenheit der Meteoriten. Facsimile reprint, with English translation, published 1964 in *Smithsonian Contributions to Astrophysics, 4,* pp. 137-239.

Urey, H.C., and Craig, H., 1953. The composition of the stone meteorites and the origin of the meteorites. *Geochim. Cosmochim. Acta, 4,* pp. 36-82.

Van Schmus, W.R., and Koffman, D.M., 1967. Equilibration temperatures of iron and magnesium in silicate meteorites. *Science, 155,* pp. 1009-1011.

Van Schmus, W.R., and Wood, J.A., 1967. A chemical-petrological classification for the chondritic meteorites. *Geochim. Cosmochim. Acta, 31,* pp. 747-765.

Wasserburg, G.J., Burnett, D.S., and Frondel, C., 1965. Strontium-rubidium age of an iron meteorite. *Science, 150,* pp. 1814-1818.

Wasson, J.T., and Kimberlin, J., 1967. The chemical classification of iron meteorites, 2. Irons and pallasites with germanium concentrations between 8 and 100 ppm. *Geochim. Cosmochim. Acta, 31,* in press.

Whipple, F.L., 1966. Chondrules: suggestion concerning the origin. *Science, 153,* pp. 54-56.

Paul D. Feldman

The Composition of Comets

New observational techniques give us a better view of what is probably the original composition of our solar system

The apparition of a bright comet belongs to the class of transient astronomical phenomena that have since the beginning of recorded history been regarded as celestial omens, as objects of mystery. They were also events of particular importance to the early astronomers. As early as the first part of the nineteenth century, the study of the celestial mechanics of comet orbits showed that comets were members of the solar system, and various hypotheses were put forward as to their origin, such as ejection from the sun or breakup of a planet. Later on, it appeared that many properties of the comets made their existence incompatible with that of the planets and asteroids, and an origin outside the solar system in interstellar space was considered more likely. The paucity of new comets with hyperbolic orbits—i.e. comets not gravitationally bound to the sun—has argued against this point of view. As we shall see below, these two views are not completely incompatible.

In recent years, partially stimulated by the beginning of an era of direct exploration of the planets by means of rocket-propelled space probes, there has been a great increase in interest and activity in the question of the origin and evolution of the solar system. While this is a basic cosmogonic problem, the inherent difficulties in an a priori approach can be simply seen by comparing the Earth's atmosphere, which is predominantly nitrogen and oxygen, with that of its neighbors. Mars and Venus both have predominantly CO_2 atmospheres, that of Mars being one hundred times less dense at the surface than that of Earth, and that of Venus being one hundred times more dense; Jupiter and Saturn have atmospheres made up predominantly of hydrogen, methane, and ammonia, and Mercury has no atmosphere at all. When one considers the differences in surface features of the planets, the presence of rings around Saturn, and the belt of minor planets, or asteroids, as well as the comets, it appears almost hopeless to try to account for the present solar system, with its widespread differentiation, by means of a unique self-consistent model. How would one even begin to attack such a problem?

The clues come from observational astronomy. We cannot observe the actual early evolution of a star, or a protostar, as the formative stages are known, because the changes to be seen are small over the course of a human lifetime—the evolutionary time scale being of the order of 10^4 to 10^5 years. However, when we look into our own galaxy, estimated to contain a total mass equal to $\sim 10^{11}$ solar masses, we see a statistical distribution of stars in different phases of evolution, with the probability of seeing a particular phase being proportional to the length of time spent in that phase. We also observe clouds of interstellar gas and dust, and these clouds are not distributed randomly in the galaxy but tend to be concentrated along the galactic plane. This is the same region in which hot, bright, massive stars (spectral type O and B) are seen, and since these stars have lifetimes of the order of 10^6–10^7 years, much shorter than the "age" of the sun (4.7×10^9 years), they must have formed relatively "recently" and, most likely, out of the interstellar clouds in the same area.

The number of observed protostars in the phase between cloud and luminous star is small, indicating a short lifetime ($\sim 10^4$ years) for this phase, but it must be in this phase that the differentiation of the primitive solar system occurs. Thus the basic model of a solar nebula would begin with a collapsing, probably rotating, disk-shaped cloud of about one solar mass and evolve along lines describable from the theory of stellar structure. As the central region collapses to form the sun, instabilities in the rotating gas cloud would lead to secondary condensations of the planets, satellites, asteroids, and comets. The exact mechanisms involved are not known and the various hypotheses remain hotly debated today (1).

The composition of the interstellar matter is also changing with time. As stars evolve toward the end of their lives they erupt, sometimes completely as in a supernova explosion, and eject material into space. This material is generally characterized by a high abundance of the heavier elements, created in the core of the star by nucleosynthesis during the main sequence period when the stellar luminosity is provided by the burning of nuclear fuel. The heavy-element component of the gas then is incorporated into the newly formed stars, which, in effect, do show considerably

After receiving his Ph.D. from Columbia University in atomic physics, Paul D. Feldman went to the Naval Research Laboratory in Washington, DC, to join the growing field of experimenters using rockets and other space platforms. He has continued in this endeavor at Johns Hopkins, where he is professor of physics, and has conducted experiments in space in such diverse disciplines as infrared astronomy, atmospheric and auroral physics, ultraviolet astronomy, and cometary physics. The work described in this article is supported by a grant from NASA. Address: Physics Department, Johns Hopkins University, Baltimore, MD 21218.

higher metal abundances than the older "late-type" stars, such as the sun.

Another continuous process also alters the composition within the interstellar clouds: the buildup of complex molecules through gas-phase reactions. In this case, the densities of both gas and dust must be relatively high, the gas to ensure a reasonable probability for the chemical reactions to take place and the dust to shield the molecules so formed from the destructive effect of ultraviolet radiation from the nearby hot stars.

One of the great achievements of radio astronomy over the past few years has been the detection in dense clouds of over thirty different molecules ranging from water and carbon monoxide to methyl alcohol, methyl cyanide, formaldehyde, and even more complex hydrocarbons (2). If we add the radicals CH, CN, and CH$^+$, previously identified optically, and H$_2$, identified in the far ultraviolet, we have a picture of molecular synthesis taking place in the middle of the dark clouds, with the basic stable atomic species H, O, C, and N as the fundamental ingredients. Thus the molecular composition of a dense interstellar cloud appears to have exactly the composition needed to produce the observed features of the visible cometary spectrum, which consists mainly of emission bands of unstable free radicals such as CH, CN, CH$^+$, OH, C$_2$, C$_3$, NH, and NH$_2$. Presumably these species are liberated by the action of sunlight as the comet approaches the sun, the parent molecules—the heavier hydrocarbons—being in some way embedded in the basic cometary structure.

We can carry this discussion to a logical conclusion. If, as has been supposed, the comets are formed near or beyond the present orbit of Jupiter (about 5 AU [astronomical unit] from the sun), far enough from the sun to ensure the retention of volatiles, and are then ejected into a cloud of some 150,000 AU in extent (the Oort cloud) where they pass most of their existence in the cold of interstellar space, then on their first appearance near the sun (the observed long-period comets are presumed to have their orbits modified by a nearby star in a way that takes them into the inner solar system) they will display the composition of the "primordial" solar

Figure 1. Comet West (1975 n) was photographed from an Aerobee rocket on 5 March 1976, in visible light (*left*) and in the light of hydrogen Lyman-α emission at 1216 Å (*right*). The ultraviolet hydrogen coma is much more extended than the coma visible to the eye. For comparison, note that the diameter of the sun is about one half of one degree. (From Opal and Carruthers (*19*), courtesy of C. B. Opal, Naval Research Laboratory.)

nebula—i.e. the composition of what the solar system was like at the time the comets were created. And, if this solar nebula is in fact derived from an interstellar cloud, the similarity in composition of the cometary coma is to be expected.

Yet there are difficulties with this picture; we do not observe the comet directly but only the coma produced by evaporation and potentially mixed up by the same gas-phase chemistry that produces the complex molecules in the interstellar clouds. Not all new comets are alike, some having much higher dust/gas ratios than others; and finally, the composition of the comets does not appear to be at all similar to that of the major planets, which it should resemble if the formation actually took place simultaneously.

We will consider first a working model (though not unanimously accepted) of cometary structure and the classical observations that have led to it. Then we will survey the new techniques that have only recently been applied to cometary observations and, finally, try to place the information derived from these observations in a proper perspective with respect to the cosmogonical problems discussed above.

Structure of a comet

The model of the comet given below, originally described by Whipple (*3*), evolved from attempts to fit a variety of different types of observations, such as visible spectra, visual appearance, and orbital characteristics, into a consistent picture while at the same time avoiding conflict with any basic physical principles. The model has not been verified by the direct observation of a cometary nucleus (such observation must await a future space flight to a comet), but recent developments in cometary spectroscopy have not offered any evidence to dispute the basic premises of the model. Also, it is necessary to remember that comets are not all alike; even if they were all created simultaneously, some have been considerably altered as a result of previous traversals of the inner solar system. The model can serve only as a rough guide in efforts to interpret the evolution of cometary behavior.

At large distances from the sun, the comet resembles a small planet or a large meteoroid, a solid body or several such bodies, not necessarily spherical, and with dimensions of the order of 1 to 10 km. Frozen gases and micron-size dust particles are the main constituents of this *nucleus,* giving rise to the name "icy conglomerate" or "dirty snowball." As the comet approaches the sun, solar heating results in the evaporation of the gas (the dust particles are carried along with the gas) and the formation of the visible *coma* of perhaps 10^4–10^5 km in extent. The *tail,* consisting of two components, one dust and the other ionized gas, which appears in the anti-sunward direction and may attain an extent of 2×10^7 km (one AU, the mean distance of the Earth from the sun, is 15×10^7 km), results from the interaction of the gas and dust in the coma with the solar radiation field or the solar wind.

The nucleus is most likely rotating, and since the sunward side becomes hotter than the anti-sunward side and hence evaporates gas at a higher rate, the rotation leads to a net momentum transfer component perpendicular to the sun-comet vector that produces deviations in the comet's orbit from the predictions of the inverse-square Newtonian gravitational force law. Since the evaporation velocity is relatively easy to determine, by considering the net momentum loss required to account for the inappropriately titled "non-gravitational forces," we can arrive at a total mass loss rate due to evaporation. For a "new" comet within 1 AU, this rate is about 10^{30} molecules per second, or roughly 30 tons per second.

Assuming the brightness of a comet to be proportional to the rate of evaporation from the nucleus, a study of the brightness variation with heliocentric distance—taking into account the solar radiant flux at the comet, the equilibrium temperature of the nucleus (assuming it to behave as a blackbody), and the latent heats of vaporization of possible cometary molecules—leads to the conclusion that the primary constituent must be water (4). Any other possible parent molecule, such as CH_4 or CO_2, is much too volatile to survive at distances close to the sun, where the comets appear at their brightest and most spectacular. However, some amount of volatile material must exist, since many comets are seen to have a diffuse appearance—suggesting the presence of a gaseous coma—as far as 5 AU from the sun, indicating the presence of a molecule with a significant vapor pressure at 125°K.

The fact that these volatiles persevere through perihelion passage led Delsemme and Swings (4) to postulate that the volatile molecules existed in the form of hydrates, in which one molecule was chemically bound in a solid lattice to six water molecules, and that the evaporation was controlled by the vaporization heat of water. That different comets show different brightness dependence on heliocentric distance and that a given comet may increase or decrease by one or two magnitudes in brightness after passing around the sun indicate that no unique composition characterizes all comets.

Spectrum of a comet

One of the most important consequences of space research to astronomy (which is still to be realized in the form of an orbiting observatory commensurate in quality and versatility with the best ground-based telescopes) is the freeing of the astronomer from the narrow confines of the electromagnetic spectrum transmitted by the terrestrial atmosphere. Traditional astronomy, which is limited to the spectral region from 3000 to about 10,000 Å, is a remarkably successful discipline given the large extent of electromagnetic phenomena not accessible to the ground-based observer.

Consider then the comet observer, who on obtaining a spectrogram of a cometary coma finds not a trace of any molecule or atom that exists as a stable gas on the Earth but only simple free radicals, now identified as OH, C_2, CN, C_3, CH, NH, and NH_2. In the case of tail spectra, ions of more familiar species—N_2^+, CO^+, and CO_2^+ (and, most recently, H_2O^+) —are observed. In sun-grazing comets, those coming within 0.1 AU of the sun, lines of various metals—Fe, Ni, Cr, etc.—similar to those found in meteor spectra appear, most likely coming from the vaporization of the dust particles themselves. But clearly, it is the presence of the free radicals observed in the coma that presents the most interesting clues to the composition of the cometary snows.

The observed radicals are all highly reactive and, if produced at densities at which collisions are frequent, will ultimately all be converted to more stable molecules. For this reason, they are extremely difficult to study in the laboratory, and, in fact, many were first identified on the basis of analysis of high-resolution cometary spectra. Because of their reactiveness, it is highly unlikely that they could exist in frozen form within the cometary snowball for the length of time since the formation of the comet, and so they are presumed to be produced by the photodissociation of "parent" molecules at some 10^3–10^4 km from the nucleus, where collisions are already negligible. Since the parents are not observed, their nature remains speculative, although in many cases the assignment seems obvious; CH from CH_4, NH from NH_3, and OH from H_2O.

It was gratifying to see the presence of CH_4 and NH_3 inferred from the observations of the radicals. Both are important constituents of the atmospheres of the giant planets, and both are likely to be present in large abundance if the comets were formed near the orbits of Jupiter or Saturn. From the nature of the band spectra of the radicals, when observed at modestly high spectral resolution, it was possible to ascertain that the emission mechanism responsible for the radiation was resonance fluorescence of solar radiation. During the lifetime of one radical (it is important here to realize that the radical itself can be photodissociated into its component atoms), it has a finite probability—determined by its quantum mechanical oscillator strength and the solar flux at the particular wavelength—of absorbing a solar photon and then reemitting another photon isotropically into space. If we know these parameters, then a measure of the intensity of a spectral line in the coma gives a measure of the density of radicals (and their spatial distribution) and, consequently, the rate at which they are produced.

Then, arguing that the production rate of radicals must accurately reflect the rate at which the parent molecules are being produced (although in some cases there are multiple dissociation paths for a given parent), the evaporation rate of the unseen parents can be deduced. If we exclude ground-based observations of the 3090 Å emission of OH, which is severely attenuated by atmospheric ozone, the production rates deduced for all the parents of the visible radicals (excepting H_2O) amount to at most 1% of the evaporation rate needed to produce the "non-gravitational" accelerations. Only the OH parent, H_2O, is present in sufficient abundance to provide the necessary momentum, and yet the above-mentioned atmospheric ozone precludes a ground-based determination of this rate. Thus the spectrum of the comet visible from the ground lets us glimpse only the relatively minor constituents of the cometary atmosphere.

Inherent in this approach to determining cometary abundances is the assumption that the daughter radicals that are observed have direct lineage with the volatile parents liberated from the cometary nucleus. Thus, the effects of gas-phase chemical reactions within the collisional zone of the coma (again 10^3–10^4 km, depending on the total gas production rate) are assumed to produce only minor perturbations of the original composition. This assumption may not be valid, and Oppenheimer (5) has shown that it is possible to start with a solid methane comet, a minor source of oxygen atoms, and a suitable choice of reaction rate coefficients (most of which have not been measured in the laboratory) and arrive at a composition of radicals and ions that closely resembles the observed cometary abundances. Caution is thus warranted in the interpretation of the visible spectrum.

We must turn to other spectral regions if we are to probe further into the composition of the cometary coma—to radio astronomy for direct observations of the complex parent molecules and to the vacuum ultraviolet for observations of the basic atomic constituents produced by successive photodissociations of parent and daughter molecules. Both techniques have only recently been applied to the observations of comets, with impressive results, and more is expected in the future.

Radio observations

In the past ten years, radio astronomers working in the centimeter and millimeter wave ranges of the spectrum have compiled a catalog of over thirty molecules detected by their emission in dusty and, consequently, cool interstellar clouds (2). Taken along with H_2 (the most abundant interstellar molecule detected in absorption in the far ultraviolet from above the Earth's atmosphere), these molecules, with a few notable exceptions, are formed of elementary combinations of four basic elements—H, O, C, and N—with CO being the second most abundant molecule, two orders of magnitude larger than any of the others. These molecules are found in large abundance only in the dusty clouds since the dust grains distributed through the cloud serve to protect the molecules from the destructive effects of the interstellar ultraviolet radiation field. Note the similarity with the observed cometary coma, in that dust grains always appear to accompany the cometary gas, although the gas/dust ratio is highly variable from comet to comet.

Naturally, the similarity in composition between dark interstellar clouds and the cometary coma led to radio searches for cometary emissions. The first successful detection of such emission was made in December 1973–January 1974 with the observation of OH, CH_3CN, HCN, and CH in comet Kohoutek (1973 XII) (6). A search for H_2O radio emission in comet Kohoutek was unsuccessful, but a positive detection of H_2O in comet Bradfield (1974 b) in May 1974 was later reported (7). Not only are the cometary emissions weak due to the limited number of emitting molecules along the line of sight, but the interpretation of the radio emission in terms of the production rates of the molecules is potentially hazardous unless the excitation mechanisms are completely understood. In the interstellar cloud, the characteristic time between collisions is shorter than the radiative lifetime of the molecule, and thermal equilibrium can be assumed in the calculation of the population of excited states.

Thermal equilibrium cannot be assumed in the cometary atmosphere, however, where the escape time from the collisional zone is short and where the distribution of quantum levels can be significantly altered by optical pumping effects of the solar radiation. To illustrate this, we cite the observations of OH emission from comet Kohoutek, which were made over a period of several weeks. Pre-perihelion, the OH lines appeared as absorption lines in the continuum radio background; then they decreased in intensity and reappeared post-perihelion as emission lines (8). A similar change in behavior in the intensity distribution of individual lines in the optical band at 3090 Å has been observed in several comets (9) and was explained on the basis of the Doppler shift of Fraunhofer lines (absorption features) in the solar spectrum, such that selected levels would absorb the solar radiation and fluoresce and the presence of a given component would depend on whether or not that component coincided with an absorption dip in the solar flux. The spectral width of these features is such that different lines would be coincident when the radial velocity had the opposite sign. It is this fluorescence that selectively populates either the upper or lower radio levels and produces the change from absorption to emission (10). Accurate knowledge of the pumping mechanism makes possible a determination of the OH production rate, and a recent observation of OH radio emission from comet West (1975 n) gives good agreement with the ultraviolet measurements discussed below (11).

However, it is not OH, readily observable in the ultraviolet, that is most important in the radio range; rather, it is the triatomic and larger molecules that are not easily accessible by optical techniques but for which the excitation mechanisms are not known. Unfortunately, no advantage was taken at the apparition of the bright comet West (1975 n) to pursue further the radio search for cometary parent molecules begun with comet Kohoutek (6). With improvements in receiver sensitivity, we may expect identifications of further molecules at the next apparition of a bright comet and perhaps, with further data from molecular spectroscopists, determination of the production rates of these molecules as well.

Ultraviolet observations

In the spectral region below 3000 Å, ozone and oxygen in the Earth's atmosphere absorb all the extraterrestrial radiation so that it is necessary to use a rocket- or satellite-borne instrument for observations in this region. Yet it is certainly worth the effort since the four basic atomic ingredients—H, O, C, and N—all have their strongest spectral emission lines (resonance transitions) between 1200 and 1700 Å, and several simple stable molecules formed from these atoms— CO, NO, and H_2—are also strongly radiating at these wavelengths. Emission from these constituents can be produced by various mechanisms, including resonance scattering, electron excitation, or photodissociative excitation by solar ultraviolet radiation. An important aspect of this spectral range is that the solar output is relatively weak: the albedo, or reflectivity, of most materials is also fairly low, and reflected sunlight will be much fainter than the gas emission. This feature permits the study of a rather tenuous planetary atmosphere against the background of the solar illuminated disk and has been successfully applied to the study of planetary atmospheres from rockets and, even more spectacularly, from interplanetary probes such as the Mariner Mars orbiters and the Pioneer Venus/Mercury flyby missions (12).

Perhaps the most important of these emission features is the Lyman-α transition of atomic hydrogen at 1216 Å. (In the spectra of single atoms, lines appear in ordered series, the intensity and spectra of the lines decreasing until a series limit is reached. The strongest line of the series, found at the longest wavelength, is denoted by α, the next by β, etc. In hydrogen the different series are named for their discoverers.) Since this hydrogen emission is particularly intense in the solar spectrum (about four orders of magnitude brighter than the adjacent continuum) and the oscillator strength for the transition is large, resonance scattering of Lyman-α is a particularly sensitive means of detecting the presence of atomic hydrogen and has been exploited extensively in the study of the hydrogen

geocorona surrounding the Earth (13), as well as that of interplanetary hydrogen thought to be swept into the solar system by the relative motion of the sun through interstellar space (14). A search for atomic hydrogen in the lunar atmosphere from Apollo 17 led to an upper limit of 10 atoms per cm^3 at the lunar surface (15), which is to be compared to a number density of N_2 and O_2 of 3×10^{19} cm^{-3} at the surface of the Earth.

The first observations of a comet in the light of hydrogen Lyman-α were made in 1970 when two bright comets, Tago-Sato-Kosaka (1969 IX) and Bennett (1970 II), were observed, first by the Orbiting Astronomical Observatory (OAO-2) (16), and then by the Orbiting Geophysical Observatory (OGO-5). In the latter case, a sensitive photometer, designed to map the distribution of geocoronal hydrogen, was able to measure the hydrogen coma of comet Bennett out to a distance of 10^7 km from the center of the comet (17). Later that year, the periodic comet Encke (with a well-known period of three years), a likely candidate for a cometary flyby mission, was also observed in Lyman-α.

Comet Kohoutek (1973 XII), which was discovered some ten months before perihelion passage, aroused great expectations in both the astronomical community and the general public. It was observed between December 1973 and January 1974 by instruments aboard a large number of spacecraft, and the first images of the Lyman-α coma were recorded by electrographic ultraviolet cameras aboard both Skylab and a sounding rocket (18). Figure 1 shows a Lyman-α photograph of the bright comet West (1975 n) taken in March 1976, together with a visible photograph recorded a few minutes later and reproduced to the same scale (19). The isophotes of constant brightness, derived by combining the data from several exposures of different duration, are shown in Fig. 2.

The intensity and distribution of the Lyman-α emission could easily be accounted for from the icy-conglomerate model and provided the first direct evidence for the mass outflow required to explain the "non-gravitational" motions seen in most comet orbits. The observations of comet Bennett by OAO-2 also provided a

measure of the unattenuated OH emission at 3090 Å, and a comparison of derived H and OH production rates seemed to justify the assumption of H_2O as the common parent molecule (20). The distortion of the Lyman-α isophotes—the elongation in the anti-sunward direction—could be qualitatively explained by the fountain model, originally invoked to explain certain features of the visible coma (21).

In this model, the atoms are assumed to originate in a point source at a rate Q per second and to flow radially outward, isotropically, with velocity v. If we consider only those atoms produced in a small time interval, as they flow outward from the source they will constitute a spherical shell and their density in the shell will be proportional to ρ^{-2}, where ρ is the radius of the shell at time t, given by $\rho = vt$. In addition, there will be a decrease in density as t increases, due to the finite probability of the atom being ionized either by solar ultraviolet radiation or by collision with solar wind protons, and this decrease can be represented by $e^{-\gamma t}$, where γ is the total loss probability. To this, the fountain model adds the effect of radiation pressure, an effect that produces an acceleration in the anti-sunward direction as a result of the transfer of momentum from solar photons during the resonance scattering process.

The particle trajectories thus resemble the paths of water droplets ejected isotropically by a fountain and then drawn downward by the Earth's gravity. Since all the atoms in a given shell are acted on by this force for the same length of time, the effect can be considered as an acceleration of the center of each shell, so that the observed Lyman-α coma can be thought of as the superposition of a large number of emitting shells, each one larger than the preceding one and with its center displaced a fixed amount in the anti-sunward direction. This can be represented schematically as in Figure 3, which can be seen to reproduce the general features of the observed isophotes quite well. The actual solution is complicated by changes in comet–sun distance and velocity, the finite size of the source region, and radiation trapping effects, but in principle it is possible to derive from the data the unknown parameters Q, v, and γ uniquely.

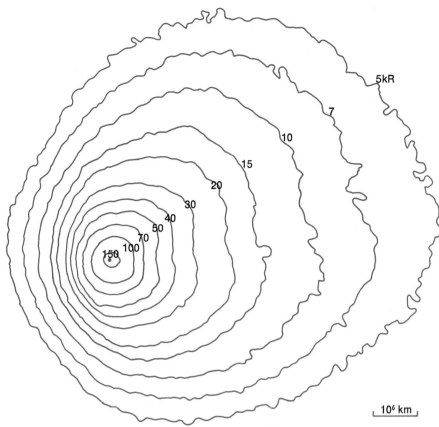

Figure 2. The hydrogen Lyman-α isophotes (equal intensity contours) for comet West (1975 n) are derived by combining the data from several exposures of different duration. The units are kiloRayleighs (kR), a measure of the surface brightness of the coma. Note the scale on the diagram; the diameter of the sun is 1.3×10^6 km. (From Opal and Carruthers (19), courtesy of C. B. Opal, Naval Research Laboratory.)

Two factors contribute to the very large extent of the observed hydrogen coma: the relatively low probability for ionization of hydrogen and a large outflow velocity. The latter results from the requirement of energy and momentum conservation in the dissociation of the parent molecule, with the lighter product having a velocity higher by a factor of the ratio of the masses of the two products. Detailed analysis of the Lyman-α isophotes

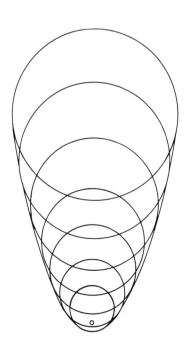

Figure 3. The elongation of the Lyman-α isophotes in the anti-sunward direction can be explained by the fountain model, schematized here. Atoms originate at a point and flow radially outward isotropically. Solar radiation pressure acts to displace each circle away from the sun. What is observed is the superposition of many such circles, each displaced by an amount depending on the time of its formation.

from comet Bennett (22) shows, in fact, that two high-velocity components—8 km sec^{-1} and 20 km sec^{-1}— are needed to fit the data. These correspond to the different mean energies available from the two sources of hydrogen:

$$H_2O + h\nu \text{ (solar)} \rightarrow OH + H$$

$$OH + h\nu \text{ (solar)} \rightarrow O + H$$

where $h\nu$ represents a solar ultraviolet photon of sufficient energy to dissociate the molecule. The velocity of the heavier product is comparable to the thermal radial outflow velocity of the parent molecules, 0.5–1.0 km sec^{-1}.

A summary of the hydrogen production rate, Q_H, for several comets that have now been observed in Lyman-α is given in Table 1. Relative to these production rates, which are of the correct order of magnitude required by the "non-gravitational" effects, the production of the radicals observed in the visible spectrum represents only about 1% of the total cometary gas production.

Note that periodic comet Encke, which has been in the inner solar system for a long time, seems to have lost most of its volatile material; this is borne out by the feebleness of its hydrogen coma. For many of the comets given in Table 1, observations of the OH emission were also made, and the derived production rates are also included. If water is the predominant source of the hydrogen in the coma, then we expect to find

$$Q_{H_2O} = Q_{OH} = \frac{1}{2} Q_H$$

and this is largely supported by the data. Of particular interest are the observations of H and OH in comet Bennett, made by OAO-2 over a period of one month during which the heliocentric distance of the comet, r, increased from 0.77 to 1.26 AU (20). The dependence of Q on r could be determined independently for the two species, and it was found that $Q \propto r^{-2.3}$ for both, suggesting a common parent. Moreover, since the solar radiation input varies as r^{-2}, the closeness of the Q dependence to solar input suggests that the H and OH parent is vaporized in direct response to solar heating and hence must be a primary constituent of the comet. Thus, the evidence of the ultraviolet observations of H and OH, while rig-

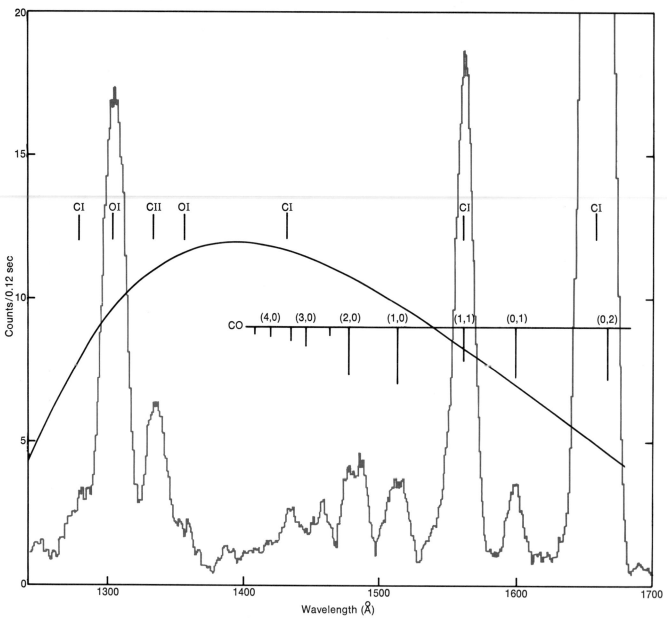

Figure 4. This ultraviolet spectrum of comet West (1975 n) from 1250 to 1700 Å was obtained during the rocket flight of 5 March 1976. The atomic species are identified by their chemical symbol and I for neutral species, II for singly ionized. For a diatomic molecule like CO, the emission is in the form of bands, each containing several (but, in this case, unresolved) lines. Though more complex than the atomic spectrum, the molecular spectrum, indicated by the superimposed scale, assures a positive identification of the emitting species. The smooth curve shows the response of the instrument to a source of constant brightness at all wavelengths. (From Feldman and Brune [24].)

orously still only circumstantial, appears to confirm the principal features of the icy-conglomerate model and, in any case, presents no new evidence in contradiction to the model.

The OAO-2 spectra of comet Bennett suggested the presence of other atomic emission features, but since the instrument was an objective grating spectrograph, these weak features were superimposed on the large Lyman-α signal recorded because of the large extent of the hydrogen coma. The first positive identifications of atomic carbon and oxygen emissions were made by two rocket experiments launched in January 1974 to observe comet Kohoutek (18, 23) and provided the first indication of significant carbon abundance relative to hydrogen or water.

These observations were repeated, with much improved results, on comet West in March 1976 (24), a comet that to the naked eye fulfilled all the expectations aroused for comet Kohoutek, and these results will be discussed briefly to illustrate some of the problems underlying the determination of production rates and, ultimately, the relative abundances.

Moderate resolution spectra of comet West are shown in Figures 4 and 5. In the range 1250–1700 Å, the resolution is 15 Å, and in the range 1800–3200 Å, it is 22 Å. The short wavelength spectrum of Figure 4 is particularly interesting in that the principal features are due to atomic carbon, atomic oxygen, and the most probable parent, carbon monoxide. The ultimate decay product of carbon, the ion C^+, is also observable by its emission at 1335 Å. Thus, all the species in-

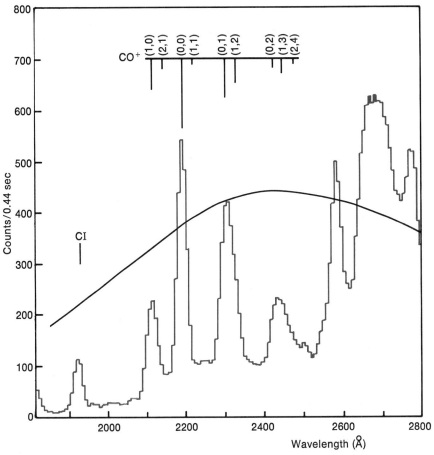
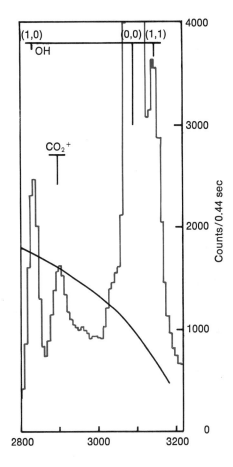

Figure 5. Ultraviolet spectrum of comet West (1975 n) for the spectral range 1800–3200 Å. Note again that whereas a single line is observed for an atomic transition, the radiation from CO$^+$ and OH is in the form of a series of bands. For OH, the (0,0) band is very intense at 55,000 counts/0.44 sec. The smooth line here, as in Figure 4, shows the response of the instrument to a source of constant brightness at all wavelengths. (From Feldman and Brune [24].)

volved in a particular photochemical chain of dissociation and ionization are observed simultaneously.

The long wavelength ultraviolet spectrum contains emissions of CO$^+$, CO$_2^+$, and OH, permitting a direct determination of the C/H ratio. At first appearance, this spectrum looks deceptively like the spectrum of Mars (12) in the same spectral region, which, though consisting predominantly of CO bands, is the spectrum of an almost pure CO$_2$ atmosphere— the CO bands being forbidden are excited indirectly during the dissociation of CO$_2$. Also, it should be noted that molecular hydrogen, if present either as an original constituent or as the result of the dissociation of methane, can be detected by fluorescence in the Lyman bands, the strongest of which appears near 1608 Å (25). The only basic species not detectable in these spectra is nitrogen, whose resonance transition at 1200 Å is masked by the intense Lyman-α line of hydrogen. (In fact, in our instrument a CaF$_2$ filter is used to reduce the Lyman-α signal by a factor of 100 to prevent instrumental scattering of Lyman-α from masking the weaker long wavelength emissions.)

Table 1. Species production rates derived for several comets

	Heliocentric distance r (AU)	Production rates (10^{29} sec^{-1})			
		Hydrogen (Q_H)	OH (Q_{OH})	Oxygen (Q_O)	Carbon (Q_C)
Tago-Sato-Kosaka (1969 IX) (30)	1.0	4.3	1.3	—	—
Bennett (1970 II) (20, 17)	1.0	5.4	2.0	—	—
	0.81	8.8	—	—	—
p/Encke (periodic comet) (17)	0.7	0.06		—	—
Kohoutek (1973 XII) (27, 23)	0.43	6.0	—	1.5	0.9
	0.34	5.4	0.8	1.2	0.5
West (1975 n) (19, 24)	0.39	32	—	—	—
	0.39	—	9.6	11	3.1

However, for hydrogen, oxygen, and carbon, the ultraviolet spectra allow a fairly complete picture of their photochemistry to be drawn.

The quantity of interest for each species is the production rate Q_i expressed in atoms or molecules per second for the ith species. If the mean lifetime of this species is τ_i, at any time the coma will contain N_i of this species, where

$$N_i = Q_i \tau_i$$

It is customary to express the efficiency of the resonance fluorescence or scattering in terms of a "g-factor," g_i, which represents the probability that in one second a given atom or molecule will absorb and reemit a solar photon at the wavelength of its resonance transition. The g-factor includes the oscillator strength and the solar flux at that wavelength. Thus, if we know g_i, and if we assume that the coma is optically thin in a given transition—i.e. the coma does not become opaque because of the large number of absorbers in it—the luminosity at the wavelength λ_i will be given by

$$L_i = g_i N_i = g_i Q_i \tau_i$$

Finally, if the whole coma is observed, the received flux at the Earth (distance Δ from the comet) is

$$F_i = \frac{L_i}{4\pi\Delta^2} = \frac{Q_i g_i \tau_i}{4\pi\Delta^2}$$

By inverting this equation, the production rate can be determined from the observed flux. Note that both g and τ depend on the solar flux, which varies as r^{-2}. The lifetime τ also can depend on the flux of solar wind particles, which also follows an inverse square law in r. Since g increases close to the sun, while τ decreases, the product $g\tau$ is independent of r and can be conveniently evaluated at $r = 1$ AU.

The heliocentric velocity of the comet produces a Doppler shift in the apparent wavelength of the exciting solar radiation, which must be taken into account in the evaluation of the g-factor (26). The maximum velocity can be written easily from orbital mechanics as

$$\dot{r}_{max} = \left(\frac{M_\odot G}{q}\right)^{1/2}$$

$$= 21\, q^{-1/2}\ \text{km sec}^{-1}$$

where $M_\odot$ is the solar mass, G the gravitational constant, and q the comet's perihelion distance in AUs. Expressed in terms of a wavelength shift,

$$\left(\frac{\Delta\lambda}{\lambda}\right)_{max} = \frac{\dot{r}_{max}}{c} = 7 \times 10^{-5}\, q^{-1/2}$$

For comet Kohoutek, with $q = 0.14$, the maximum Doppler shift at 1302 Å—the oxygen resonance transition—is 0.24 Å.

The most favorable viewing geometry occurs when the comet has nearly its maximum radial velocity. Thus for both comets Kohoutek and West, the ultraviolet spectra were recorded when the Doppler shift was sufficient to shift the solar line by more than its halfwidth—in the case of oxygen, 0.20 Å—and hence no resonance scattering should occur since the underlying continuum in the solar spectrum is extremely weak. Yet a strong signal at 1302 Å is observed in the spectrum of Figure 4, and a monochromatic image of the coma at this wavelength was recorded for comet Kohoutek (27). The large extent of this image also implied that the source of excitation could not be collisional, which would have confined the emission to the inner coma, but rather must derive from the incident solar radiation.

The resolution of this problem lies with a mechanism first proposed in the 1930s to account for the emission of certain lines in gaseous nebulae that occurs when there are accidental coincidences between spectral lines of different elements. In this particular case, the "Bowen fluorescence mechanism" utilizes the coincidence between the hydrogen Lyman-β line at 1025.72 Å and an oxygen line at 1025.77 Å. Absorption of the solar Lyman-β photon is followed by cascade, as shown in Figure 6, with a significant fraction of the initial photons emerging at 1302 Å. The shape of the solar Lyman-β line must be considered in calculating the effective fluorescence g-factor, but since the line width is of the order of 0.5 Å, the shift due to the comet's velocity is not as significant.

Not surprisingly, when this g-factor is used to evaluate the production rate of oxygen, as given in Table 1, the value of Q_O is found to be comparable to that of Q_H and Q_{OH}. One doubt about the proposed mechanism remains, though, in that the cascade transition at 8447 Å, which should have the same flux as the ultraviolet

1302 Å line, has not been observed in any ground-based cometary spectra—but this may be the result of the rather extended oxygen coma, which would escape detection by standard cometary spectroscopy. One can imagine that near perihelion, as the heliocentric velocity decreases (and the tangential component increases) and the solar line comes into coincidence with the cometary oxygen atoms, that the comet would appear extremely bright at 1302 Å.

The case of the oxygen transition represents the extreme effect of a Doppler shift completely off a resonance line, due to the comet's heliocentric velocity component. For the carbon resonance transition, which is actually a multiplet of six individual lines spread out over 2 Å, the velocity change produces a factor of 2 variation in g-factor, which can easily be evaluated (26). It is slightly more difficult to obtain accurate g-factors for molecular transitions, because the detailed structure of the molecular ground state must be included, and this requires knowledge of the "temperature" of the molecule. High-resolution visible spectra, such as those of OH, suggest very low temperatures, of the order of 200°K, which greatly eases the problem. Again, attention must be paid to the details of the solar spectrum, both for emission lines and, at wavelengths longward of 1700 Å, absorption features (10).

The lifetimes of the various species are also uncertain, in some cases by perhaps a factor of two. Many of the laboratory measurements that are needed to evaluate the probability of ionization or dissociation by solar ultraviolet radiation or by collisions with solar wind particles have not yet been made, and there are still uncertainties in the absolute value (and the variation with time) of the solar extreme ultraviolet flux. Monochromatic images or slit spectrograms can be used to determine a *scale length* from the decrease in intensity with distance from the center of the coma, but to extract the lifetime, a model that assumes a radial outflow velocity must be used. Despite these limitations, a fairly consistent picture of atomic production is beginning to emerge, based largely on the data from comet West. Future comets will show whether or not it is correct.

In Table 2 we summarize the pro-

Table 2. Production rate of common species in the coma of comet West

Species	Production rate Q $(10^{29}\ sec^{-1})$
H	32
OH	9.6
O*	14
C*	4.2
CO	4.2
CO_2	$\leqslant 2.3$
H_2	$\leqslant 1.0$

*Corrected for limited field of view of spectometer.

duction rates of the molecular species observed in comet West, as well as those for the atoms already included in Table 1. For CO_2 and H_2, these are upper limits from the data. The results are remarkable and must be considered fortuitous to the degree of consistency. Clearly,

$$Q_{H_2O} = Q_{OH} = \frac{1}{2} Q_H$$

$$Q_{CO} = Q_C$$

$$Q_O = Q_{H_2O} + Q_{CO}$$

Because of the uncertainties in g and τ, the upper limit for CO_2 must be treated cautiously, but the implication is that CO is a major constituent of the cometary snow. Further evidence for CO as the ultimate parent comes from the 1931 Å line observed, which originates on a metastable state of atomic carbon. This state is populated mainly by recombination of CO^+ ions and electrons, which requires a large source of CO^+ near the nucleus, and consequently suggests direct evaporation of CO rather than evaporation of CO_2 followed by photodissociation.

The upper limit on H_2 production is also an upper limit on the production rate of methane that would be photodecomposed into H_2 and CH_2. Thus methane is no more than 10% as abundant as water, and this is well in keeping with the relatively low production rates of the free radicals observed in the visible. Neglecting methane and the other hydrocarbons that may be present in a total abundance of about 1%, we find the observed abundances to be consistent with a cometary composition that is 70% H_2O and 30% CO or CO_2 (there is some evidence from comet Bennett that CO_2 was more abundant than

water [28], but each comet must be considered by itself, and the corroborating evidence from the ultraviolet spectra of comet Bennett is not very convincing). With these two major constituents, the formation of the two dominant cometary ions, CO^+ and H_2O^+, can be explained without the need to invoke complex gas-phase chemistry, even though it does appear that H_2O^+ ions are more readily destroyed by reactions with H_2O in comets with large gas-production rates.

Assuming that the production rates Q accurately reflect the total composition of the comet (remember that we observe only what evaporates from the surface), the relative abundance of oxygen to carbon is 3.4. A similar, though less certain, result was obtained from comet Kohoutek (27). This can be compared with the most recent evaluation of the relative O/C solar abundance as 2 (29), which is derived from analyses of meteorites of the carbonaceous chondrite class. These meteorites are presumed also to be representative of the undifferentiated composition of the primitive solar nebula, although there is the possibility that they are the residue left by the passage of earlier comets through the inner solar system. Hydrogen, which is cosmically a thousand times more abundant than oxygen, could have existed in the comet (or in the precursor nebula) primarily in the form of H_2, which is extremely volatile even at the temperatures of interstellar space, and would not have survived to the present time. The agreement between the O/C ratio found in comet West with the solar abundance ratio suggests that this comet at least does in fact represent the primordial composition of the solar system.

The large carbon abundance relative to water in comets Kohoutek and West requires some revision in the details of the cometary nucleus. In the clathrate model (4), water is the dominant species and controls the rate of evaporation of all the other species, which are supposed to exist as hydrated molecules. For instance, a molecule of CO would be chemically bound to a cluster of six water molecules and the entire cluster would be sited in the ice lattice. However, six is the minimum number of H_2O molecules needed to form a bound hydrate, and thus a relative abundance

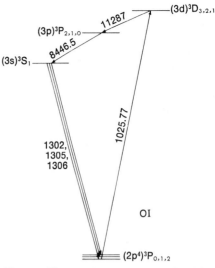

Figure 6. This atomic oxygen energy level diagram shows the fluorescence mechanism responsible for cometary 1302 Å emission. The solar Lyman-β line at 1025.72 Å coincides closely with the oxygen line at 1025.77 and excites the oxygen atoms to the ^{3}D state. About 65% of the time the atoms reradiate at this same wavelength, but in the remaining 35% the cascade indicated by the arrows results, with ultimate emission of a 1302 Å photon. (From Feldman et al. [26].)

of a carbon-bearing molecule to water of 3/7 cannot be accommodated by this model. Before this question can be resolved, it is necessary to know how the rate of carbon production varies with the heliocentric distance of the comet relative to the production rate of water. The necessary observations have so far not been made but are clearly within the capabilities of the next generation of satellite astronomical observatories.

Despite great progress made in recent years, comets still retain a measure of their former mystery. Although much has been learned about particular comets through recent advances in observational techniques, the basic questions of origin and structure still lack definitive answers. Each new comet provides new opportunities for additional observations, but each answered question seems to be accompanied by two new unanswered ones. Perhaps our state of knowledge is limited by selection effects (since we can only observe those few comets that pass near the sun) and the limited observing time available, but then again, perhaps we can make up for this by means of a direct mission of exploration to a comet—a project not possible in the foreseeable future for stars, nebulae, or galaxies.

References

1. H. Reeves, ed. 1974. *On the Origin of the Solar System.* Paris: CNRS.

2. W. D. Watson. 1976. *Rev. Mod. Phys.* 48:513; see also G. H. Herbig. 1974. Interstellar smog. *Am. Sci.* 62:200.

3. F. L. Whipple. 1950. *Astrophys. J.* 111:375; 1951. 113:464.

4. A. H. Delsemme and P. Swings. 1952. *Ann. Astrophys.* 15:1.

5. M. Oppenheimer. 1975. *Astrophys. J.* 196:251.

6. W. F. Huebner, D. Buhl, and L. E. Snyder. 1976. *Astron. J.* 81:671.

7. W. M. Jackson, T. Clark, and B. Donn. 1976. In *The Study of Comets,* ed. B. Donn et al., p. 272. NASA SP-393.

8. F. Biraud, G. Burgois, J. Crovisier, R. Fillit, E. Gerard, and I. Kazès. 1974. *Astron. and Astrophys.* 34:163.

9. A. L. Lane, A. N. Stockton, and F. H. Mies. 1975. In *Comet Kohoutek,* ed. G. A. Gary, p. 87. NASA SP-355.

10. F. H. Mies. 1974. *Astrophys. J. (Letters)* 191:L145.

11. L. E. Snyder, J. C. Webber, R. M. Crutcher, and G. W. Swenson, Jr. 1976. *Astrophys. J. (Letters)* 209:L49.

12. C. A. Barth, A. I. Stewart, C. W. Hord, and A. L. Lane. 1972. *Icarus* 17:457.

13. J. L. Bertaux and J. E. Blamont. 1973. *J. Geophys. Res.* 78:80.

14. J. L. Bertaux and J. E. Blamont. 1971. *Astron. and Astrophys.* 11:200.

15. W. G. Fastie, P. D. Feldman, R. C. Henry, H. W. Moos, C. A. Barth, G. E. Thomas, and T. M. Donahue. 1973. *Science* 182:710.

16. A. D. Code, T. E. Houck, and C. F. Lillie. 1972. In *The Scientific Results from Orbiting Astronomical Observatory (OAO-2),* ed. A. D. Code, p. 109. NASA SP-310.

17. J. L. Bertaux, J. E. Blamont, and M. Festou. 1973. *Astron. and Astrophys.* 25:415.

18. C. B. Opal, G. R. Carruthers, D. K. Prinz, and R. R. Meier. 1974. *Science* 185:702.

19. C. B. Opal and G. R. Carruthers. In press. *Icarus.*

20. H. U. Keller and C. F. Lillie. 1974. *Astron. and Astrophys.* 34:187.

21. L. Haser. 1957. *Bull. Acad. Roy. Sci. Belgique* 43:740; 1966. *Cong. Coll. Univ. Liège* 37:233.

22. H. U. Keller and G. E. Thomas. 1975. *Astron. and Astrophys.* 39:7.

23. P. D. Feldman, P. Z. Takacs, W. G. Fastie, and B. Donn. 1974. *Science* 185:705.

24. P. D. Feldman and W. H. Brune. 1976. *Astrophys. J. (Letters)* 209:L45.

25. P. D. Feldman and W. G. Fastie. 1973. *Astrophys. J. (Letters)* 185:L101.

26. P. D. Feldman, C. B. Opal, R. R. Meier, and K. R. Nicolas. 1976. In *The Study of Comets,* ed. B. Donn et al., p. 773. NASA SP-393.

27. C. B. Opal and G. R. Carruthers. 1977. *Astrophys. J.* 211:294.

28. A. H. Delsemme and M. R. Combi. 1976. *Astrophys. J. (Letters)* 209:L149.

29. A. G. W. Cameron. 1973. *Space Sci. Rev.* 15:121.

30. H. U. Keller and C. F. Lillie. Submitted to *Astron. and Astrophys.*

Elbert A. King

The Origin of Tektites: A Brief Review

Pieces of the Moon? Some researchers thought so, but the lunar samples have completely dismissed this claim

During the past forty or so years, few topics have been characterized by such disagreement and acrimonious debate in the scientific community as the origin of a group of curious, natural glassy objects called tektites. The answer is now clear—tektites are produced from terrestrial rocks melted by hypervelocity impacts of large, extraterrestrial objects. Now that we have resolved the question of origin, it seems in order to review the steps that have led most of us to this conclusion (a few diehards still tout an extraterrestrial origin), as well as to point out some of the remaining uncertainties.

Tektites were described in the literature as early as A.D. 950 (Barnes 1969) and have been known and utilized as implements since prehistoric times (Suess 1916). They are pieces of natural glass, ranging in size from microscopic grains to masses weighing tens of kilograms. Most tektites are dark brown in transmitted light (many are so dark brown that they appear black in reflected light), but some are olive-brown and others are various shades of green. Tektites are found scattered about the surface of the ground or mixed with geologically young sediments in four major geographic areas (Table 1). However, the location and distribution of tektites have not been easy to relate to local geologic features or phenomena, and this has led to much of the speculative debate about origins. Tektites sometimes are found in shapes that suggest fluid splash forms; some have been reworked by streams into abrasion-rounded pebbles; also, tektites commonly have finely sculptured surfaces (Figs. 1 and 2) that result primarily from the etching and partial dissolution of the glass by groundwater and soil acids.

The difficulty of the problem of the origin of tektites is attested to by the large number of published papers that suggest (or dogmatically defend) radically different origins for them, only a *very few* of which are listed here. Suess (1900), who coined the word *tektite* from a Greek root meaning melted or molten, suggested that they were glass meteorites. His work was widely read, and his well-known name and reputation caused most universities and museums to house collections of tektites with meteorites; many institutions still do so. Berwerth (1917) believed them to be artificial glass; Darwin (1844) considered them terrestrial volcanic glass; Easton (1921) thought them dried siliceous gels; Vogt (1935) and Barnes (1939), lightning-fused dust or surface soil; Hardcastle (1926) and Hanus (1928) viewed them as meteoritic ablation droplets; Ehmann and Kohman (1958), as meteorites from outside the Solar System; Kahn (1947), as the fall of antimatter; Cassidy (1956), Stair (1956), and Barnes (1951) suggested fused sedimentary rocks or crustal rocks from a destroyed planet.

Some of the more popular theories have claimed a lunar origin for tektites. La Paz (1938), Fenner (1938), and later O'Keefe (1958, 1961, 1963) proposed lunar or other material in Earth's orbit as the answer. Nininger (1940, 1952), Chapman (1960, 1964, 1971), Chapman and Larson (1963),

Dr. King, Professor of Geology at the University of Houston, joined the faculty in 1969 as Chairman of the Geology Department and served in that capacity until 1975. He previously worked at the NASA Johnson Space Center from 1963 until after the first return of lunar samples to the Lunar Receiving Laboratory, of which he was Curator in 1967–69. Dr. King's professional activities and research have included astronaut training, lunar samples from all Apollo Missions, tektites, meteorites, and Mars and Mercury photogeologic mapping. His recently published book, Space Geology, *covers many of the geological aspects of the exploration of space. He presently is on research leave as a Guest Professor at the Mineralogical Institute of the University of Tübingen. Address: Department of Geology, University of Houston, Houston, TX 77004.*

Table 1. The major groups of tektites

Group name	Localities	Age[1]
Australasian tektites	Australia, Indochina, Philippines, Billiton, Thailand, Sumatra, and numerous other sites	~0.7 m.y.
Ivory Coast tektites	Ivory Coast	~1.3 m.y.
Moldavites	Czechoslovakia (Bohemia and Moravia)	~15 m.y.
North American tektites	Texas, Georgia, and Martha's Vineyard	~34 m.y.

1. Age in millions of years before present, as determined from potassium-argon and fission track dates from a number of different laboratories

Figure 1. These Czechoslovakian tektites (moldavites) from southern Bohemia are transparent green and so attractive that they commonly have been used in jewelry. The fine surface sculpture is caused by partial dissolution of the glass in groundwater and soil acids. The drop-shaped specimen at left is approximately 6 cm long. (Reprinted from King 1976, by permission of John Wiley & Sons.)

Chao et al. (1962), and O'Keefe (1969) have believed them lunar impact glass. Early in the debate, Verbeek (1897), Linck (1928), and, most recently, O'Keefe (1976) have suggested ejecta of lunar volcanoes.

Those on the right trail, who considered tektites to be terrestrial impact glass, have included Spencer (1933), Urey (1955, 1963), Barnes (1961, 1971), Faul (1966), Taylor (1962, 1965), King (1968), and King et al. (1970). This bewildering array of ideas for the origin of tektites, in addition to his own experience, led Faul (1966) to state: "To anyone who has worked with them, tektites are probably the most frustrating stones ever found on earth."

Inclusions and internal structures

A number of relatively recent observations indicate that tektites originate from the shock melting of surface or near-surface rocks. Coesite, a dense form of silica, was identified in tektites from Southeast Asia by X-ray diffraction and optical methods by Walter (1965); it occurs in frothy silica glass inclusions associated with quartz. We know that coesite is formed at the surface of the Earth only by very high pressure accompanying the hypervelocity impacts of large meteorites, and possibly other cosmic bodies, at impact sites such as Meteor Crater, Arizona, and the Ries Basin, Germany. In 1956, H. H. Nininger suggested that coesite might be found in the silica-rich sandstone at Meteor Crater; however, the mineral was not actually identified there until

Figure 2. Dumbbell and tear-drop fluidal shapes of tektites from Indochina. Tektites from Southeast Asia and Australia typically are very dark brown glass and may appear to be completely opaque. The dumbbell is approximately 10 cm long. (Reprinted from King 1976, by permission of John Wiley & Sons.)

several years later (Chao et al. 1960).

Baddeleyite, a monoclinic form of zirconium dioxide, was identified in tektites from North America by Clarke and Wisonski (1967) and King (1966). Baddeleyite is produced in tektites and other impact glass, for example in the Ries Basin and at Aouelloul Crater, Mauritania (El Goresy 1965), by the thermal decomposition of zircon (tetragonal zirconium silicate) at very high temperature. The occurrence of baddeleyite in tektite glass, together with the better-known silica glass inclusions (Barnes 1939), such as those shown in Figure 3, indicates a very high temperature origin for tektites. Nickel-iron spherules containing schreibersite (iron-nickel phosphide) and troilite (iron sulfide) were found in Southeast Asian tektites by Chao et al. (1962); they are common minerals in most meteorites.

Spencer (1933) had already reported the occurrence of magnetic metallic spherules in Australian tektites and noted their similarity to the nickel-iron spherules in impact glass from the Wabar Craters, Saudi Arabia, but he did not have modern tools such as the electron microprobe analyzer at his disposal. Spencer did, however, conclude correctly that tektites were a variety of terrestrial impact glass long before the heated discussions of the 1950s and 1960s. His conclusion was based on the general similarities of impact glasses and tektites under the microscope. Tektites and other impact glasses were observed to have very contorted flow structure and schlieren, most of them much more complex than similar features in volcanic rocks, and also many tiny spherical bubbles. Thus, the inclusions and internal structures of tektites established rather well their mode of origin, and arguments were reduced to the more important question—Are tektites impact-melted Earth rocks or are they rare samples splashed to Earth from the surface of some other body?

Shapes

Much of the discussion regarding the possible extraterrestrial origin of tektites focused on the shapes of many Australian tektites (Fig. 4). Their general form was first illus-

Figure 3. These silica glass inclusions in a North American tektite from Dodge County, Georgia, show silica glass in rather equant grains (*top*) and in thin filaments in the tektite glass (*center and bottom*). Dark circles are small spherical bubbles. Photomicrograph in plane polarized transmitted light, length of field of view is approximately 2 mm. (Reprinted from King 1976, by permission of John Wiley & Sons.)

trated by Charles Darwin (1844), who believed that tektites were a peculiar form of obsidian, a common silica-rich volcanic rock. These "buttons" and some other shapes have flanges that apparently are produced by a second period of melting, almost certainly caused by aerodynamic heating and flowage. D. R. Chapman (1960, 1964) convincingly reproduced this morphology in a series of highly successful experiments at the NASA Ames Research Center arc jet facility. Chapman even succeeded in reproducing the ring wave spacing and amplitudes on the ablation surface of some tektites. He argued very forcefully that his experiments and calculations supported an extraterrestrial origin for tektites. Later, he refined his arguments to state that his data were uniquely compatible with a lunar origin for tektites, in agreement with the idea of H. H. Nininger, who suggested in 1940 and later papers that tektites might be fused lunar rocks splashed off the Moon by meteorite impacts.

Chapman's conclusions were not accepted by all. There was some speculation that the "button" shape of many Australian tektites might be caused by a peculiar history of some of these tektites, and that the shape need not be explained in any pro-

posed general solution to the origin of tektites. However, aerodynamically sculptured specimens were recognized from other groups of tektites (King 1964; Chao 1964), and it eventually became clear that aerodynamic sculpturing is a general feature of tektites that simply is better preserved on the youngest tektites. Chapman's experiments and observations demonstrated that at least some tektites from each of the four major groups had two periods of melting: one that formed the body of glass of the tektite and a second period of aerodynamic melting on the anterior surface of the tektites as they passed downward, at high velocity, through the Earth's atmosphere.

Chemistry

It has been well established that the chemistry of tektites closely resembles that of terrestrial rocks. Taylor and Kaye (1969) showed that the compositions of the Australian tektites are similar in their major and some minor element contents to common types of terrestrial sedimentary rocks, particularly clayey sandstones. Similar conclusions that tektites are chemically very like terrestrial sedimentary rocks were reached by Barnes (1939), Urey (1958), and Taylor et al. (1961), based primarily on previously published analyses but also on new analytical work. Other workers, such as Greenland and Lovering (1963), compared tektites to silica-rich terrestrial rocks, particularly granites. The rare Earth element content of tektites was studied by Haskins and Gehl (1963), who found that the pattern is identical to that of terrestrial rocks. Numerous other chemical and isotopic studies have further documented the similarities between terrestrial rocks and tektites (e.g. Taylor 1962, 1968; Setser and Ehmann 1964). At least one group of tektites, the bediasites from Texas, were shown to be associated stratigraphically with rocks of similar major element chemistry (King 1962, 1968a).

However, there are two obvious differences between the chemistry of tektites and terrestrial rocks: water content and the oxidation state of iron. Most terrestrial rocks contain water ranging from approximately 0.5 to 2.0 weight percent. Tektites are extremely dry, with water content close to 0.01 weight percent. Also,

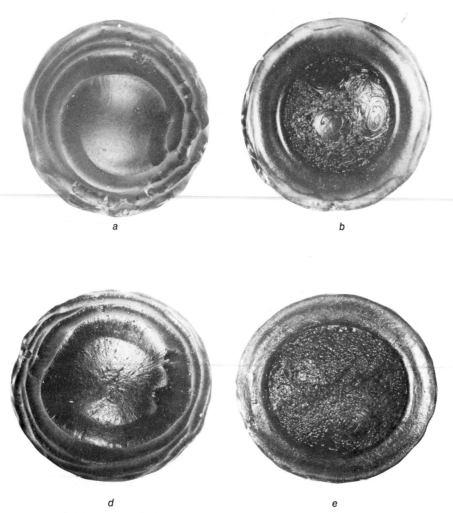

Figure 4. Australian tektites show the "button" shape that results from a second period of melting during hypervelocity passage through the Earth's atmosphere. (*a through c*) Anterior, posterior, and side view of a specimen from near Princetown, Victoria. (*d through f*) Anterior, posterior, and side view of a tektite from near Port Campbell, Victoria. Note that ring waves have formed on the remelted layer on the anterior surface, and that some of the remelted glass has flowed back from aerodynamic pressure to form a "flange." The ablation is similar to that observed on some missle nosecones. Both specimens are approximately 25 mm in diameter. (Reprinted from King 1976, by permission of John Wiley & Sons.)

there is a substantial amount of ferric iron in virtually all terrestrial rocks, but the amount in tektites is 0.08 weight percent or less, based on the most careful analyses. Both of these features of tektite chemistry, we now realize, are produced by the very high-temperature origin of tektites, which is estimated to be several thousand degrees centigrade. Water, of course, is highly volatile at the high temperatures developed by the passage of the shock wave due to a large impact event. Also, it is well known from the steel industry and from experimental petrology that the ratio of ferrous to ferric iron in silicate melts increases with temperature for melts of similar compositions in contact with air.

Viste and Anders (1962) made an important isotopic analysis of tektites for the possible presence of Al^{26}, which is produced in meteorites by the relatively high flux of cosmic rays in space. The absence of detectable Al^{26} in tektites meant that tektites had to have spent less than 90,000 years in space as small or unshielded bodies. Fleischer et al. (1965) searched for cosmic-ray tracks in tektites and found none. Based on the volume of tektites examined and on the assumption that the tektites were small bodies in space, they concluded that an upper limit for the time tektites had spent in space could only have been a few thousand years. The probability is extremely minute that impact ejecta from any other source in the Solar System besides the Moon would be swept up by the Earth in such a short time. Thus, from dynamical and probability arguments, the *only* possible extraterrestrial source of tektites became the Moon.

The search on Earth

There were, of course, attempts to locate source craters for tektites on Earth. It was gradually realized that the Ries Basin in southern Germany was an impact crater 24 km in diameter. Although this origin for the Ries Basin was suggested as early as 1904, coesite and other undeniable petrographic evidence of shock metamorphism was not found in the Ries structure until much later (Shoemaker and Chao 1961; see also von Engelhardt and Stöffler 1968). The center of the Ries structure is located approximately 256 km from the closest known occurrences of moldavites in Bohemia, and it was not until 1961 that Cohen suggested that the moldavites might have originated as fusion products from this crater. There was considerable uncertainty because the compositions of impact glasses in the Ries Crater ejecta have not been found to resemble the compositions of moldavites (von Engelhardt and Hörz 1965; von Engelhardt 1967). Nevertheless, the basement rocks underlying the Ries Crater are very complex and the sampling and comparison problem is extremely difficult; thus the coincidence in age between the moldavites and the Ries Crater (Gentner et al. 1967) is the most direct correlation, and it con-

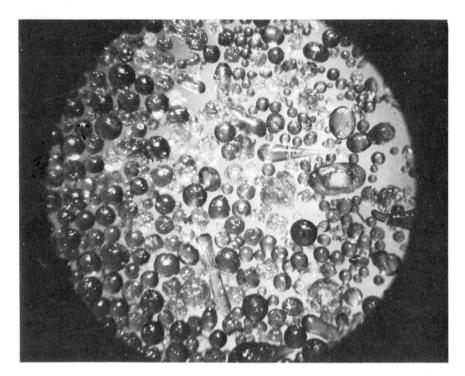

Figure 5. These microtektites, from marine sediment cores taken off the west coast of Africa, presumably were formed by the same impact that produced the larger Ivory Coast tektites and the Bosumtwi Crater, Ghana. Notice the abundance of fluid splash-form shapes: spheres, ellipsoids, dumbbells, rods, and tear drops. Width of field of view is approximately 5 mm, reflected light. (Photograph courtesy of Dr. Billy Glass, Department of Geology, University of Delaware.)

analyzed these minute glassy objects and showed them to be related to the Southeast Asian and Australian group of tektites. He later documented the occurrences of microtektites in marine cores from the African (Figs. 5 and 6) and North American impact events (Glass 1968; Glass et al. 1973). The age of microtektites in ocean bottom cores could be well dated by magnetic reversal stratigraphy—the dates agree rather well with the previously determined tektite ages. The recognition of microtektites served mainly to extend the geographic limits and to revise upward the estimates of the total masses of tektites in the different occurrences. They also provided evidence of a greater range in chemical compositions than had been found in larger tektites.

The investigation of the strontium and rubidium isotopic compositions of Bosumtwi Crater rocks and the Ivory Coast tektites by Schnetzler et al. in 1966 was of special significance. These investigators found that the strontium and rubidium isotopic composition of the target rocks at the Bosumtwi Crater were rather distinctive for terrestrial rocks, and that the composition could be directly related to the isotopic contents in the Ivory Coast tektites. They concluded that "the evidence available at present suggests that the Ivory Coast tektites are most probably the fusion products of meteorite impact at the Bosumtwi crater site." This was a dramatic reversal of opinion for this research group, which had previously supported an extraterrestrial origin. The strontium/rubidium isotopic data, the geographic proximity, and the age coincidence of the crater glass and the Ivory Coast tektites left little doubt in most of the scientific community that tektites must be fusion products of terrestrial rocks resulting from the hypervelocity impacts of

vinced many scientists that this crater was the source of the moldavites.

Fortunately, the target rocks in the vicinity of the Bosumtwi Crater in Ghana are more homogeneous than those at Ries. The suggestion was made by Cohen (1961, 1963) that the Ivory Coast tektites could have originated as impact-fused rocks from the Bosumtwi Crater, because the closest

known Ivory Coast tektites are approximately 250 km from the crater. The Bosumtwi Crater had been considered a possible impact site for a number of years, but the work of Littler et al. (1962) and El Goresy (1966) left no room for doubt. From potassium-argon and fission track measurements, the age of the Bosumtwi Crater was determined by Gentner et al. (1963, 1967) to be identical, within experimental uncertainties, to the age of the Ivory Coast tektites.

An age coincidence of an impact crater and a nearby occurrence of tektites in not one but *two* cases was quite convincing to many scientists, but there were still other remote possibilities. It was suggested that the craters were localities where large lunar fragments had impacted, and that the tektites nearby were simply ablation droplets from the outside of the mass as it passed through the Earth's atmosphere. The lunar origin of tektites continued for several years to be vigorously supported by Chapman, O'Keefe, and a few others, and O'Keefe *still* argues that tektites originated as volcanic ejecta from deep within the Moon! (If curious, see O'Keefe 1976.)

Although very small tektites were previously known, "microtektites" were first reported in cores from the Indian Ocean by Glass (1967), who

Figure 6. This scanning electron microscope image of an Ivory Coast microtektite shows the deilicate elongated dumbbell shape. Length of the specimen is approximately 905 μm. (Photograph courtesy of Dr. Billy Glass, Department of Geology, University of Delaware.)

large meteorites, asteroids, and/or comets.

Extraterrestrial evidence and final questions

The elegant aspect of the last ten years of the tektite debate is that all workers in this field knew that the Space Program would surely provide a conclusive answer—and it did! The Surveyor Program made the first in situ chemical analyses of the lunar surface by means of an alpha particle backscattering device (Turkevich et al. 1967, 1969). The results showed that the lunar surface rocks did not resemble tektites in composition; there was, nevertheless, some uncertainty about the accuracy of remote analyses performed under adverse conditions. The lunar samples returned by the Apollo Program provided definitive results and demonstrated that the alpha backscatter analyses were quite accurate! No lunar sample returned by the Apollo or Luna missions is a suitable parent material for tektites (King et al. 1970; Taylor 1973; King 1976). Although O'Keefe claimed that tektite glass had been discovered in the Apollo 12 samples (O'Keefe 1970), this was quickly refuted (King 1971; Taylor 1973; Barnes and Barnes 1973). The fact is that the chemistry of lunar samples differs in a number of fundamental ways from terrestrial rocks and tektites. These differences include distinctive rare Earth element distribution patterns, chromium contents, potassium to uranium ratios, aluminum and silicon contents, and lead and oxygen isotopic ratios. These and other chemical differences have been convincingly summarized by Taylor (1973).

Thus the major problem of the origin of tektites is solved—it is the Earth. Nevertheless, a number of unanswered or only partially answered questions remain. We have a source crater (Bosumtwi) for the Ivory Coast tektites, and at least a possible source (Ries Crater) for the moldavites. The source crater for the North American tektites is no particular problem because, owing to the age of the tektites, we would expect the source to be buried under a great thickness of sedimentary rocks somewhere in the area of the coastal plain of the Gulf of Mexico. However, the lack of a candidate crater for the source of the largest and most recent tektite group, the Australasian tektites, is particularly annoying. Various possibilities have been suggested, including a possible large crater under the ice in Wilkes Land, but none can be strongly supported with the evidence at hand. Urey (1957) suggested that tektites might be formed by the impact of a comet and that, because of the diffuse, low-density nature of the impacting comet, a large crater might not be formed. However, our knowledge of possible cometary impacts is poor indeed, and the validity of this mechanism to produce the layered tektites described by Barnes (1971), who also supports a cometary impact origin for at least some tektites, is difficult to evaluate.

Another group of remaining problems involves the mechanics and time sequence of events that can accelerate fused droplets to the velocities required to distribute tektites hundreds to thousands of kilometers from the source crater. The velocities, which should range from approximately 2.4 km/sec to as much as 6 km/sec (O'Keefe 1976), do not appear to be such a serious problem themselves because large meteorites, asteroids, and comets impact the Earth at velocities ranging from 11 to 72 km/sec, and it has been well established from experimental impact mechanics that some of the target material may be accelerated to a velocity that equals or even exceeds the velocity of the impacting projectile. The problem for tektites is that molten droplets, or even solid bodies, cannot be propelled through the atmosphere at velocities that could account for the observed geographic distributions. They must escape the Earth's atmosphere into near-Earth space. But even here there are still uncertainties, because the mechanics and sequence of events by which the atmosphere is blown aside, or otherwise permits the ejection of tektites into near-Earth space, is not well understood.

Tektites have, in retrospect, probably received much more attention than they deserve. They should now be relegated to a rather minor topic within shock metamorphism. It is apparent that more than a little confusion was caused by an overly restrictive application of the term *tektite*. Many later workers assumed that true tektites should have all of the properties of the moldavites as described by Suess (1900). L. J. Spencer (1933) saw the problem and the solution clearly, but because of the primitive state of knowledge of shock metamorphism and cratering mechanics at that time, his prescient conclusions were not widely accepted. It remained for us to reach the same conclusions, after very much more work, in only slightly more quantitative terms.

References

Barnes, V. E. 1939. North American tektites. *Bur. Eco. Geol. Univ. Texas Pub.* 3945:477–656.

———. 1951. New tektite areas in Texas. *Geol. Soc. Amer. Bull.* 62:1422.

———. 1961. Tektites. *Sci. Amer.* 205:58–65.

———. 1969. Petrology of moldavites. *Geochim. Cosmochim. Acta* 33:1121–34.

———. 1971. Description and origin of large tektite from Thailand. *Chem. Erde* 30.13–19.

Barnes, V. E., and M. A. Barnes, eds. 1973. *Tektites.* Benchmark Papers in Geology. Stroudsburg, PA: Dowden, Hutchinson & Ross.

Berwerth, F. 1917. Können die Tektite als Kunstprodukte gedeutet werden? (Eine Bejahung). *Centralbl. Min., Geol. Palaeon.* pp. 240–54.

Cassidy, W. A. 1956. Australite investigations and their bearing on the tektite problem. *Meteoritics* 1:426–37.

Clarke, R. S., Jr., and J. F. Wosinski. 1967. Baddeleyite inclusion in Martha's Vineyard tektite. *Geochim. Cosmochim. Acta* 31:90–114.

Chao, E. C. T., E. M. Shoemaker, and B. M. Madsen. 1960. First natural occurrence of coesite. *Science* 132:220.

Chao, E. C. T., I. Adler, E. J. Dwornik, and J. Littler. 1962. Metallic spherules in tektites from Isabela, Philippine Islands. *Science* 135:97–98.

Chao, E. C. T. 1964. Spalled, aerodynamically modified moldavite from Slavice, Moravia, Czechoslovakia. *Science* 146:790–91.

Chapman, D. R. 1960. Recent reentry research and the cosmic origin of tektites. *Nature* 188:353.

———. 1964. On the unity and origin of Australasian tektites. *Geochim. Cosmochim. Acta* 28:841–80.

———. 1971. Australasian tektite geographic pattern, crater and ray of origin, and theory of tektite events. *J. Geophys. Res.* 76:6309–38.

Chapman, D. R., and H. K. Larson. 1963. On the lunar origin of tektites. *J. Geophys. Res.* 68:4305–58.

Cohen, A. J. 1961. A semiquantitative hypothesis of tektite origin by asteroid impact. *J. Geophys. Res.* 66:2521 (abstract).

———. 1963. Asteroid- or comet-impact hypothesis of tektite origin: Moldavite strewn-fields. In *Tektites*, ed. J. A. O'Keefe, pp. 189–211. Univ. Chicago Press.

Darwin, C. 1844. Geological observations on coral reefs, volcanic islands, and in South America, pp. 38–39. London: reprinted 1851, Smith, Elder & Co., Cornhill.

Dunn, E. J. 1911. *Pebbles.* pp. 30–34. Melbourne: Robertson & Co.

———. 1912. Australites. *Geol. Surv. Victoria, Bull.* 27:3–23.

Easton, N. W. 1921. The billitonites, an attempt to unravel the tektite puzzle. *K. Akad. Wetensch. Amsterdam Verh., Tweede Sectie,* 22:1–32.

Ehmann, W. D., and T. P. Kohman. 1958. Cosmic-ray-induced radio-activities in meteorites II. ^{26}Al, ^{10}Be, and ^{60}Co, aerolites, siderites and tektites. *Geochim. Cosmochim. Acta* 14:364–79.

El Goresy, A. 1965. Baddeleyite and its significance in impact glass. *J. Geophys. Res.* 70:3453–56.

———. 1966. Metallic spherules in Bosumtwi Crater glasses. *Earth Planet. Sci. Letters,* 1:23–24.

Engelhardt, W. v. 1967. Chemical composition of Ries glass bombs. *Geochim. Cosmochim. Acta.* 31:1677–89.

Engelhardt, W. v., and F. Hörz. 1965. Riesglaser und Moldavite. *Geochim. Cosmochim. Acta* 29:609–20.

Engelhardt, W. v., and D. Stöffler. 1968. Stages of shock metamorphism in the crystalline rocks of the Ries Basin, Germany. In *Shock Metamorphism of Natural Materials,* ed. B. M. French and N. M. Short, pp. 159–68. Baltimore: Mono Book Corp.

Faul, H. 1966. Tektites are terrestrial. *Science* 152:1341–45.

Fenner, C. 1938. Australites, III. A contribution to the problem of the origin of tektites. *Roy. Soc. S. Aust., Trans.* 62:192–216.

Fleischer, R. L., C. W. Naeser, P. B. Price, R. M. Walker, and M. Maurette. 1965. Cosmic ray exposure ages of tektites by the fission track technique. *J. Geophys. Res.* 70:1491–96.

Gentner, W., H. J. Lippolt, and O. A. Schaeffer. 1963. Argon-Bestimmungen an Kaliummineralien, XI. Die Kalium-Argon Alter der Gläser des Nördlinger Rieses und der boehmischmaehrischen Tektite. *Geochim. Cosmochim. Acta* 27:191–200.

Gentner, W., B. Kleinman, and G. A. Wagner. 1967. New K-Ar and fission-track ages of impact glasses and tektites. *Earth Planet. Sci. Letters* 2:83–86.

Glass, B. P. 1967. Microtektites in deep-sea sediments. *Nature* 214:372–74.

———. 1968. Glassy objects (microtektites) from deep sea sediments near the Ivory Coast. *Science* 161:891–93.

Glass, B. P., R. N. Baker, D. Storzer, and G. A. Wagner. 1973. North American microtektites from the Caribbean Sea and their fission track ages. *Earth Planet. Sci. Letters* 19:184–92.

Greenland, L. P., and J. F. Lovering. 1963. The evolution of tektites: Elemental volatilazation in tektites. *Geochim. Cosmochim. Acta* 27:249–59.

Hanuš, F. 1928. O moldavitech cili vltavinech z Cech a Moravy. *Česk. Akad. Ved a Uměni,* Rozpr., Třida II, Ročnik 37, Číslo 24, pp. 1–83; England trans., Washington, DC: NASA, TT F-111.

Hardcastle, H. 1926. The origin of australites—plastic sweepings of a meteorite. *New Zeal. J. Sci. Tech.* 8:65–75.

Haskins, L., and M. A. Gehl. 1963. Rare earth elements in tektites. *Science* 139:1056–58.

Kahn, M. A. R. 1947. Atomic bombs, the tektite problem, and "contraterrene" meteorites. *Meteorit. Soc., Contrib.* 4:35–36.

King, E. A. 1962. Possible relation of tuff in the Jackson Group (Eocene) to bediasites. *Nature* 196:569–70.

———. 1964. An aerodynamically sculptured bediasite. *Jour. Geophys. Res.* 69:4731–33.

———. 1966. Baddeleyite inclusion in a Georgia tektite. *Amer. Geophys. Union, Trans.* 47:145 (abstract).

———. 1968. Recent information on the origin of tektites. In *Shock Metamorphism of Natural Materials,* ed. B. M. French and N. M. Short, Baltimore: Mono Book Corp., (abstract).

———. 1968a. Stratigraphic occurrence of bediasites. *Geol. Soc. Amer.,* Prog. with Abstrs., 1968 Ann. Mtg., Mexico City, pp. 160–61 (abstract).

———. 1976. *Space Geology: An Introduction.* NY: Wiley.

King, E. A., M. F. Carman, and J. C. Butler. 1970. Mineralogy and petrology of coarse particulate material from the lunar surface at Tranquillity Base. *Science* 167:650–52.

King, E. A., R. Martin, and W. Nance. 1971. Tektite glass *not* in Apollo 12 sample. *Science* 170:199–200.

La Paz, L. 1938. The great circle distribution of tektites. *Pop. Astron.* 46:194–200.

Linck, G. 1928. Oberfläche und Herkunft der Meteorischer Gläser. *Neues Jahrb.* 57:223–36.

Littler, J., J. Fahey, R. S. Dietz, and E. C. T. Chao. 1962. Coesite from the Lake Bosumtwi Crater, Ashanti, Ghana. *Geol. Soc. Amer., Spec. Pap 68,* p. 218.

Nininger, H. H. 1940. The Moon as a source of tektites. *Geol. Soc. Amer., Bull.* 51:1936 (abstract); 1941. *Amer. Mineral.* 26:199.

———. 1952. *Out of the Sky.* Denver: Univ. Denver Press.

O'Keefe, J. A. 1958. Origin of tektites. *Nature* 181:172–73.

———. 1961. Tektites as natural Earth satellites. *Science* 133:562–66.

———. 1963. Origin of tektites. In *Tektites,* ed. J. A. O'Keefe, pp. 167–88. Univ. Chicago Press.

———. 1969. The microtektite data: Implications for the hypothesis of the lunar origin of tektites. *J. Geophys. Res.* 74:6795–804.

———. 1970. Tektite glass in Apollo 12 sample. *Science* 168:1209–10.

———. 1976. *Tektites and Their Origin. Developments in Petrology 4.* Amsterdam: Elsevier.

Schnetzler, C. C., W. H. Pinson, and P. M. Hurley. 1966. Rubidium-Strontium age of the Bosumtwi Crater Area, Ghana, compared with the age of the Ivory Coast tektites. *Science* 151:817–19.

Setser, J. L., and W. D. Ehmann. 1964. Zirconium and hafnium abundances in meteorites, tektites and terrestrial materials. *Geochim. Cosmochim. Acta* 28:769–82.

Shoemaker, E. M., and E. C. T. Chao. 1961. New evidence for the impact origin of the Ries basin, Bavaria, Germany. *J. Geophys. Res.* 66:3371–78.

Spencer, L. J. 1933. Origin of tektites. *Nature* 131:117–18, 876.

Stair, R. 1956. Tektites and the lost planet. *Sci. Monthly* 83:3–12.

Suess, F. E. 1900. Die Herkunft der Moldavite. *Jahrb. K. K. Reichanst.,* Vienna, 50: 193–382.

———. 1916. Können die Tektite als Kunstprodukte

gedeutet werden? *Centralbl. Min. Geol. Palaeon.* 569–78.

Taylor, S. R. 1962. Consequences for tektite composition of an origin by meteorite splash. *Geochim. Cosmochim. Acta* 26:915–20.

———. 1965. Tektites: Origin of parent material. *Science* 149:658–59.

———. 1968. Geochemistry of Australian impact glasses and tektites (australites). In *Origin and Distribution of the Elements,* ed. L. H. Ahrens, pp. 533–41. Oxford: Pergamon.

———. 1973. Tektites: A post-Apollo view. *Earth-Sci.Rev.* 9:101–23.

Taylor, S. R., M. Sachs, and R. D. Cherry. 1961. Studies of tektite composition, I. Inverse relation between SiO_2 and other major constituents. *Geochim. Cosmochim. Acta* 22:155–63.

Taylor, S. R., and M. Kaye. 1969. Genetic significance of the chemical composition of tektites: A review. *Geochim. Cosmochim. Acta* 33:1083–100.

Turkevich, A. L., E. J. Franzgrote, and J. H. Patterson. 1967. Chemical analysis of the Moon at the Surveyor V landing site. *Science* 158:635–37.

———. 1969. Chemical composition of the lunar surface in Mare Tranquillitatis. *Science* 165:277–79.

Urey, H. C. 1955. On the origin of tektites. *Nat. Acad. Sci. Proc.* 41:27–31.

———. 1957. Origin of tektites. *Nature* 179:556–57.

———. 1958. Origin of tektites. *Nature* 181:1457–58.

———. 1963. Cometary collisions and tektites. *Nature* 197:228–30.

Verbeek, R. D. M. 1897. Glaskogels van Billiton. *Jaarb. Mijnwes. Ned. Indie* 235–72.

Viste, E., and E. Anders. 1962. Cosmic-ray exposure history of tektites. *J. Geophys. Res.* 67:2913–19.

Vogt, T. 1935. Notes on the origin of the tektites I. Tektites as aerial fulgarites. *K. norske vidensk. selsk. Forh., Trondhjem* 8:9–12.

Walter, L. S. 1965. Coesite discovered in tektites. *Science* 147:1029–32.

"Actually, they all look alike to me."

PART 5 *Earth Viewed from Space*

William R. Muehlberger
Verl R. Wilmarth

The Shuttle Era: A Challenge to the Earth Scientist

From Earth-orbiting spacecraft, astronauts will be able to identify and analyze unique phenomena, thereby adding a potent new technique to the study of Earth's dynamic systems

Nearly a decade and a half ago, TIROS satellites provided man's first synoptic view of weather systems that encircle the Earth. These spectacular views of Earth increased interest in space exploration as a valuable technology for expanding man's understanding of the Earth and its dynamic systems. By the mid-1960s space photographs from the manned Gemini and Apollo programs led to serious consideration of remote sensing as an important tool for Earth exploration. With the launch of Landsat in 1970, the ability to obtain repetitive multispectral data for large areas of the Earth for resources analysis was established. Landsat was joined by Landsat 1 in 1972 and by Landsat 2 in 1975, and the era of analysis of the Earth using space data had begun.

Skylab, the nation's first manned space station, was launched in 1973 and, as part of the multidiscipline scientific programs, carried the Earth

William R. Muehlberger, Professor of Geological Sciences at the University of Texas at Austin, was a coinvestigator of the Visual Observations Experiment for Skylab 4 and the Apollo-Soyuz mission. His research in the fields of structural geology and tectonics includes studies on the Moon, major fault zones of Earth using space imagery, and ground studies in west Texas, New Mexico, Guatemala, and Honduras. This paper is an outgrowth of talks given while on tour as a Distinguished Lecturer for the American Association of Petroleum Geologists. Verl R. Wilmarth, Acting Chief, Science Payloads Division of NASA, served as Project Scientist for the Earth Resources Experiment Package (EREP) on Skylab and for the Skylab 4 Visual Observations Project. His present major task is developing scientific payloads for shuttle and planning scientific operations to optimize the use of shuttle for Earth and outer space experimentation. Address: Professor Muehlberger, Department of Geological Sciences, University of Texas, Austin, TX 78712.

Resources Experiment Package (EREP), which viewed the Earth with a system of optical, infrared, and microwave sensors. Skylab also included a program of visual observation and photography by the astronauts, and similar experiments were conducted during the joint USA-USSR ASTP (Apollo-Soyuz Test Project) mission, July 15–24, 1975 (astronauts Slayton, Stafford, and Brand). Although these data are only partially analyzed, they extended the use of remote sensing in the study of the land, sea, and atmosphere.

Concurrently with the expansion of Earth resources studies came the development of Earth-orbiting weather satellites by 1970 and the Synchronous Meteorological Satellites (SMS) of the mid-1970s. As the end of this decade approaches, scientists have expanded the technology to study the Earth and increased their understanding of the complexity of the Earth's systems.

In planning for future Earth science programs from space that will integrate the capabilities of man and the repetitive overflights of satellites, consideration must be given to what manned flights can contribute. Both Skylab and ASTP missions have clearly demonstrated that trained crewmen can analyze and obtain useful and sometimes unique data on Earth processes and phenomena. For instance, sunglint—the reflection of the sun off the sea—was typically avoided during aircraft photographic missions. However, the photographs returned by the astronauts showed the distinct advantage of sunglint in the study of water surface patterns. New data on the location of volcanoes, sand-dune fields, and the distribution and growth of sea ice were

recorded by the crew during these missions. From the Skylab and ASTP missions, man in space provided the capability to detect new features and phenomena, document them, and extend this information to related processes in other parts of the world. Some of this information must be gathered in off-nadir position or when the sun angle highlights the features. In contrast, unmanned satellites view the Earth only at nadir and at high sun angles.

The 1980s will bring the advent of the manned shuttle program and its large (60,000 pounds) science payloads, along with the next generation of improved unmanned satellites. It is imperative that the various approaches to study of the Earth be carefully integrated so that we can further our understanding of the Earth, its energy sources, environment, and resources. It has been demonstrated that both manned and unmanned satellites are important in repetitive data-gathering roles. A major challenge for Earth scientists is to develop useful programs that incorporate the unique skills of man and the capability of satellites. It is the purpose of this paper to report how astronauts were used during Skylab and ASTP to discover new Earth features and phenomena and to outline program elements that will incorporate the use of astronaut observers on the shuttle to supplement the wide variety of satellites proposed for study of the Earth.

Visual observation experiment

Skylab, the nation's first space station, was a gigantic experiment designed to see how long man could

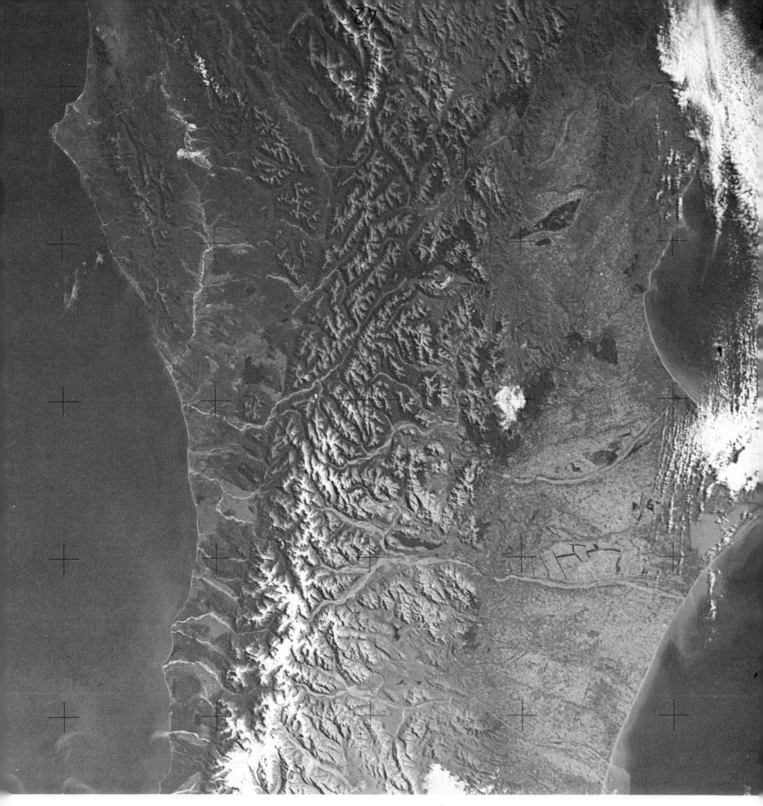

Figure 1. In this photograph of most of north-central South Island, New Zealand, taken during the Skylab 4 mission, the shadows beautifully display the striking linearity of the Alpine fault for over 100 miles of its length, its gigantic kink to the left (near top center of the photo), and the system of branch faults developing as a result of the kinking. These branch faults are visible as linear valleys in a wedge-shaped region to the right of the Alpine fault (point of the wedge is along the Alpine fault near the center of the photograph). Mt. Cook is on snow-covered ridge near bottom left center. Christchurch is under clouds at right center edge. Field of view of photograph is approximately 150 miles on a side. North is toward upper left. (SL4-137-3700)

survive in space, to explore the utility of Earth-orbiting platforms for scientific and engineering purposes, and to enlarge man's role as an active participant in scientific and engineering studies. It was launched into Earth orbit on May 14, 1973, and was occupied by three three-man crews during the period May 25, 1973, through February 8, 1974 (Table 1). During the 171 days the astronauts lived in Skylab they conducted a wide variety of experiments in physical science, medicine, and technology. The preliminary results appear in a series of NASA documents (1).

During the second manned mission (Skylab 3) the crew requested additional experimental work, and a sequence of observational and photographic tasks for selected sites and features were teletyped up to the spacecraft. The results led to the initiation of a Visual Observations Experiment for the Skylab 4 mission. The experiment team of 19 scientists was responsible for lectures to the Skylab 4 astronauts and mission support on a wide variety of topics on the dynamic processes that affect the Earth's atmosphere, oceans, and land surface (2). In addition to studies of solar dynamics, it was the "Vis Obs" experiment that demonstrated that man was not just an extension of a machine but was an integral part of the studies being made.

The last mission (Skylab 4) lasted for 84 days and flew through the months of December 1973 and January and February 1974—the winter of the northern hemisphere and the summer of the southern. The orbits of Skylab covered the Earth from 50°N to 50°S at an altitude of 235 nautical miles (450 km). Every five days it would repeat an earlier orbit; thus repetitive observations could be made with changing sun angle and season. The astronauts took more than 2,000 hand-held photographs to document the many visual observations they made.

Training was brief—too brief considering the bewildering variety of subject matter—but time was short, and the demands on the astronauts' time limited each scientist to about one hour of lecture. This was supplemented with a loose-leaf book that flew with the crew; it contained brief descriptions of each site, what to look for, why it was important, and suggestions on how to photograph the scene. The transcripts for each five-day cycle were reviewed by the science team members, and new instructions were relayed to Skylab via the teletype or by voice.

An important objective of the Vis Obs experiment was to determine what features and phenomena the crew could identify and describe. Consequently a wide variety of features were selected, such as ocean currents, plankton blooms, atmosphere-ocean interactions, storm development, snow cover patterns, lake and sea ice growth, sand-dune patterns, the Sahelian drought area, air pollution—both natural and manmade—active volcanism, world cereal crop plantings, geology of southwestern North America, and major fault zones of the world. For each of these topics the astronauts contributed many observations and spec-

Table 1. Skylab program

Mission no.	Crew	Lift-off date	Splashdown date	Days in space
1	unmanned[a]	May 14, 1973	—	—
2	Conrad Kerwin Weitz	May 25, 1973	June 22, 1973	28
3	Bean Garriott Lousma	July 28, 1973	Sept. 25, 1973	59
4	Carr Gibson Pogue	Nov. 16, 1973	Feb. 8, 1974	84

[a] Skylab workshop operated unmanned between manned missions.

tacular photographs, and for some they acquired data that have made significant improvements in our understanding of the processes involved.

Because of the short, 8-day flight, our astronauts in the joint USA-USSR Apollo-Soyuz mission (ASTP) had only a single opportunity to observe any given area on Earth. As a result of the time constraints and experience gained in Skylab, the astronauts were intensely briefed on a limited set of target areas. Further, their flight altitude was about half that of Skylab, and thus comparisons could be made between the two altitudes (e.g. naked-eye resolution of specific objects such as the Egyptian pyramids) and between different types of spacecraft. Even with the time and altitude limitations, the Apollo-Soyuz mission added considerable new knowledge, including determining the "true" color of desert and ocean areas by using the first color wheel ever carried into space.

Deserts and desertification

The Sahel, the transition zone between the Sahara Desert and tropical Africa, extends across Africa from the Atlantic Ocean to the Nile River. Here, empires that flourished and fell were isolated for centuries from European knowledge or conquest by the broad expanse of the Sahara to the north and the malarial jungles to the south. For the past several years, this region of grasslands and fertile fields along the main rivers has suffered a severe drought. As their grass and

forage failed, the nomadic tribes and their herds moved south into the already heavily populated farming region. The failure of the crops and the extra population produced widespread starvation, with untold deaths.

A focal point of study for the Skylab astronauts was the inland delta of the Niger River, a well-watered, heavily cultivated and populated region. The Niger River flows for nearly 2,600 miles across west Africa. Its headwaters are in the low mountains a few hundred miles from the Atlantic Ocean in the southwestern corner of west Africa, and from there it flows northeasterly to the edge of the Sahara Desert near Timbuktu, Mali, where it turns easterly and southeasterly to the Atlantic Coast at the southeastern corner of west Africa.

Upstream from Timbuktu is the inland delta of the Niger River. The "delta" (an alluvial fan) is the result of a decrease in stream gradient due to the blocking of the river by large sand dunes and downcutting by the river through the cover of sedimentary rocks onto the harder rocks of the African shield. Here the annual flood spreads across the inland delta through distributary channels, flooding the low areas between the dunes and filling the lakes. This annual irrigation and addition of new soil makes this an important cropgrowing region.

Standard practice in the Sahel, as in many parts of the world, is to burn the fields to clear out the stubble, noxious insects, and other pests. It was generally believed by agricultural experts of this region that most of the burning was done just before the rainy season, but this conflicted with the observations of the astronauts who saw extensive smoke plumes during the day and widespread range fires during night orbits. They found that much of the burning took place *after* the rainy season, just before the hot, dry season when strong winds blow south out of the Sahara Desert. Thus instead of new grass to protect the soil from wind erosion and leaves on the bushes for forage for the animals, there is no food for the animals or protection for the soil. This sequence leads to progressive drying of the region, an increase in wind erosion, and a gradual conversion of once-fertile country to desert. In observing the Sahel the crew noted a general "blandness," which from field studies was found to be the effect of an increase in atmospheric dust, a direct result of soil erosion. The Skylab astronauts' observations were important clues to furthering our understanding of how cultural activities can affect arid lands, not only in the Sahel but potentially throughout the world. The contrast in observations is a result of the very limited area that can be studied by ground or aircraft studies versus the synoptic view observed from spacecraft.

The wide influence of this burning as a climate modifier is now beginning to be recognized. For example, the astronauts described gray cirrus clouds over the Indian Ocean, which aroused their curiosity because cirrus clouds are made of ice crystals and should appear white (our weather satellites do not recognize shades of color in clouds). On successive (90-minute) orbits as Skylab moved westward the crewmen continued to track the clouds. Over 1,500 miles to the west of where they first observed them, the crew was able to identify the source of these clouds as fires covering many hundreds of square miles in east Africa. Clouds covering such a wide area as this change the surface heating of oceans and land, decrease convection in the atmosphere and the possibility of rainfall, and thus alter the climate toward aridity for at least the length of the burning season.

Many dust storms were also seen in the desert regions of the world. One that the astronauts commented upon

was seen leaving Africa and passing westward out into the Atlantic Ocean (Fig. 2). This cloud was identifiable as a faint smudge on the SMS weather satellite imagery and was tracked across the Atlantic Ocean and into the Caribbean Sea before it was lost to sight. African dust may make up the nuclei of raindrops that fall on the United States and be part of the bottom sediments of the Caribbean Sea.

Another area of man-induced desertification that the astronauts could easily identify is in the northern Sinai Desert (see the cover of this issue). Here the fence that separated Egypt and Israel prior to the 1967 war divided the region into two contrasting cultures. To the west, on the Egyptian side, Bedouins with their herds of goats progressively overgrazed the sparse desert vegetation and virtually eliminated it. On the Israeli side, the absence of herds of goats has left the desert vegetation intact and, in fact, has allowed it to increase. The boundary as seen on the photographs appears to fade southward into the desert, because the vegetation decreases away from the Mediterranean Sea. The prevailing onshore breezes bring moisture that the plants absorb, and at night dew forms under the bushes, thus helping their growth. However, where there are no plants, there is no dew; thus, the area may never recover from overgrazing.

The present worldwide trend toward cooling and drying is a cause for concern for the future. If the trend continues, our present abilities to produce food cannot be maintained, and malnutrition and starvation may soon become commonplace for an increasing part of the world's population. If man's present agricultural practices are accelerating this process, then changes need to be made as soon as possible.

Oceans

The general locations of major ocean current patterns are well known, but how they change positions, the evolution of eddies, the location of internal waves, and the interactions between ocean currents and atmosphere are not well known. The repetitive coverage of Skylab, particularly in the high southern latitudes, paid off handsomely in the observa-

tions and photography of these phenomena, whereas traditional aerial photography cannot cover a large enough area to see these major relationships.

Recognition of ocean current boundaries from space turns out to be relatively easy. Looking into the sunglint, current boundaries can be seen as sharp lines separating regions of differing wave patterns. These patterns can be seen for only a very short time because of the rate of movement of the spacecraft and because of the constantly changing sun angle; thus a man's eye, brain, voice, and camera are necessary to describe and record it.

Elsewhere, plankon blooms mark the main current and frequently outline eddies within the current. Plankton, an important factor in the food chain of the oceans, provides food for larger organisms and ultimately for the fish we harvest. The size of the blooms observed by the crewmen are truly enormous—it is difficult to imagine the billions of organisms that make up this profligate outburst of life. Study of the plankton distribution and the changes over time will help in understanding the quantity of life in the sea, where it is, and how much can be safely harvested.

The eddies that spin off from an ocean current generate a vertical cylinder of water unlike the water around it and alter the normal temperature gradient both in the ocean and in the overlying atmosphere. Hot-water eddies evaporate more moisture, causing a cap of cloud blanket to form over them. Cold-water eddies bring cool water up from deep in the ocean and along with it a large quantity of nutrients, which, in turn, increases the fish population in that region (Fig. 3). Further, the cooler water chills the air passing over it, which produces a ring of heavier clouds, with a clear center, around the eddy. Thus we have a new clue as to where the schools of fish are likely to be on the day you are fishing! Eddies were identified in the photographs taken on the first manned mission, and their characteristics were transmitted to the crew then in space. As a result eddies were recognized and photographed in many regions of the ocean where they had gone unnoticed before.

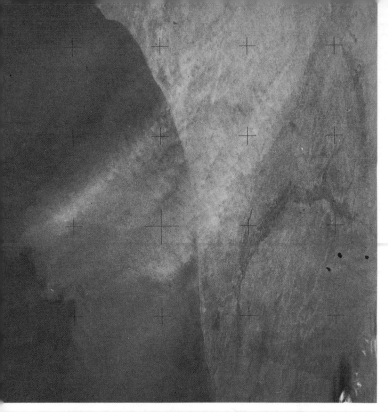

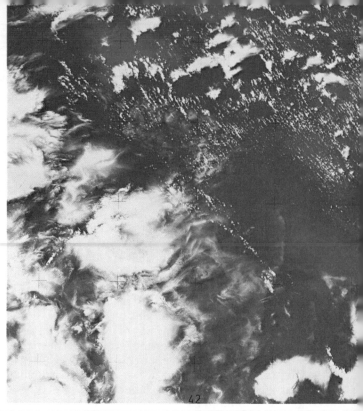

Figure 2. In this duststorm photographed by the Skylab 4 crewmen over Mauritania, west Africa, the wind is carrying dust westward over the Atlantic Ocean. (SL4-140-4227)

Figure 3. Study of photographs taken by the Skylab multispectral camera system led oceanographers to recognize that ocean eddies were visible from space and that the slight (few degrees) difference in water temperature caused characteristic cloud formations. These eddies, in the Falkland current off the coast of Argentina, (*bottom left*), are in one of the many areas where the astronauts recognized such patterns and photographed them with hand-held cameras. The wind and the Falkland current are northerly. (SL4-137-3608)

Internal waves in the ocean were identified in many areas, some in places where they were not expected. They develop along the boundary between the surface current and an underlying current, and they appear as long dark lines when the crew's viewing direction is into the sunlight (Figs. 4 and 5).

Plate boundaries

Plate tectonics has revolutionized our understanding of the mechanism that governs the evolution of the Earth. The apparent absence of plates on the other planets supports the notion that their evolution terminated early in the history of our solar system. In contrast, our Earth is still dynamic, with destruction of crustal segments counterbalanced by regeneration of new material elsewhere—one of nature's recycling systems within which many other interdependent subcycles are interwoven.

Although the rate of movement along plate boundaries is slow—a few centimeters per year—over millions of years this movement generates complicated patterns of deformed regions within and adjacent to the plate margins. Most maps and diagrams of these boundaries are oversimplified, resulting in the impression that it is easy to understand what is going on and how it developed, when actually the geologist has before him a bewildering array of deformed rocks that do not lend themselves to such easy analysis. The astronauts' observations and photographs along the plate margins beautifully illustrate these complexities and present views from many vantage points that give the trained interpreter new insights how the region evolved.

The astronauts can see many features that can't be easily seen from the surface. Most zones where plates are pulling apart end up being covered by oceans and thus are not visible from space, but exceptions are Iceland (too far north for viewing from either Skylab or ASTP), the Red Sea rift zone, and the Rio Grande rift that extends from southern Colorado, through New Mexico and southward along the Texas-Mexico border, to near El Paso, Texas. Spectacular examples of extension faulting (stretching) are visible in the Afar triangle at the southwest corner of the Red Sea fault zone. This region is new crust resulting from basalt upwelling along the rifting zone that has kept most of this region above sea level. Fault patterns similar to this are found in southern Oregon and northeastern California. Plates sliding by each other are well displayed in many areas of the world: excellent examples are the San Andreas fault system in California, the Dead Sea–Lebanese fault zone, and the Alpine fault zone in New Zealand (Fig. 1). At large bends along the length of these faults there is major subsidiary faulting and folding. The geologic history and the types of geologic units differ in each region, but comparison of the structures for similarities and differences helps interpret the complexities found in the more highly deformed areas.

The astronauts were asked to take oblique photographs along the major fault trends, to highlight the linearity or any subtle bends. When possible, they took pictures when the sun was low so that the shadows would emphasize the relief features. The eye and brain are remarkable integrators

Figure 4. This oblique view toward the Straits of Gibralter shows an apparently uniform ocean surface. However, when the sun's reflection travels with the spacecraft and the crew looks into the light, a different phenomenon.(Fig. 5) appears. (AST-27-2365)

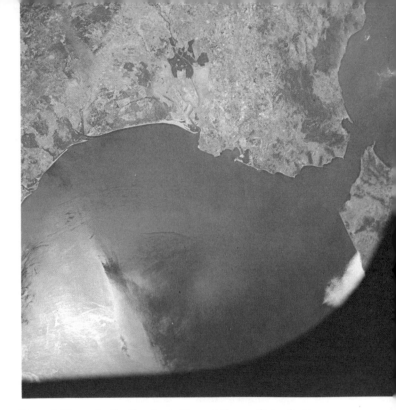

Figure 5. Moments after the view in Figure 4, the sunglint reflections show the magnitude and occurrence of unexpectedly large internal waves (*left*)—50 to 60 km long—in this region. Internal waves were anticipated here because of the salt-rich underflow leaving the Mediterranean (*upper right*) under the surface influx of normal sea water from the Atlantic Ocean (*bottom half*) through the Straits of Gibralter. (AST-27-2367)

of segments of lines into long continuous features that may not be recognizable from conventional aerial photography but may show up in space imagery because of the large area covered by a single frame. Oblique photographs from space along these alignments can assist in the analysis that is currently a subject of intense interest to structural geologists.

Many oblique, near-vertical, and vertical photographs were taken along each of these fault zones. The photographs on the cover and in Figure 1 picture segments of the Dead Sea fault zone and the New Zealand Alpine fault zone that illustrate the complexity of the deformation near a band and the relative simplicity of the straight segments. The east side of the Dead Sea fault has moved north about 100 km (60 miles) relative to the west side. The New Zealand Alpine fault zone has had the opposite sense of movement, has moved about 580 km (350 miles), and has a much longer, more complex history of movement. Both have developed a similar diamond-shaped fracturing pattern at large bends along the major

fault. These patterns are easily recognized north of the Sea of Galilee and in north-central South Island, New Zealand, in these Skylab photographs but not from existing geologic maps.

Satellite data application

Earth studies from manned Earth-orbiting spacecraft are productive because the crew selects the scene, analyzes its information content, and obtains discrete data pertinent to study of the feature or phenomenon. In contrast, the nadir-seeking sensors on Landsat satellites and on Skylab are programmed by ground operations and provide a different set of information, including measurements in the visual, infrared, and microwave portions of the electromagnetic spectrum. For example, each composite image by a Landsat multispectral scanner contains information in the visible to near-infrared on an area about 160 km on a side; similarly, one scene from the six-camera system carried by Skylab covers an equal area in color, color infrared, and on various black-and-white films. The

advantages of the orthophotographic and repetitive data collecting systems on satellites such as Landsat are readily apparent and offer the scientist a new perspective in Earth science.

An excellent example of the use of space-required data in geologic mapping is the application of Skylab and Landsat data in studies, by Houston et al. (1975), of the Boysen Reservoir area, a semiarid, well-exposed region of contrasting red-bed colors in north-central Wyoming. In the Boysen area a series of uplifted blocks have displaced rocks ranging in age from Precambrian to Quaternary. The large east-trending South Owl Creek Mountain thrust fault marks the southern limit of the east-trending uplift of the Owl Creek Mountains. The steep-dipping south flank of the uplift is complexly faulted into a series of diversely tilted blocks.

A comparison of data used by Houston et al. from the Skylab EREP S-190A cameras and the Landsat composite imagery indicates that, although the Skylab photograph did have improved resolution, the marker beds were not as easily recognized as on the Landsat images. For example, all red beds appear as greenish-yellow on the Landsat images and on the S-190A infrared transparencies. However, on the S-190A color photograph only the deep red of the Permo-Triassic is recognized, whereas the Tertiary Wind River Formation red beds are easily identified on the Landsat images. The comprehensive study by these investigators has shown that neither Skylab EREP data nor Landsat data alone are sufficient for overall geologic mapping; using both data sets produces better results. Houston et al. recommend that future spacecraft that view the Earth for geologic studies, whether manned or unmanned, include sensors with resolution of 10–30 meters; produce stereoscopic images; provide spectral data in visible, infrared, and thermal infrared bands; and use an orbit giving repetitive coverage of the same area 4 to 6 times a year.

Shuttle era challenge

With the approach of the shuttle era and the greatly expanded capability to carry large payloads into space either as satellites or as an integral part of the manned orbiter, study of the Earth from space will stretch the imagination of most scientists. An integrated manned or unmanned orbital Earth science system must be designed to minimize cost and to maximize knowledge of Earth.

First we must begin with the identification of the dynamic systems of the land and ocean surfaces, the atmosphere, and the land/air interfaces. Once defined, the next phase is to assess what types of features and phenomena can best be remotely sensed by onboard instruments, by man, or by a combination of man and sensors. We have briefly illustrated here the types of data that can be obtained by these methods. The final decisive phase defines the role of man, the type of sensors, the data sequences, and the application of the data to the space system study. In total this represents an enormous effort by dedicated scientists of all disciplines.

One of the striking accomplishments of the astronauts on Skylab and Apollo-Soyuz was their ability to discern unexpected or unique patterns. The new features observed and the new concepts generated from these observations have made the missions well worthwhile. When we can send professional scientists into space we should learn even more. This is one of the challenges of the shuttle era.

References

1. Skylab earth resources experiment package investigative summary, NASA SP-399. In press; Skylab explores the earth, NASA SP-380. In press; Skylab biomedical experiments, NASA SP-377. In press.

2. Skylab 4. Visual observation project report, NASA TMX-58142, June 1974.

3. Third Earth Resources Technology Satellite Symposium, NASA SP-356, vols. 1–3. May 1974.

4. Proceedings of the NASA Earth Resources Survey Symposium, NASA TMX-58168, vols. 1–2. June 1975.

5. Houston, R. S., R. W. Marrs, and L. E. Bergman. 1975. Final report—Multidisciplinary study of a Wyoming test site. NASA contract 9-13298, December 1975.

Lawrence C. Rowan

Application of Satellites to Geologic Exploration

Recent experiments in two spectral regions, the visible and near-infrared and the thermal-infrared, confirm the value of satellite observations for geologic exploration of the earth

Remote-sensing has played a key role in geologic exploration because it is the most practical method of measuring many pertinent physical properties of large, generally inaccessible areas. The information derived provides a means for studying regional features, for extrapolating local measurements to regional scales, and for identifying critical areas for subsequent detailed studies. Ideally, analysis should begin with large regions and then progress to successively smaller key areas. Such a multilevel approach, used very successfully in lunar and Martian exploration, has become one of the basic concepts of terrestrial remote-sensing (Colwell 1973). The sequence of study, it should be noted, is essentially the reverse of classical geologic investigations of the past, in which detailed information was gathered for many small areas before regional features could be delineated.

After receiving his Ph.D. from the University of Cincinnati in 1964, Lawrence C. Rowan joined the Branch of Astrogeology of the U.S. Geological Survey in Flagstaff, Arizona. His principal research was in lunar geologic mapping using small-scale satellite images, and he also worked on the Geological Survey's Lunar Orbiter Program, coordinating phases of geologic analysis of the possible Apollo landing sites. In 1969, he joined a small group of geologists and geophysicists to form the Remote Sensing Geophysics Section, which now constitutes the Geological Survey's main research effort in this field. As Staff Geologist for Remote Sensing from 1972 to 1974, Dr. Rowan coordinated the geologic remote-sensing program during the Landsat-1 and Skylab experiments. His current research focuses mainly on spectral and spatial analysis of Landsat, Skylab, and aerial photographs and images, especially as applied to mineral exploration. Publication of this article authorized by the Director of the U.S. Geological Survey. Address: Geological Survey, U.S. Department of the Interior, Reston, VA 22092.

The earliest indications of the potential value of satellite photography for terrestrial studies were revealed in the geologic experiments using photographs obtained from the Viking and Aerobee rockets (Lowman 1965). Later, geologic evaluation of Mercury, Gemini, and especially Apollo photographs reaffirmed this potential (Lowman 1972) and provided much of the impetus for subsequent satellite experiments specifically designed for earth-resource investigations. The first satellite devoted to earth-resources evaluation, Landsat-1 (previously ERTS-1), is now widely recognized to be an important new tool for geologic exploration because of the proliferation of applications of its data. Their number and variety, reported in Freden and Becker (1973, 1974) and in Williams and Carter (in press), fully demonstrate the suitability of this new approach to resolving many geologic problems. Moreover, Landsat data are unencumbered by political or security restrictions and are available to all at reasonable cost.

The Landsat Multispectral Scanner (MSS) records reflected radiation in two near-infrared (0.7–0.8 μm and 0.8–1.1 μm) as well as two visible (0.5–0.6 and 0.6–0.7 μm) wavelength bands, a response range that represents only a narrow part of the entire electromagnetic spectrum. Other, commonly more diagnostic properties can be measured by analyzing radiation reflected or emitted in other spectral regions: the gamma-ray (10^{-7}–10^{-5} μm), the remainder of the reflective near-infrared (to 2.5 μm), the thermal-infrared (8–14 μm), and the microwave (1 cm–3 m) regions are presently being exploited for terrestrial remote-sensing. Of these spectral regions, the thermal-infrared has the broadest, most clearly defined geologic potential from satellite platforms.

Although Landsat-1 and Landsat-2, launched on 23 January 1975, do not record thermal-infrared radiation, the third satellite anticipated in this series, Landsat-C, will have a 10.0–12.5 μm band as well as the two visible and two near-infrared wavelength bands. In addition, NASA is considering another satellite, the Heat Capacity Mapping Mission, designed specifically for thermal-infrared geologic applications. Three aspects of the MSS system and the Landsat-1 orbital geometry combine to make the images unique and especially useful for geologic studies: (1) the synoptic view of large areas, (2) the spatially registered multispectral radiometry, and (3) the repetitive imaging, which provides information over time. In addition, data analysis is facilitated because the information is available on computer-compatible tapes. Although not unique to Landsat, another important data-transmission subsystem, the Data Collection Platform, records and relays data from earth-based instruments. A few geologic applications depend on only one of the three aspects of the MSS data, but most result from a complex relationship between two or among all three of them.

Uses of spatial information

Images displaying progressively larger areas show that different scales and types of information are revealed by changing the perspective. For example, major crustal breaks or faults are rarely exposed continuously for great distances; in fact, major faults are usually zones of many smaller faults and related deformational features rather than a continuous single fracture. Either segments are directly exposed in scattered outcrops or, more commonly, the major breaks appear as a combination of erosional escarpments, linear streams and valleys, and tonal changes owing to differences in rock type. Because conventional aerial photographs of small areas show only segments of a major fault zone, the full magnitude of the feature might escape the notice of the interpreter. Individual aerial photographs can be combined to show larger areas, but the time and expense required to make these mosaics as well as their nonuniform illumination and inferior quality are usually formidable problems.

In contrast, Landsat-MSS images are especially useful for regional landform studies because effective mosaics showing entire continents at nearly uniform illumination can be compiled from these already synoptic, essentially planimetric images. One of the most important observations made from examination of these images and image mosaics is the presence of a surprisingly large number of linear features, many of which are hundreds of kilometers long but do not appear on geologic maps. Although evaluation of these large-scale features is· far from complete, the results show that some are previously undetected zones of crustal weakness and many others almost certainly have similar origins (Collins et al. 1974; Isachsen et al. 1974; Gold 1974; Rowan and Wetlaufer 1975). The discovery of these features is already having a significant impact on regional geologic interpretations and, therefore, on many applications, including mineral, petroleum, and ground-water exploration; power

plant and dam sitings; and earthquake studies.

In Nevada, where the origin and distribution of ore deposits have been debated for many decades, analysis of Landsat-1 images combined with synthesis of geophysical data indicates the presence of seven major lineament systems, six of which appear to have influenced the occurrence of ore bodies (Rowan and Wetlaufer 1975). One of these lineaments, the Midas Trench system, extends nearly 500 km from near Lake Tahoe to the northeastern corner of the state (Fig. 1). Morphologic evidence seen in Landsat-1 mosaics of the coter-

minous United States prepared by the U.S. Soil Conservation Service and in aeromagnetic data (Zeitz 1974) suggests that this feature may continue northeast to and possibly beyond the Yellowstone Park volcanic province in Wyoming. The other six major systems identified in Nevada are 75 to 250 km long (Rowan and Wetlaufer 1975).

These seven lineament systems have several geologic and geophysical characteristics in common. First, judging from the density of locally mapped faults, they represent fracture zones which have substantial vertical as well as horizontal extent. Two of the lineaments

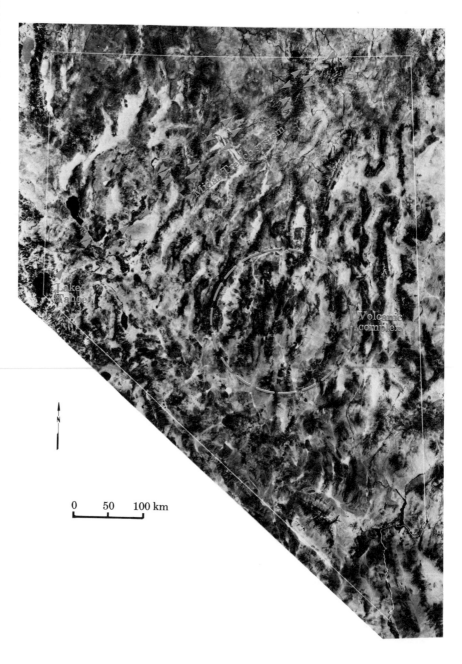

Figure 1. This Landsat-1 MSS image mosaic of Nevada shows the location of the Midas Trench lineament system and the central Nevada volcanic complex. Arrows locate the end points of individual lineaments. (Mosaic prepared by Aerial Photographers of Nevada, Reno, from band-5 images recorded during September and October 1972. From Rowan and Wetlaufer 1975.)

were previously documented as important crustal zones, but the other five resulted from extension of locally mapped faults along major lineaments seen in Landsat-1 images and from synthesis of geophysical data. Second, because all of the zones probably originated in the Precambrian, over 600 m.y. ago, the complex pattern of movement found along them is not surprising. Third, most of the major lineament systems are generally coincident with the trends of total intensity magnetic anomalies.

These characteristics indicate that these lineaments are old, deep-seated zones of crustal weakness which

provided the main conduits for episodically rising magma. These zones also appear to have resulted in the concentration of volcanism between about 30 and 19 m.y. ago within a large, previously undefined volcanic complex (see Fig. 1). The horizontal extent of the central Nevada volcanic complex, as seen on the images, and its postulated vertical extent are supported by regional aeromagnetic, gravity, and seismic refraction data (Rowan and Wetlaufer 1975).

The influence of these features on the distribution of ore deposits in Nevada is indicated by comparing their areal distributions (Fig. 2). In

general, the mining districts are concentrated along the major lineament systems, especially the Midas Trench. The generally lower areal density of ore-producing deposits within the central Nevada volcanic complex appears to be caused by the superposition of a thick pile of essentially barren volcanic rocks on older ore deposits followed by only very limited subsequent volcanic activity.

In Alaska, landform analysis of Landsat-1 images substantiated by geophysical data persuaded Lathram and his colleagues (1973) to propose a new area for petroleum exploration. As seen on the images, lakes in the Arctic coastal plain are predominantly elongate, with their long axes trending about N9°W. Northwest of the Umiat oil field, an additional strong east-trending regional lineation (Fig. 3), not previously recognized on aerial photographs or in field study, is expressed by elongation of some lakes, alignment of others, and by linear interlake areas. This second lineation is parallel both to the trend of deflections in contours of the magnetic and gravity fields in the area and to westerly deflections in the northwest ends of northwest-trending folds mapped to the south. In addition, the alignment of many small lakes forms a large ellipse superimposed on the regional lineation. Sparse seismic profiles show periodic reversals in dip and regional arching in shallow strata beneath the lineated area.

These data suggest that structures may be concealed beneath the Quaternary-mantling Gubik Formation in the area of Figure 3. Although the shallow-fold strata of the Gubik Formation are younger than those tapped by the oil wells of the Umiat field to the south, they may contain favorable reservoir beds. Similarly integrated landform and geophysical studies in Oklahoma convinced a commercial petroleum firm to reduce estimates of exploration costs by 20–50 percent because

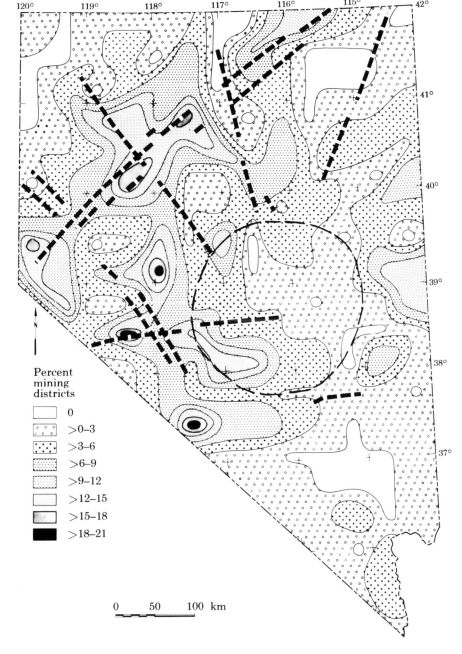

Percent mining districts

	0
	>0–3
	>3–6
	>6–9
	>9–12
	>12–15
	>15–18
	>18–21

0   50   100 km

Figure 2. Contour map showing the areal distribution of metal mining districts in Nevada in relation to the major lineaments (dashed lines) and a large circular feature derived from analysis of Landsat-1 images and synthesis of geophysical evidence. The contour interval is 3 percent. (Compiled from Rowan and Wetlaufer 1975.)

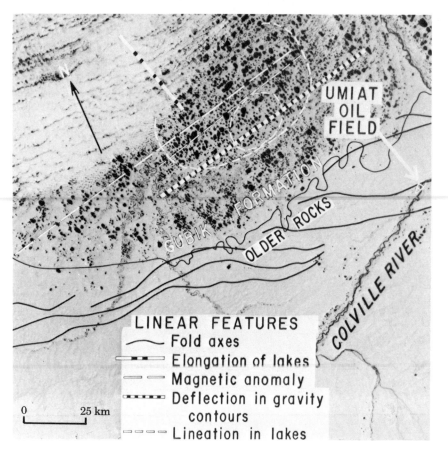

LINEAR FEATURES

Fold axes

Elongation of lakes

Magnetic anomaly

Deflection in gravity

contours

Lineation in lakes

0   25 km

Figure 3. Landsat-1 MSS image of an area northwest of Umiat, Alaska, showing elongation and lineation of lakes, locations of magnetic anomaly, deflection of gravity contours, and orientation of fold axes. (Image No. E1004–21395. From Lathram et al. 1973.)

Figure 5. Color-ratio composite prepared from a Landsat-1 MSS image of an area in south-central Nevada. Blue represents ratio image 4/5; yellow, 5/6; and magenta, 6/7. *A*, mineralogically altered area; *B* and *C*, felsic rocks with variable albedo. (Image No. E1072–18001, recorded 3 Oct. 1972. From Rowan et al. 1974.)

The spectral reflectance of rocks and soils in the MSS response range is determined mainly by the oxidation-reduction state and coordination of the constituent transitional metal ions, commonly iron. Electronic transitions in these metal ions cause broad absorption bands which, although commonly conspicuous in spectra for individual iron-bearing minerals, are generally subdued in rock spectra where ferrous and nonferrous minerals are combined. These broad bands are largely responsible for the shapes of rock spectra in the MSS response range. For example, in Figure 4, the differences between spectra for predominantly quartz-feldspar, or felsic, rocks and for predominantly ferromagnesian, or mafic, rocks are due mainly to the prominence of broad absorption bands in the spectra of the latter. However, visible and near-infrared laboratory spectra for a large number of rocks ranging from felsic to ultramafic have a large variance (Hunt et al. 1973a, b; 1974). In nature, the diagnostic value of these spectra is reduced even further by the effects of the surface coating and impurities.

These factors limit the value of spectral radiance data recorded by the MSS for making determinations of composition. In contrast to absolute *identification* of composition, however, *discrimination*, or differencing, is far less demanding because it depends only on relative spectral radiance differences. Several Landsat studies have now shown that the spectral radiance of many rock units is surprisingly uniform and that the differences among them are sufficient to permit discrimination.

While the spectral reflectance dif-

the use of Landsat data would reduce the amount of geophysical work required (Collins et al. 1974).

It must be emphasized that the geologic significance of the lineaments discussed here—and most of those described elsewhere—have not yet been fully evaluated, mainly because of their horizontal and apparent vertical magnitude. Additional systematic geologic and geophysical field studies should be used to determine the origin of these features, because studies conducted thus far clearly indicate their potential mineral and petroleum importance.

Uses of spectral information

One of the main objectives of most geologic remote-sensing studies is to discriminate among different materials and ultimately to identify them, thereby aiding the mapping process fundamental to many investigations. In the visible part of the electromagnetic spectrum, some geologic materials are reasonably distinctive owing to brightness, color, or textural differences and can therefore be mapped on conventional black-and-white or color photographs. Discrimination among many rock and soil units, however, can be improved substantially by also measuring radiation reflected in the near-infrared region.

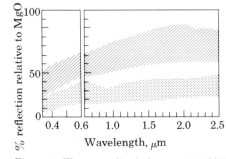

Figure 4. The generalized shapes of visible and near-infrared reflectance spectra for felsic (diagonal lines) and mafic (stippling) rocks. Grain size of samples is 0–75 μm. (From Hunt et al. 1974.)

Figure 6. The standard band-7 image showing the same area of Nevada represented in Figure 5. *A*, mineralogically altered area; *B* and *C*, felsic rocks with variable albedo. (Image No. E1072–18001, recorded 3 Oct. 1972. From Rowan et al. 1974.)

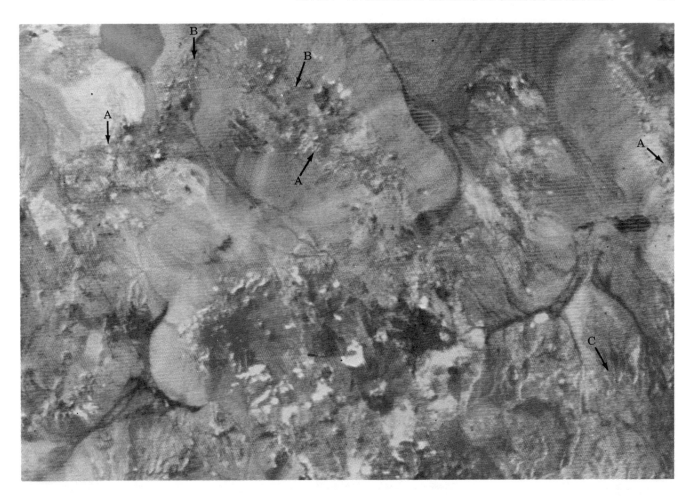

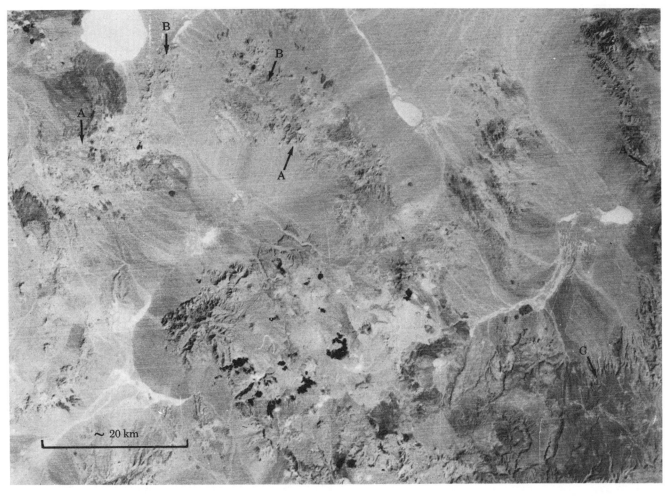

~ 20 km

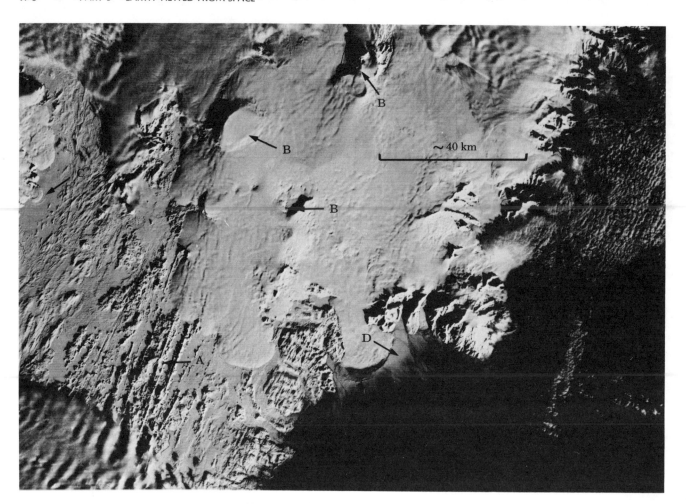

Figure 7. Landsat-1 composite image of the Vatnajökull area, Iceland, taken in January 1973. Note the full snow cover and the detailed relief afforded by the low illumination angle (7°) of winter. A, volcanic ridges, craters, and graben; B, volcanoes and calderas; C, recessional moraines; D, outwash plains. (Image No. E1192–12084. From Williams et al. 1974.)

ferences between a few rock types are great enough to be detected by visually comparing the MSS images, the differences among most rocks are so small that some form of image enhancement is needed to distinguish between them. Although many enhancement methods are presently in use, one of the most effective is a combination of two widely employed techniques: digital computer processing (Billingsley and Goetz 1973) and color compositing. During the computer processing, each of the digital picture elements which constitute an individual MSS band is divided by the spatially registered elements of another band. The contrast of a selected part of the ratio values is enhanced, and the resulting values are used to record new black-and-white ratio images. The gray tones in these images show spectral radiance differences; variations due to albedo differences and topography are minimized. Color composites can be prepared by using either an additive or subtractive process.

Figure 5 shows a Landsat-1 image of an area in south-central Nevada which is made up of three digitally processed ratio images, each presented in a different diazo, or subtractive, color. This color-ratio composite displays very subtle spectral-radiance differences, thus permitting many distinctions among rock units which cannot be seen in an unenhanced image (Fig. 6) or in color and color-infrared images (Rowan et al. 1974). The numerous patterns of green and brown shading to red-brown are of particular interest to the exploration geologist because they represent chemically and mineralogically altered areas in which metal-ore

Figure 8. Landsat-1 color infrared composite image of the same area of Iceland as in Figure 7, taken in July 1973. A large portion of the snow cover has melted, and an outwash plain is visible at D. The high illumination angle (42°) suppresses topographic details visible in Figure 7. (Image No. E1372–12080. From Williams et al. 1974.)

deposits are often found. The orange-pink areas represent compositionally similar rocks, even though the albedo within the unit is highly variable (B and C, Figs. 5 and 6). Therefore, the visible and near-infrared spectral radiance recorded by the four broad-wavelength bands of the MSS is reasonably uniform, despite the variable albedo.

Similar computer studies of MSS data are now being used to explore poorly known areas for ore deposits. In one of the few published studies, Schmidt (1975) used digital computer processing to determine the spectral radiance of a known copper-porphyry deposit as a frame of reference to classify other unknown altered areas in Pakistan. Field examination indicates that approximately 17 percent of the potential areas have mineral characteristics worthy of further evaluation.

Uses of temporal information

Many geologic phenomena are manifestations of dynamic processes. Some, such as volcanic and earthquake activity, are relatively short-lived catastrophic events, while others, including erosion and deposition, movement of crustal plates, and periodic seasonal and annual thermal variations, operate continuously over a longer period of time. Still other features are not dynamic in themselves, but their detection takes place over time. For example, the appearance of many landforms, especially linear structures, is influenced by seasonal variations in vegetation, soil moisture, and snow cover as well as by the shorter-term effects of illumination and atmospheric conditions. Owing to these factors, information gathered over time is essential for many remote-sensing geologic studies, although the time span may range from ten minutes or hours to several years. While aircraft can be used for monitoring predictable periodic variations of small areas, satellites are the most practical means of collecting information over time for large areas and for phenomena that are unpredictable or operate over long periods.

For an impressive example of en-

hancement of a variety of morphologic features, compare Figures 7 and 8, which are composite images of the Vatnajökull area of Iceland. This region, which is near the center of Iceland, sits astride the Mid-Atlantic Ridge, a major crustal feature fundamental to the volcanic development of the island. The series of closely spaced linear ridges which shows so well in the southwestern part of the winter image (Fig. 7) consists of a combination of volcanic ridges, volcanic crater rows, and down-faulted linear blocks (graben) that lie along the projection of the Ridge. Unfortunately, the projection of this important zone passes beneath Vatnajökull, the largest icecap in Iceland, where the volcanic features become subglacial or at least partially subglacial. In the summer view (Fig. 8), little evidence of these features can be seen because of their low topographic relief and the high sun angle. In contrast, the winter view shows several elliptically shaped volcanoes and calderas that were either previously unknown or poorly mapped.

While both images are useful for study of the glacial features, morphologic detail shows best in the winter image—for example, the recessional moraines—whereas the truly dynamic characteristics, such as meltwater outwash plains, are portrayed in the summer view. Repetitive imaging of this complex area over a long period will permit analysis of glacial movement, sedimentation rates, geothermal activity, development of new volcanic features, and many other temporal aspects.

In Iran, populated areas are concentrated around the margins of playas—large flat areas that are natural collection basins for seasonal runoff from adjacent mountains. Because the large volume of water which collects in the basins evaporates, most of the water needed for the population centers is obtained from deep wells and small reservoirs. Although economically important salt deposits result from the evaporation process, evaluation of the salt and water resources has been limited by the inaccessibility of these areas and by the inadequacy of a few scattered ground measurements.

In southwestern Iran, the Neriz Playa is one of the principal drainage basins in the Zagros Mountains Watershed. Using analysis of limited ground measurements and color-infrared composite Landsat-1 images acquired between 2 September 1972 and 28 August 1973, Krinsley (1974) has developed an important new method for monitoring the seasonal water fluctuations. His method shows that the areal extent of the lake occupying the playa ranges from a few percent to nearly 100 percent (Fig. 9), depending on precipitation and snow-melt runoff rates and the impact of man's activites in the area. From February through May 1973, at least 600×10^6 m^3 of water was available in the lake at Neriz, but by August 1973 only 19×10^6 m^3 remained because of evaporation losses. Obviously an enormous resource was lost and will continue to be lost until water-storage facilities are constructed in the mountains.

These conditions are found elsewhere in Iran and in many other arid regions of the world now undergoing economic development. Krinsley (1974) has also used Landsat-1 images to propose a new highway route across the Great Kavir, the largest playa in Iran, and, through digital processing, to detect subtle spectral reflectivity differences among the playa deposits.

Thermal-infrared techniques

Thermal-infrared images recorded from orbital platforms have been used by several workers to observe high-intensity thermal radiation associated with active volcanism (Friedman and Williams 1968). The value of thermal-infrared satellite data for mapping geologic materials, however, was not demonstrated until Pohn and his colleagues (1974) used a quantitative approach, referred to as thermal-inertia mapping, for analyzing Nimbus III and IV images of Oman. This technique, developed through analysis of aircraft images (Rowan et al. 1970) and theoretical considerations (Watson et al. 1971), takes advantage of the responses of materials to diurnal solar heating and cooling (Fig. 10). Materials of low thermal inertia, such as sand, have large-amplitude radiative flux vari-

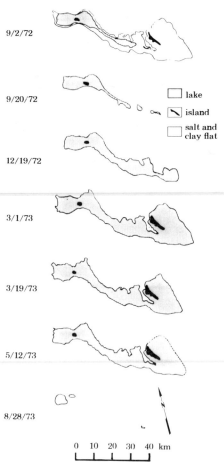

9/2/72

9/20/72

☐ lake
⌒ island
☐ salt and clay flat

12/19/72

3/1/73

3/19/73

5/12/73

8/28/73

0 10 20 30 40 km

Figure 9. Lake fluctuations at the Neriz Playa, 1972–73, as determined from Landsat-1 images. (After Krinsley 1974.)

ations, whereas high thermal-inertia materials, such as nonporous mafic rocks, have relatively small variations. Several factors other than thermal inertia influence the diurnal flux, including physical properties of the materials, albedo and emissivity, topographic configuration (slope, orientation, and elevation), and atmospheric conditions. The computer model developed by Watson and his colleagues accounts for these factors and

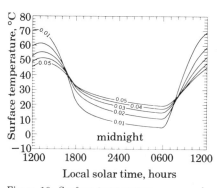

Figure 10. Surface temperature versus solar time for thermal inertias. (From Pohn et al. 1974.)

thereby provides a method for making quantitative determinations of thermal inertia. This technique is significant because it permits discrimination of many geologic materials on the basis of a fundamental physical property.

The results of the Oman study may be understood by comparing the thermal-inertia map (Fig. 11) with a regional geologic map of the same area (Fig. 12). The agreement between areas of high thermal inertia and geologic units consisting of high-density materials, a parameter which correlates strongly with thermal inertia, is remarkably good, especially in light of the 8-km spatial resolution of the Nimbus data. The coincidence of the two highest thermal-inertia levels with the ophiolite suite, mainly mafic rocks of high density, is most noteworthy. In addition, limestone is also reasonably accurately portrayed as areas of moderately high thermal inertia. In contrast, areas of low thermal inertia are generally underlain by relatively low-density materials, such as wadi deposits, sand, silt, and gravel. Some of the differences between the two maps later proved to be inaccuracies in the regional geologic map (Pohn et al. 1974).

This analysis, based on a thermal model developed from earlier aircraft experiments, provided the basis for the proposed Heat Capacity Mapping Mission (HCMM), an imaging system which will record twice daily the reflected solar-infrared (0.5–1.1 μm) and thermal-infrared (10.5–12.5 μm) radiation of an area. Readings are recorded at approximately 1400 and 2400 hrs, near the time of maximum thermal-inertia differences; anticipated thermal and spatial resolutions are 1°C and 0.5 km, respectively, the latter being more than an order of magnitude higher than the resolution achieved by Nimbus III and IV. Although the thermal channel planned for Landsat-C has an expected spatial resolution of 0.25 km, the times of overflight—09:40 and 21:40 at the equator—are not well suited for thermal-inertia mapping because they nearly correspond to the time of minimum contrast (Fig. 10). These data should be useful for many other purposes, however, including monitoring of

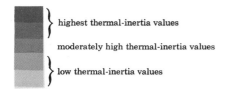

highest thermal-inertia values

moderately high thermal-inertia values

low thermal-inertia values

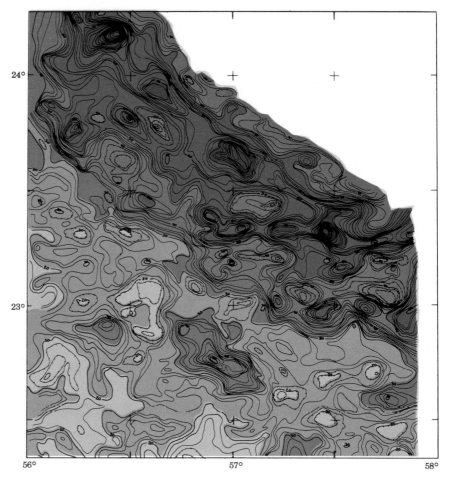

Figure 11. In this thermal inertia map of part of Oman, the contour interval is 2 × 10^2 cal cm -2 s $-1/2$. (After Pohn et al. 1974.)

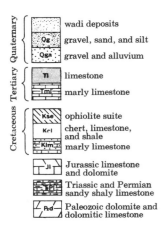

Quaternary
wadi deposits
Qg   gravel, sand, and silt
Qga   gravel and alluvium

Tertiary
Tl   limestone
Tml   marly limestone

Cretaceous
Kse   ophiolite suite
Kcl   chert, limestone, and shale
Klm   marly limestone

Jl   Jurassic limestone and dolomite

⊤Pl   Triassic and Permian sandy shaly limestone

P₂d   Paleozoic dolomite and dolomitic limestone

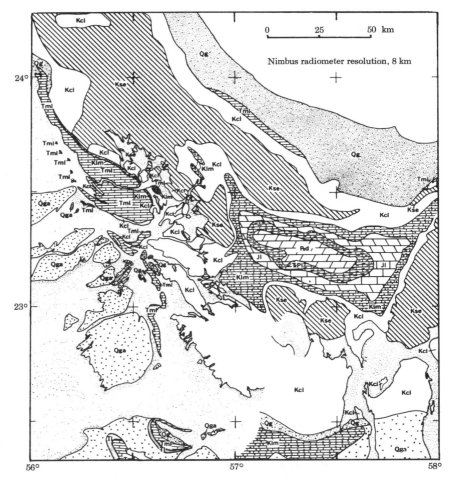

Figure 12. Geologic map of the same part of Oman represented in Figure 11. (After Pohn et al. 1974.)

volcanic activity, soil moisture conditions, glacial movement, coastal erosion and sedimentation, thermal pollution, and detection of some fault and fracture zones. Together, HCMM and Landsat-C should be an important asset in geologic investigations.

Although the mathematical model developed during these studies has direct application to detection and analysis of geothermal areas, it may be limited to airborne techniques, because extraction of the geothermal flux from satellite-recorded thermal-infrared data is complicated by many factors that tend to overwhelm the relatively low fluxes typical of most geothermal areas (Watson 1974). A detailed discussion of this and other geologic applications of thermal-infrared images is given by Watson (1975).

Another thermal-infrared technique deserves comment—even though it has not been demonstrated from orbital platforms—because it is a potential source of information about the composition of rocks. Spectroscopic measurements of minerals and rocks (Lyon 1964; Vincent et al., in press) show that the emissivity of some rocks departs significantly from black-body behavior in the thermal-infrared. For example, silicates have characteristic emissivity minima between 8 and 12 μm which generally occur at progressively longer wavelengths as the composition of the rocks becomes more mafic (Fig. 13). These spectral emission features, or reststrahlen bands, are caused by changes in the fundamental vibrations of the Si-O bond in tetrahedral coordination as the metal at the secondary coordination site changes from aluminum to iron.

A method has been developed for detecting the presence of reststrahlen bands, for monitoring their wavelength positions, and for displaying images that relate this spectral information to the spatial distribution of the surface materials (Vincent and Thomson 1972; Vincent et al. 1972). The principle is similar to that used in the ratio processing of Landsat MSS bands. The most straightforward application permits discrimination between silicate and nonsilicate rocks —for example, certain granites in

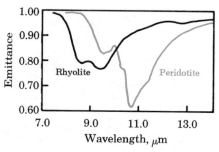

Figure 13. The emission spectra of two silicates, rhyolite and peridotite, show the shift of the reststrahlen bands toward longer wavelengths as the composition becomes more mafic. The vertical axis represents the ratio of black-body emission to sample emission. (From Vincent et al., in press.)

contrast to limestone and dolomite —which may appear similar in visible and near-infrared images. Although limestone and dolomite exhibit reststrahlen bands, these bands are weaker and narrower than those of silicate rocks, and they are only partially coincident (Lyon 1967).

This method has also been used to discriminate among silicate rocks, a more difficult problem than distinguishing silicates from nonsilicates. In the example shown in Figure 14, two analog thermal-infrared images of the Pisgah Crater area in California were ratioed to produce a third image, which displays the spectral information in tones of gray and permits distinction of two compositionally different silicate rocks, dacite and basalt. Other materials, alluvium and playa deposits, can also be identified. In another experiment, near Halloran Springs, California, using visible and near-infrared ratio images as well as thermal-infrared ratio images, Dillman and Vincent (1974) classified the predominantly silicate rock units with an estimated overall accuracy of 77.9 percent.

The capability of obtaining even gross information about composition from satellite images would obviously benefit regional geologic mapping in most parts of the world. The emissivity-ratioing technique, however, faces several formidable problems, the most important being atmospheric absorption between 9.5 and 9.9 μm owing to ozone. In addition, textural and mineralogic variations can result in apparent differences in composition, but the magnitudes of these effects are not yet well defined. A systematic program is needed to evaluate this technique fully, beginning with intensive laboratory studies and analysis of high-altitude aircraft data.

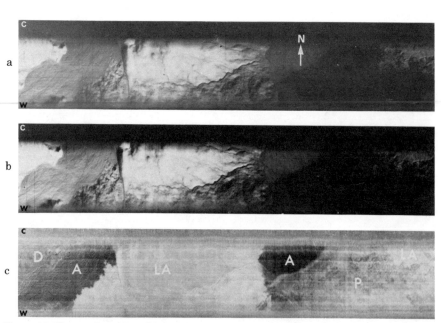

Figure 14. Two analog thermal-infrared images, a and b, of the Pisgah Crater, California, area have been ratioed to produce the third image, c. Image a was recorded at 8.2–10.9 μm and b at 9.4–12.1 μm; in both, the bright areas are warm. In c, the varying tones of gray permit the distinction between dacite (D) and basalt (LA), which are two compositionally different silicate rocks. A and P represent alluvium and playa deposits. C and W mark edges of cold and warm calibration plates. 1 cm $\simeq$ 315 m. (From Vincent and Thomson 1972.)

In spite of the success of Landsat-1 and the potential of thermal-infrared techniques, geologic and geophysical remote-sensing from satellite platforms is still in its infancy. Landsat-1 and -2 are experimental satellites, and their four broad MSS bands were selected about 6 years ago, largely on the basis of agricultural considerations and scanty geologic information. Analysis of spectral reflectance measurements in the 0.4–2.5 μm region (Goetz and Rowan, in press) and preliminary study of Skylab/Earth Resources Experiment Package images (Pohn 1974) indicate that recording in at least two bands between 1.0 and 2.5 μm would significantly improve our ability to distinguish rock units. In addition, stereoscopic images and a higher spatial resolution than that possible with the MSS are needed for many investigations. To date, a thermal-infrared imaging system that has both high spatial and thermal resolutions and repetitive viewing times determined mainly by geologic considerations has not been operated successfully from orbit.

Although I have confined this discussion to satellite platforms, it must be emphasized that the spawning grounds for such techniques are nearly always systematic laboratory and field experiments using airborne instruments. For example, analysis of energy emitted in the microwave region is a promising area of research, but because of the complex relationships among several factors, this method has not progressed from the theoretical modeling and aircraft experiment stage to the operational satellite phase, although some experimental satellite data are being studied (England, in press). Other proven remote-sensing techniques are either impractical or impossible from orbital altitudes: airborne gamma-ray surveys, for example, are used very extensively in some mineral exploration programs, but useful measurements from satellites are precluded by atmospheric absorption. Therefore, airborne platforms will continue to be essential for development of new techniques and for the application of certain proven ones.

References

Billingsley, F. C., and A. F. H. Goetz. 1973. Computer techniques used for some enhancements of ERTS images. In S. C. Freden and M. A. Becker, eds. and compilers. 1973. Vol. 1, sec. B, pp. 1159–68 (see below).

Collins, R. J., F. P. McCown, L. P. Stonis, G. Petzel, and R. E. John. 1974. An evaluation of the suitability of ERTS data for the purposes of petroleum exploration. In S. C. Freden and M. A. Becker, eds. and compilers. 1974. Vol. 1, sec. A, pp. 809–22 (see below).

Colwell, R. N. 1973. Remote sensing as an aid to the management of earth resources. American Scientist 61(2):175–83.

Dillman, R. D., and R. K. Vincent. 1974. Unsupervised mapping of geologic features and soils in California. Proc. 9th International Symposium on Remote Sensing of Environment, Ann Arbor, Mich.,15–19 Apr. 1974, vol. 3, pp. 2013–25.

England, A. W. In press. The effect upon microwave emissivity of volume scattering in snow, in ice, and in frozen soil. Proc. 1974 URSI Conf. on Microwave Emission and Scattering from the Earth.

Freden, S. C., and M. A. Becker, eds. and compilers. 1973. Symposium on Significant Results Obtained from the Earth Resources Technology Satellite-1, New Carrollton, Md., 5–9 Mar. 1973. Washington, D.C.: NASA SP-327, vol. 1, sec. A, pp. 1–976; sec. B, pp. 977–1994.

Freden, S. C., and M. A. Becker, eds. and compilers. 1974. Third Earth Resources Technology Satellite-1 Symposium. Washington, D.C.: NASA SP-351, vol. 1, sec. A, pp. 976.

Friedman, J. D., and R. S. Williams, Jr. 1968. Infrared sensing of active geologic processes. In Proc. 5th Symposium on Remote Sensing of Environment, Ann Arbor, Mich., 16–18 Apr. 1968, pp. 787–815.

Goetz, A. F. H., and L. C. Rowan. In press. The realities of using satellite spectral data for lithologic identification (abs.). In Proc. Annual Meeting Soc. Photogram., Washington, D.C., March 1975.

Gold, David. 1974. Oral communication.

Hunt, G. R., J. W. Salisbury, and C. J. Lenhoff. 1973a. Visible and near-infrared spectra of minerals and rocks-VII: Acidic igneous rocks. Modern Geol. 4:217–24.

Hunt, G. R., J. W. Salisbury, and C. J. Lenhoff. 1973b. Visible and near-infrared spectra of minerals and rocks-VIII: Intermediate igneous rocks. Modern Geol. 4:237–44.

Hunt, G. R., J. W. Salisbury, and C. J. Lenhoff. 1974. Visible and near-infrared spectra of minerals and rocks-IX: Basic and ultrabasic igneous rocks. Modern Geol. 5:15–22.

Isachsen, Y. W., R. H. Fakundiny, and S. W. Forster. 1974. Evaluation of ERTS imagery for spectral geological mapping in diverse terranes of New York State. In S. C. Freden and M. A. Becker. 1974. Vol. 1, sec. A, pp. 691–718.

Krinsley, D. B. 1974. The utilization of ERTS-1 generated images in the evaluation of some Iranian playas as sites for economic and engineering development. NASA Type-III report, pts. 1 and 2.

Lathram, E. H., I. L. Tailleur, and W. W. Patton, Jr. 1973. Preliminary geologic application of ERTS-1 imagery in Alaska. In S. C. Freden and M. A. Becker. 1973. Vol. 1, sec. A, pp. 257–64.

Lowman, P. D., Jr. 1965. Space photography: A review. Photogrammetric Engineering 31(1): 76–86.

Lowman, P. D., Jr. 1972. Apollo 9 multispectral photography: Geologic analysis. Greenbelt, Md.:NASA, Goddard Space Flight Center X-644-69-423. 53 p.

Lyon, R. J. P. 1964. Evaluation of infrared spectrophotometry for compositional analysis of lunar and planetary soils: Rough and powdered surfaces. Final report, pt. 2, NASA report CR-100.

Lyon, R. J. P. 1967. Infrared absorption spectroscopy. In Physical Methods in Determinative Minerology, J. Zussman, ed. N.Y.: Academic Press, pp. 371–403.

Pohn, H. A. 1974. Near-infrared reflectance anomalies of andesite and basalt in southern California and Nevada. Geology 2(11):547–50.

Pohn, H. A., T. W. Offield, and K. Watson. 1974. Thermal inertia mapping from satellite-discrimination of geologic units in Oman. U.S. Geol. Survey J. Res. 2(2):147–58.

Rowan, L. C., T. W. Offield, K. Watson, P. J. Cannon, and R. D. Watson. 1970. Thermal infrared investigations, Arbuckle Mountains, Oklahoma. Geol. Soc. America Bull. 81(12):3549–61.

Rowan, L. C., P. H. Wetlaufer, A. F. H. Goetz, F. C. Billingsley, and J. H. Stewart. 1974. Discrimination of rock types and detection of hydrothermally altered areas in south-central Nevada by the use of computer-enhanced ERTS images. U.S. Geol. Survey prof. paper 883, 35 p.

Rowan, L. C., and P. H. Wetlaufer. 1975. Iron-absorption band analysis for the discrimination of iron-rich zones. NASA Type-III report, 160 p.

Schmidt, R. G. 1975. Exploration for porphyry copper deposits in Pakistan using digital processing of ERTS-1 data. U.S. Geol. Survey open-file report, no. 75–18, 26 p.

Vincent, R. K., and F. J. Thomson. 1972. Rock-type discrimination from ratioed infrared scanner images of Pisgah crater, California. Science 175:986–88.

Vincent, R. K., F. Thomson, and K. Watson. 1972. Recognition of exposed quartz sand and sandstone by two-channel infrared imagery. J. Geophys. Res. 77:2473–77.

Vincent, R. K., L. C. Rowan, R. E. Gillespie, and C. Knapp. In press. Thermal-infrared spectra and chemical analyses of twenty-six igneous rock samples. Modern Geol.

Watson, K. 1974. Geothermal reconnaissance from quantitative analysis of thermal infrared images. In Proc. 9th International Symposium on Remote Sensing of Environment, Ann Arbor, Mich., 15–19 Apr. 1974, vol. 3, pp. 1919–32.

Watson, K. 1975. Geologic application of thermal infrared images. J. Inst. Electrical and Electronic Engineers 63(1):128–37.

Watson, K., L. C. Rowan, and T. W. Offield. 1971. Application of thermal modeling in the geologic interpretation of IR images. In Proc. 7th International Symposium on Remote Sensing of Environment, Ann Arbor, Mich., 17–21 May 1971, vol. 3, pp. 2017–41.

Williams, R. S., Jr., A. Böovarsson, S. Frioriksson, G. Pálmason, S. Rist, H. Sigtryggsson, K. Saemundsson, S. Thorarinsson, and I. Thorsteinsson. 1974. Environmental studies of Iceland with ERTS-1 imagery. In Proc. 9th International Symposium on Remote Sensing of Environment, Ann Arbor, Mich. 15–19 Apr. 1974, vol. 1, pp. 31–81.

Williams R. S., Jr., and W. D. Carter. In press. ERTS-1: A new window on our planet. U.S. Geol. Survey prof paper 929.

Zietz, Isidore. 1974. Written communication.

Ralph Bernstein
George C. Stierhoff

Precision Processing of Earth Image Data

Landsat images of the earth can be corrected by digital techniques to yield more precise information

Fourteen times a day each of the two National Aeronautics and Space Administration's Landsat satellites orbits the earth collecting resources information from the surface. Sensors aboard the spacecraft convert the hue and intensity of the sunlight reflected from the earth below into numerical data that are transmitted to receiving stations in North America. This information can be used either in digital or visual form, in fields as diverse as agriculture, oceanography, cartography, and pollution monitoring (see Table 1). Each of the four spectral bands of the Landsat system can be processed to produce an image in shades of gray. Three of the gray images, which may be compared to color separations, can be combined into a three-color image called a color composite. Before the integration can be done, however, the numerical data should be corrected to remove errors such as those caused by rotation of the earth and motion of the satellite. Precise corrections of Landsat data are useful for generating land-use maps, detecting various crops and determining their acreage, and detecting changes. Each user of Landsat

Mr. Bernstein, who joined IBM in 1956, is a senior engineer and manager of the Advanced Image Processing Analysis and Development Department. Since 1967 he has had important roles in analyzing and designing systems for the acquisition and processing of satellite and aerial imagery, including serving as a principal investigator on the NASA Landsat program. Mr. Stierhoff, Associate Editor of the IBM Systems Journal, joined the IBM Research Division in 1952. At present he is also Adjunct Professor of Geophysics at Briarcliff College. The authors are grateful to C. W. Niblack for his help in preparing the images that accompany this article. The work reported here was done under contract with NASA. Addresses: Mr. Bernstein, IBM Corp., 18100 Frederick Pike, Gaithersburg, MD 20760. Mr. Stierhoff, IBM Corp., Armonk, NY 10504.

imagery may visually extract information related to his own discipline. The purpose of this paper is to describe computer-processing of Landsat data so that users, by understanding techniques better, can get more information from the imagery.

When enhancement and correction reach a level of accuracy equal to that of United States Geological Survey maps at a particular scale, the data processing may be called *precision processing*. The method yields precise images of areas of the earth's surface that can be used by scientists in many fields. A corrected image has the advantage of being both a cartographic product and one that contains the desired disciplinary information. Because of the digital nature of the data and the vast amounts of information involved, computers have been used for such processing, and we discuss here various computer-aided techniques that have been developed for the precision processing of data collected by Landsat (*1, 2*). Precision data-processing methods are expected to be used by NASA to process data from a third Landsat spacecraft, as well as from future earth-observation satellites (*3*).

Precision processing of Landsat data has led us to experiment with applications of the accurate imagery that is produced. We have of course not attempted to enter professionally into such disciplines as metropolitan planning, land use, marine-resource management, and the many other fields in which Landsat data can be used. Rather, we have worked on developing the precision-processing techniques so that those who are engaged in such studies may extract

more useful information for their purposes.

The decision of NASA to build and launch Earth Resources Technology Satellite-1 (ERTS-1, now renamed Landsat-1) has roots in NASA's own studies and those of the Department of the Interior as well as in the Summer Study on Space Applications by the National Academy of Sciences during 1967 and 1968. At the Woods Hole Oceanographic Institution in July 1968, the Academy's study group completed its deliberations on technological and economic issues concerning remote sensing of earth resources by satellite as part of the United States' overall space applications program. Their report (*4*) drew on all previous experience in space, including the World Weather Watch, the Application Technology Satellite, and the Nimbus program. The experience of manned spaceflights was extensively analyzed, and reports and photographs returned by the Gemini astronauts (*5, 6*) were used to evaluate the practicality of studying weather and climate, monitoring pollution, surveying earth resources, and studying earth and ocean physics. Aircraft and Apollo 9 experimented with optical filters and cameras to establish the spectral bands that would later be used with the Landsat sensors (*7, 8*). The multispectral photographic data were processed by computer to estimate the information-extraction potential of the Landsat system.

The National Academy of Sciences study group evaluated the entire proposed Landsat program—including design, construction, launching, guidance, orbit, and instrumentation—for its potential benefits to mankind. The group con-

Figure 1. A section of a Landsat image of Manhattan Island was enhanced by computer to display the square *picture elements,* or *pixels,* which usually cannot be distinguished visually on images. The Hudson River is the dark vertical column at the left of the photograph. Surprisingly, the narrow West Side Drive is clearly visible to the right of the river. This is because, in the area shown, the highway runs along a grassy slope, and the reflected green provides the contrast that delineates the road. The East Side Drive, at the right edge of the figure, blends in with the concrete and steel of the city and is less visible in the image because of lack of relative contrast. (Data in the image are from Band 4 only.)

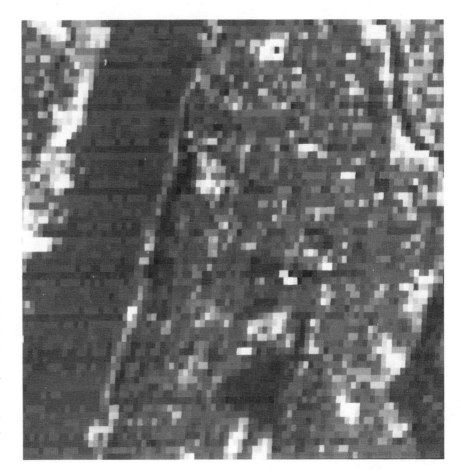

cluded that a digital-processing system would be necessary to manage the information flow, that digital or analog methods could be used for correcting some errors, that provision for complete digital processing should be included in the program for a selected sample of images, and that users would find the data most convenient in visual image form (9). NASA launched Landsat-1 on 23 July 1972 and Landsat-2 on 22 January 1975.

The sensing system

Two earth-sensing systems are installed on board Landsat-1 and -2—a three-band vidicon (television-type) system and a four-band multispectral scanner. The vidicon system on Landsat-1 was shut down early in the program because of a power switching malfunction, and thus applications for vidicon data have not been fully developed. However, the four bands of the multispectral scanners have supplied a large and continuing flow of information for several hundred varied and valuable applications (10–12).

Landsat circles the earth at an altitude of 915 kilometers (500 nautical miles) in a nearly circular, nearly polar orbit that has been designed to coordinate with the sun and the earth to record certain spectral (hue) and intensity information in the sunlight reflected from the earth below. Landsat hue signals are produced in the four spectral bands with wavelengths from 0.5 to 1.1 micrometers (Bands 4, 5, 6, and 7). Within each band, the multispectral sensor resolves 64 intensity levels. The sensors record the world in four color dimensions, which include the infrared. Man views the earth in the three

visible color dimensions of the cones of the retina (the photoreceptors responsible for color vision). Sometimes at dusk after a shower—when the last of the sunlight is diffused by clouds—we experience the intensified vision of both the cones and the rods (the sensory bodies in the retina responsive to faint light). This is the evening glow—*l'heure bleu.*

Because of the overlapping sensitivity of the cones, multiple hue and intensity signals are transmitted to the visual cortex, and human beings can detect a continuous spectrum of colors over the range of sensitivity of the eye. Landsat sensors, however, do not distinguish hues within a band because the multispectral sensor bands do not significantly overlap. Although Landsat Bands 4 and 5 extend from green to red, which we can see, Bands 6 and 7 are in the infrared region of the spectrum to which our eyes are not sensitive, and thus, when the bands are combined to form a visible color composite to aid our visualization of the data, colors we can see are substituted for colors we cannot see. The resultant image is called a "false-color" composite.

Table 1. Applications of Landsat data

Agriculture, forestry, rangeland
   Crop census
   Crop yield
   Identification of vegetational disease
   Land-use inventory

Oceanography, marine-resources
   Fish production
   Ship routing
   Sea and ice conditions

Hydrology
   Water-resource inventory
   Identification of freshwater sources
   Flood monitoring
   Health monitoring of lakes
   Pollution monitoring

Geology
   Identification of tectonic features
   Geologic and physiographic mapping
   Mineral and field exploration
   Earthquake studies
   Time-rate studies: glaciers, volcanos, erosion sites

Geography
   Thematic mapping for land use
   Physical studies for land improvement

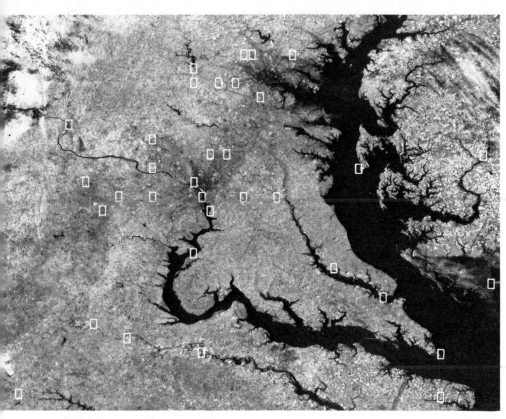

Figure 2. This unprocessed Landsat image contains raw—or uncorrected—data for the Chesapeake Bay region. The Potomac River branches to the left of the bay. The dark gray area at the point where the Potomac narrows and veers west is Washington, D.C., and the similar dark area to the northeast of Washington, near the head of the bay, is Baltimore. Very light areas in the upper left corner and to the right of the bay are clouds. White rectangles indicate *ground control points,* accurately located features on the earth used to correct the image geometrically. Although all errors are still present in the image, only those caused by the earth's rotation and by oversampling are prominent enough to be visible. As the spacecraft traveled from north to south (top to bottom) scanning the scene, the earth rotated 12.88 km from west to east (left to right); thus the lower left corner of the image is 12.88 km west of the upper left corner. The image is rectangular because of oversampling: more points were taken in the along-scan (west to east) than the cross-scan direction. Corrected images are roughly square, reflecting the 185-km^2 scene they depict. (Data are from Band 7.)

Atmospheric scattering of light occurs primarily in the blue (0.4–0.5 μm wavelength) region of the spectrum. The Landsat-1 and -2 sensors do not detect blue, and thus the data are not significantly affected by the sky-blue haze that was first explained by Lord Rayleigh. The 0.8–1.1 μm infrared Band 7 of Landsat-1 and -2 does not detect thermally generated radiation; however, Band 8 of Landsat C, planned for a 1977 launch, will provide a thermal infrared sensing capacity in the 10.4–12.6 μm range.

Like machines precisely wrapping yarn on a ball, the Landsat spacecrafts wrap their orbits around the earth, while our planet turns daily on its axis and revolves yearly around the sun. The Landsat orbits have been designed so that the multispectral data are recorded under near-constant conditions of solar illumination for each geographic location throughout the year. Called "sunsynchronous orbits," they provide

Figure 3. A color composite—called "false-color" because bands in the infrared region of the spectrum, to which the human eye is not sensitive, are represented by colors the eye can see—was produced by combining data from 3 of the 4 Landsat bands. The scene, imaged in September 1972, is the same Chesapeake Bay region for which uncorrected data are shown in Fig. 2. This scene was processed to conform geometrically to a corrected October 1972 scene to within 0.25 pixel.

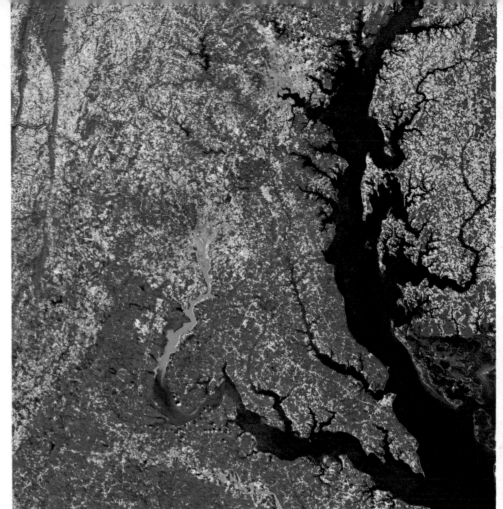

Figure 4. A false-color composite of the Chesapeake Bay area in Fig. 3 was processed from data collected 18 days later, in October 1972, during the satellite's next track over the area. The scene was precision-processed to conform to within 0.4 pixel accuracy of USGS maps.

essentially constant sunlight conditions for the many experiments that use Landsat data, and are affected only by seasonal variations in illumination.

During the 14 orbits a day made by each Landsat, the spacecraft views a 185 km-wide (100 n mi) strip as it traverses the sunlit face of the earth from north to south. In this orbit, each spacecraft provides global data between 80 degrees north and south latitudes every 18 days—offset in time so that together they provide 9-day coverage. Every equatorial crossing of the sunlit face occurs at approximately 9:30 a.m. local time. To accomplish this, the sun-synchronous orbit plane precesses westward at a rate of about one degree—170 km, or 92 n mi—per day, which maintains a constant orbit-plane angle relative to the sun-earth line.

At the equator, successive Landsat "tracks" (center lines of the 185 km-wide strips) are spaced 2,800 km (1,500 n mi) apart. Thus 14 such strips are swept across the earth each day. On the North American continent, three United States and two Canadian Landsat receiving stations record the data images. Depending on cloud cover and requirements of NASA investigators, approximately 200 *scenes* are recorded and transmitted every day. A scene, the basic unit of Landsat imagery, represents

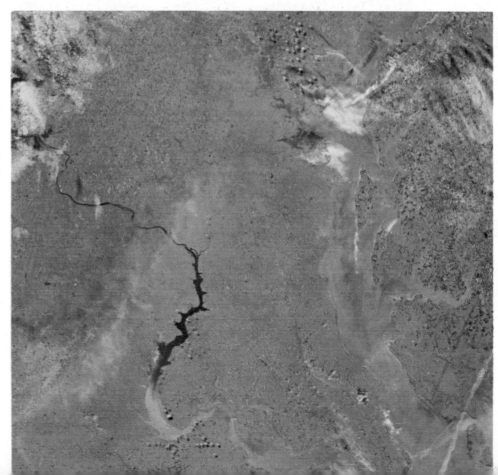

Figure 5. Precision processing made it possible to match the data in the scenes in Figs. 3 and 4 to within 0.25 pixel, accurate enough to compute differences in intensity values for each pair of pixels. The result is the "difference" scene shown here, which reveals changes that took place in the 18 days between first and second images. The Potomac River is conspicuous because of suspended sedimentation caused by a rainstorm shortly before the October scene was imaged. Clouds (*white areas*) and a jet aircraft contrail (*whitish streak in the lower right corner*), present in September but not October, are also clearly visible. Since Baltimore, Washington, and the Chesapeake Bay changed very little during the intervening 18 days, they are practically invisible on the "difference" scene.

Figure 6. A "shade-print" image, like this one of the section of Washington, D.C., where the Anacostia River joins the Potomac, is created by overstriking the set of characters shown in Fig. 7. Not only does shade printing provide a convenient way to produce Landsat images, it also permits the accurate matching of pixels. A shade print similar to the one shown was used, for example, to attain the precise pixel-to-pixel registration needed to produce the "difference" scene in Fig. 5. (Data are from Band 7.)

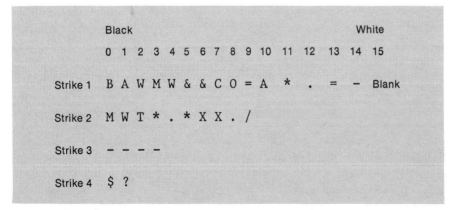

Figure 7. Overstriking these characters yields the 16 distinguishable shades of gray used to create shade prints like the one in Fig. 6. To produce the densest shade of black, first B is struck, then M, –, and $. The 64 levels of gray sensed by the Landsat system are reduced to 16 shades by arbitrarily deciding what range (0–3, 4–7, etc.) will be represented by which set of characters.

all spectral bands within an area on the ground 185 km on a side.

The multispectral sensors scan each scene in about 27 seconds. The raw data in Band 7 for the scene visualized in Figure 2 show, as the most prominent features, Chesapeake Bay, Baltimore, and Washington. While scanning this scene, the spacecraft was moving from north to south (top to bottom), during which time the earth turned 12.88 km (7 n mi) from west to east (left to right). Thus, although the reference frame of the sensors is rectangular, geographically the lower left corner of the scene is 12.88 km west of the upper left corner. The earth's rotation is but one of a number of errors in the data that must be corrected by precision data-processing techniques. Although Figure 2 contains all the errors, only the earth's rotation and oversampling effects are visible; the other errors are too subtle to be noticed.

Precision data-processing

The elementary unit of data in each band of each scene is the integrated value of intensity of reflected light detected in the field of view by each sensor. Called *instantaneous field of view,* the source of hue-intensity data is a 79-meter square on the ground. In the along-scan direction oversampling occurs, that is, each sample overlaps the previous sample by a factor of 1.4. A *picture element,* or *pixel,* is the numerical value of the energy radiated by an instantaneous field of view in each of the multispectral sensor bands. Numerically, a pixel is a radiant energy point, the value of which is currently normalized to the range 0 to 63. Graphically, a pixel is a photographic or printed image of an instantaneous field of view on a 64-level gray scale for each band. In color imagery, a pixel is the image of an instantaneous field of view in which the color saturation, or intensity, is a function of the numeric value of the energy reflected from the ground on a 64-level intensity scale.

There are approximately 8 million instantaneous fields of view or picture elements in each band, and about 32 million in all four bands transmitted from the spacecraft to the ground receiving stations. No ground feature within an instantaneous field of view can be resolved. Precision processing of earth image data involves the geo-

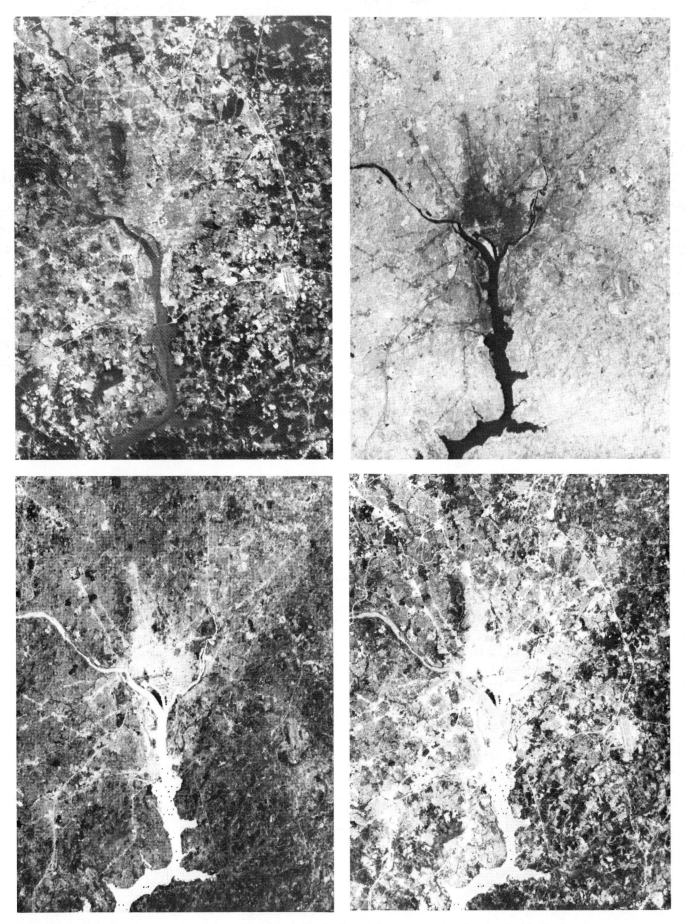

Figure 8. *Top left,* data from Band 5 of the Landsat multispectral scanner show concrete roads in and around Washington, D.C., in white; because of the poor contrast ratio, asphalt roads are all but invisible in this band. *Top right,* in Band 7 data of the area, asphalt roads are clearly visible in black, whereas concrete roads are almost invisible. *Bottom left,* Band 7 data are inverted to visualize the asphalt roads in white, and, *bottom right,* are interlaced with Band 5 data to reveal the combined concrete-asphalt road network as a white pattern. The view of Washington is extracted from the larger scene of the Chesapeake Bay area in Fig. 2.

Figure 9. Images derived from scenes comparable to those in Figs. 3 and 4 show Landsat false-color composites of (*left to right*) Washington, D.C., New York City, London, and San Jose, Calif. Each "map" is a square 34.26 km (18.5 n mi) on a side, which has been extracted from a 185-km² scene. Data-processing permits the display of similar features, and the images were enhanced to emphasize transportation networks, buildings, airports, industrial centers, recreational areas, and hydrological features.

metric and radiometric correction of the positioning and the intensity of the numeric values of the picture elements. The primary product of the processing is a computer output tape, and the Landsat figures in this article might be thought of as elaborate and beautiful graphs—pictorial representations of the data on the tapes.

Figure 1, a subimage of Band 4 of a scene that includes Manhattan Island, was especially enhanced to display the pixels, which are not usually resolvable by the eye. Individual pixels are visible, and some areas are seen to be composed of arrays of pixels. Many different intensity levels in Band 4 are visible in the scene as gradations on a gray scale. Close examination of the figure reveals faint stripings, especially across the dark channel of the Hudson River, which flows from north to south at the left of this subscene. The stripes are artifacts of sensor bias and gain differences that have not yet been completely removed. There are six pixel stripes per cycle; this is because of the design of the sensor matrix, which does not consist of one bank of four detectors that scans one pixel track in four spectral bands. Rather, there are twenty-four detectors, matrically arrayed to scan six

parallel pixel tracks in four bands simultaneously. This design has optimized data-sensing, because as a scan is completed and the scanning mirror returns to begin the next scan, the spacecraft has moved along its ground track the width of six fields of view. Radiometric errors of gain and bias, which are visualized as the striping in the figure, are removed by using preflight calibration tables that translate measured intensity signals to calibrated values from the tables. As a further precaution against drift of the sensors and their amplifiers, intensity signals are recalibrated by a standard light during the return scan of the scanning mirror.

Another interesting observation about Figure 1 is that West Side Drive (along the right side of the Hudson River in the image) is clearly visible although it is less than 79 m wide. This is because, in the area of Manhattan Island shown, the roadway borders a grassy and wooded hill, and in the integration of reflected spectral information, the contribution of reflected green provides the contrast needed to delineate the road. East Side Drive (or F.D.R. Drive, at the right edge of the image) is more nearly like the surrounding city and is much less distinguishable.

Numerous analyses of features, such as the contrast just noted, have parallels in the extraction of abstract features by the human visual cortex—detection of color, contour, and contrast, and analysis of linear features and texture. At any instant, the eye receives more data than the person can use. Landsat similarly records and transmits more data than can be used in a single application. When photographic and electro-optical techniques are used to extract Landsat information, some of the information is randomly and irretrievably lost. When Landsat information is processed by computer, potentially no information is lost. Rather, specific data can be extracted for specific applications, and the rest can be retained for other uses.

One of our objectives is to reduce all geometric errors in the Landsat data to such a degree that the images conform to USGS map standards. Since information about the precise attitude, position, and altitude of Landsat is unavailable in the present state of the art, mathematical methods have been developed for correcting the scene data that use geodetic points in the data themselves (*13, 14*)—*ground control points*, which serve much the same function as the

network of benchmarks and position control points used by the cartographers of the USGS. The ground control points used to develop a mapping function for correcting the data for Figure 2 are shown as white rectangles in that figure. Typical ground control points are landing-strip intersections, highway intersections, points of land in water, and field patterns whose positions and elevations are known precisely. A catalog of ground control points has been prepared and stored in the computer. Nine to sixteen points uniformly distributed over a scene are usually adequate for the precision processing of Landsat data. (Computationally efficient techniques for locating ground control points in digital data arrays are discussed in references *13* and *15*.)

To develop the basic mathematical approach for the precision correction of the data, differences between positions of ground control points on the reference map and the observed control points in the scene are used to evaluate the coefficients of cubic time functions of roll, pitch, and yaw, and a linear time function of altitude deviation from normal height above local earth's surface. The resultant equation, termed a *mapping func-*

tion, corrects the warped data image into one that approximates the reference map.

When the positions of pixels have been located in the uncorrected data image by means of the mapping operation, one of three methods is used to compute the intensities of the pixels. In the nearest-neighbor method, the intensity of the pixel closest to the computed position of the output point is assigned. The nearest-neighbor method requires the least data-processing of the three methods, but it does not compensate for the fine sawtooth effect on straight lines and edges due to sensor sampling. Another method is bilinear interpolation, in which four neighboring input values are used to compute the intensity of each pixel by a process of two-dimensional interpolation that removes the sawtooth but produces a slight fuzziness. A more complex method of locating intensity values, which provides the sharpest image, is cubic convulation (*13, 15*), in which sixteen neighborhood values of a given input intensity are used to compute the intensity of each pixel.

To confirm the accuracy of the Landsat system and the precision data-processing, absolute and relative

checks are made on ground control points in the images produced by the methods just described. Absolute accuracy, determined by comparing the precision-processed ground control points with those in USGS maps, has been within one pixel for 90% of the ground control points. Relative accuracy, which is determined by comparing the positions of ground control points in precision-processed scenes that were recorded at different times, has been within 0.5 pixel.

When two scenes have been corrected to conform geometrically to within less than 0.5 pixel, the difference between intensity values of each corresponding pixel pair can be computed to show the changes that have taken place between the times the two scenes were imaged. Figure 3 is a color-composite scene of the Chesapeake Bay area derived from data sensed in September 1972; Figure 4 is a color composite of the same scene in October 1972, corrected to achieve a 0.4 pixel accuracy relative to a USGS map of the area. The September scene was corrected to conform to the October data to within 0.25 pixel, the two scenes were digitally "differenced," and the resultant, or "difference," scene is shown in Figure 5. Information contained in one scene but not in the other—such as the clouds and jet aircraft contrail in the September image, plus the more subtle differences in vegetation—are clearly visible. In particular, the Potomac River stands out because of suspended sedimentation caused by rain that fell shortly before the October scene was imaged.

Techniques and applications

Applications in urban studies serve as an example of how precision processing can be used. The enhancement, extraction, enlargement, and data management methods are, however, broadly applicable.

Shade prints. Shade printing is a method of creating by computer printout image data. Shade prints, like the one in Figure 6, can give a visual representation of building density or hydrological features. They are useful for quick looks and for locating features precisely. The technique, which is analogous to viewing the data with a typographical microscope, consists of overstriking a set of characters (shown in Fig. 7) that yields sixteen distinguishable levels of gray. Since each pixel is generated separately, the intensity value of any pixel can be visually estimated and plotted on the image. In a shade print, the elementary unit of length of one pixel is determined by the point size of the typeface. Thus, an approximately 25-fold magnification occurs in translating from a 185-km^2 scene to type of the size in Figure 6.

Extraction of road features. When the subimage that included Washington, D.C., was extracted and enlarged from the data image shown in Figure 2, we observed that concrete roads are readily detectable as a white pattern in the visible Bands 4 and 5, whereas asphalt roads are virtually invisible because of their poor contrast ratio (see Fig. 8). Conversely, asphalt road patterns are readily discerned as a black pattern in the infrared Bands 6 and 7, and concrete roads are almost invisible. This observation suggested an experiment in which we extracted and combined the two road networks in one radiometrically enhanced composite image. First we inverted the Band 7 data image, which yielded the complement of each of the 64 intensity values and, of course, caused the black road pattern to become white. The inverted Band 7 data were then interlaced with Band 5, and the result was the combined concrete-asphalt road network visible as a white pattern.

Atlas of cities. Another experiment used precision-processed data to compile an atlas of selected cities of the world, on the assumption that images derived from a comparable data base would bring out similarities and differences directly, whereas they are usually inferred or deduced. Each city has a distinctive personality, as the traveler knows from experience. Particular aspects of the personalities of four cities are visible in the Landsat data images in Figure 9. These urban areas were selected for display because of their size, geographic separation, and historic differences. The images were processed to enhance the same characteristics, with emphasis on the aspects most often cited by specialists as key indices of urban environments: transportation networks, buildings, airports, industrial centers, recreational areas, and hydrological features. Areas of vegetation—parks, woods, fields, and farms—which appear as false-color red in the images, are a way of estimating the quality of life in each metropolitan area.

We have discussed the precision processing and several visualization techniques for Landsat data images. Reference *12* describes over 200 applications, and new research continues to be reported (*16*). We hope that precision data-processing techniques will contribute to the imaginative use of the data by scientific specialists. Implied, but largely unspecified here, is the fact that our methods are mathematical operations on each packet of data—each pixel. This makes possible a range of experiments of a statistical and quantitative nature in which visualization is not necessary. Some of these applications have already been explored, but many other uses of precision-processed Landsat data remain to be investigated.

References

1. R. Bernstein. 1976. Digital image processing of earth observation sensor data. *IBM J. Res. Dev.* 20(1).
2. R. Bernstein and D. J. Ferneyhough. 1975. Digital image processing system. *Photogrammetric Engineering and Remote Sensing (J. Am. Soc. Photogrammetry)* 41:1465.
3. Master Data Processor. NASA Request for Proposal. 1975. Greenbelt, MD: NASA Goddard Space Flight Center.
4. *Useful Applications of Earth-Oriented Satellites.* 1969. Report of the Central Review Committee, Summer Study on Space Applications. Washington, DC: National Academy of Sciences.
5. *Earth Photographs from Gemini III, IV, and V.* 1967. Scientific and Technical Information Division, Office of Technology Utilization, NASA. Washington, DC: U.S. Government Printing Office.
6. *Earth Photographs from Gemini VI through XII.* 1968. Scientific and Technical Information Division, Office of Technology Utilization, NASA. Washington, DC: U.S. Government Printing Office.
7. G. A. Theodore. 1971. ERTS A and B—The engineering system. *Astro. & Aero.* 41–51.
8. L. Jaffe and R. A. Summers. 1971. The earth resources program jells, *Astro. & Aero.* 24–40.
9. Pers. comm., Arthur G. Anderson, Chairman, Panel 8, Systems for Remote-Sensing Information and Distribution (Summer Study on Space Applications).
10. R. N. Colwell. 1973. Remote sensing as an aid to the management of earth resources. *Am. Sci.* 61:175–83.
11. E. P. Mercanti. 1974. Widening ERTS applications. *Astro. & Aero.* 28–39.
12. S. C. Freden, E. P. Mercanti, and M. A. Becker, eds. 1974. *Third Earth Resources Technology Satellite-1 Symposium*, Vol. 1: *Technical Presentations*, Sections A and B. Washington, DC: NASA.
13. R. Bernstein. 1975. All-digital precision processing of ERTS images. IBM final report to NASA, Contract NAS 5-21716.
14. R. Bernstein. 1974. Scene correction (precision processing) of ERTS sensor data using image processing. In Ref. *12*. Also see ref. *1*.
15. S. S. Rifman and D. M. McKinnon. 1974. Evaluation of digital correction techniques for ERTS images. TRW Corporation final report, TRW 20634-6003-TU-00. Greenbelt, MD: NASA Goddard Space Flight Center.
16. L. C. Rowan. 1975. Applications of satellites to geologic exploration. *Am. Sci.* 63: 393–403.

Index